Advances in Mathematical Fluid Mechanics

Series Editors

Giovanni P. Galdi
School of Engineering
Department of Mechanical
Engineering
University of Pittsburgh
3700 O'Hara Street
Pittsburgh, PA 15261
USA
e-mail: galdi@engrng.pitt.edu

John G. Heywood
Department of Mathematics
University of British Columbia
Vancouver BC
Canada V6T 1Y4
e-mail: heywood@math.ubc.ca

Rolf Rannacher
Institut für Angewandte Mathematik
Universität Heidelberg
Im Neuenheimer Feld 293/294
69120 Heidelberg
Germany
e-mail: rannacher@iwr.uni-heidelberg.de

Advances in Mathematical Fluid Mechanics is a forum for the publication of high quality monographs, or collections of works, on the mathematical theory of fluid mechanics, with special regards to the Navier-Stokes equations. Its mathematical aims and scope are similar to those of the *Journal of Mathematical Fluid Mechanics*. In particular, mathematical aspects of computational methods and of applications to science and engineering are welcome as an important part of the theory. So also are works in related areas of mathematics that have a direct bearing on fluid mechanics.

The monographs and collections of works published here may be written in a more expository style than is usual for research journals, with the intention of reaching a wide audience. Collections of review articles will also be sought from time to time.

Analysis and Simulation of Fluid Dynamics

Caterina Calgaro
Jean-François Coulombel
Thierry Goudon
Editors

Birkhäuser Verlag
Basel · Boston · Berlin

Editors:

Caterina Calgaro
Jean-François Coulombel
Thierry Goudon
Laboratoire Paul Painleve, UMR 8524
CNRS-Université des Sciences et Techno.
F-59655 Villeneuve d'Ascq cedex
France

e-mail: Caterina.Calgaro@math.univ-lille1.fr
jfcoulom@math.univ-lille1.fr
thierry.goudon@math.univ-lille1.fr

2000 Mathematical Subject Classification 76-06, 65Cxx, 68U20

Library of Congress Control Number: 2006937472

Bibliographic information published by Die Deutsche Bibliothek
Die Deutsche Bibliothek lists this publication in the Deutsche Nationalbibliografie;
detailed bibliographic data is available in the Internet at <http://dnb.ddb.de>

ISBN 978-3-7643-7741-0 Birkhäuser Verlag, Basel – Boston – Berlin

© 2007 Birkhäuser Verlag, P.O. Box 133, CH-4010 Basel, Switzerland,
Part of Springer Science+Business Media
Printed on acid-free paper produced from chlorine-free pulp. TCF ∞

ISBN 10 3-7643-7741-0   ISBN 3-7643-7742-9 (eBook)
ISBN 13 978-3-7643-7741-0   ISBN 978-3-7643-7742-7
                                             www.birkhauser.ch
9 8 7 6 5 4 3 2 1   www.birkhauser.ch

Contents

Foreword

This volume collects the contributions of a Conference held in June 2005, at the laboratoire Paul Painlevé (UMR CNRS 8524) in Lille, France. The meeting was intended to review hot topics and future trends in fluid dynamics, with the objective to foster exchanges of various viewpoints (e.g. theoretical, and numerical) on the addressed questions.

The content of the volume can be split into three categories:

– A first set of contributions is devoted to the description of the connection between different models of fluid dynamics. An important part of these papers relies on the discussion of the modeling issues, the identification of the relevant dimensionless coefficients, and on the physical interpretation of the models. Then, making rigorous the connection between the different levels of modeling and justifying the validity of some simplifications become an asymptotics question, and an overview of the modern tools of mathematical analysis that allow to treat such kind of problems is given. The paper by L. Saint-Raymond describes how the equations of fluid dynamics (Euler or Navier-Stokes equations) can be derived from the Boltzamnn equation. In the latter, the gas is described statistically through the evolution of the particles density function assuming that particles are subject to a binary collision dynamics. This derivation of the fluid dynamics equations is actually part of the program addressed at the International Congress of Mathematics, Paris, 1900, and it is often referred to as the 6th Hilbert's problem. T. Alazard offers a survey on the justification of reduced models through asymptotics arguments: starting from the full gas dynamics equation, he shows how other classical models can be obtained: low Mach number limit, rotating fluid models, geostrophic equations and relaxation. D. Bresch, B. Desjardins and G. Métivier focus on the derivation of the shallow water equation from the Navier-Stokes equations, a problem that is motivated by application to atmospheric flows. A common feature of these problems is the role played by dissipation mechanisms and the crucial use of "entropy dissipation" properties that lead to a prori estimates. These a priori estimates have to be combined with delicate compactness arguments.

– A second set of contributions considers the question of the stability of particular structures in fluid mechanics equations. P. Secchi investigates the stability of contact discontinuities in the multi-dimensional Euler equations. The argument is based on energy methods and relies on a suitable adaptation of the Nash-Moser theorem. Y. Trakhinin investigates the stability of vortex-sheets for the compressible and incompressible Magnetohydrodynamics equations, which is motivated by applications in astrophysics. C. Mascia and F. Rousset show the nonlinear stability of stationary steady states for the viscous Saint Venant equation, obtaining sharp decay estimates. The argument is based on a fine analysis of the properties of the Green function associated to the linearized operator. M. Hillairet considers the interaction of a viscous fluid with rigid bodies. The difficulties stem from the possible collisions between the scatterers, where the model itself might become questionable. However a careful analysis allows to exhibit particular situations in which such collisions are prohibited.

– The third set of contributions is concerned with numerical issues; it is indeed a crucial challenge to design numerical schemes that are able to capture the complex features generated by fluid flows. A. Rousseau, R. Temam and J. Tribbia are interested in simulations of the primitive equations describing the evolution of the ocean. The idea consists in mixing a spectral method in the vertical direction with a finite difference approach for treating the time variable and the horizontal directions. This strategy is coupled to the design of transparent boundary condition in order to reduce the size of the computational domain. C.-H. Bruneau develops a multigrid algorithm in order to simulate the evolution of 2D incompressible Navier-Stokes equations at high Reynolds number. The flows in the lid-driven cavity and in a channel behind arrays of cylinders are considered. Then, the presence of an attractor in 2D flows and the turbulent cascades are exhibited on numerical grounds. R. Donat and P. Mulet describe shock capturing numerical methods designed for systems of hyperbolic conservation laws. The abilities of the scheme are illustrated with applications to supersonic relativistic jet and multifluid flows. The work of F. Coquel and C. Chalons studies the approximation of nonconservative hyperbolic system, and is motivated by various turbulence models. The authors give a precise definition of admissible shock solutions, and derive an approximate Riemann solver by means of a prediction-correction method. V. Dolean and F. Nataf introduce a domain decomposition method for the 2D compressible Euler equations. Optimal parameters of general interface conditions are estimated for a classical Schwarz method and a new preconditioning one. A convergence rate analysis is given in the subsonic case, in order to obtain the best convergence rate with respect to the number of subdomains. The contribution of A.E. Lovgren, Y. Maday and E. Ronquist is an overview on the so-called reduced basis element method. This method is specifically designated to solve PDEs in complex geometries, the whole domain being split into smaller elementary domains; it couples reduced basis method, domain decomposition strategies and a posteriori estimates. Developments of such a method are particularly motivated by the simulation of air flow in the lung.

Lille, May 2006   The editors
                                                           C. CALGARO
                                                           J.-F. COULOMBEL
                                                           T. GOUDON

Analysis and Simulation of Fluid Dynamics

Advances in Mathematical Fluid Mechanics, 1–13

© 2006 Birkhäuser Verlag Basel/Switzerland

Some Recent Asymptotic Results in Fluid Mechanics

Thomas Alazard

Abstract. The general equations of fluid mechanics are the law of mass conservation, the Navier-Stokes equation, the law of energy conservation and the laws of thermodynamics. These equations are merely written in this generality. Instead, one often prefers simplified forms. To obtain reduced systems, the easiest route is to introduce dimensionless numbers which quantify the importance of various physical processes. Many recent works are devoted to the study of the classical solutions when such a dimensionless number goes to zero. A few results in this field are here reviewed.

1. Low Mach number limit

The target of the low Mach number limit is to justify some simplifications that are made when discussing the fluid dynamics of highly subsonic flows. For a fluid with density ρ, velocity v, pressure P, internal energy e and temperature T, the equations, written in a non-dimensional way, are

$$
\begin{cases}
\partial_t \rho + \operatorname{div}(\rho v) = 0, \\[4pt]
\partial_t(\rho v) + \operatorname{div}(\rho v \otimes v) + \dfrac{\nabla P}{\varepsilon^2} = \mu \operatorname{div} \tau, \\[4pt]
\partial_t(\rho e) + \operatorname{div}(\rho v e) + P \operatorname{div} v = \kappa \operatorname{div}(k \nabla T) + \varepsilon^2 \mu \tau \cdot Dv.
\end{cases}
\tag{1.1}
$$

The parameters $\mu \in [0,1]$ and $\kappa \in [0,1]$ are the inverses of the Reynolds and Péclet numbers, they measure the importance of viscosity and heat-conduction. The Mach number ε is the ratio of a characteristic velocity in the flow to the sound speed in the fluid.

Furthermore,

$$
\tau := 2\zeta Dv + \eta(\operatorname{div} v)I_d,
$$

with $2Dv := \nabla v + (\nabla v)^t$. By assumption, ζ, η, k are C^∞ functions of (P, T), satisfying $k > 0$, $\zeta > 0$ and $2\zeta + \eta > 0$.

It is also convenient to use the evolution equation for the entropy S,

$$\rho T(\partial_t S + v \cdot \nabla S) = \kappa \operatorname{div}(k \nabla T) + \varepsilon^2 \mu \tau \cdot Dv.$$

The nature of the low Mach number limit strongly depends on the size of the entropy variations. One can distinguish three cases: the almost isentropic regime where the entropy is constant except for perturbations of order of the Mach number; the almost adiabatic case where the entropy of each fluid particle is almost constant; the general non-adiabatic case. More precisely,

Almost isentropic: $\nabla S = O(\varepsilon)$,

Almost adiabatic: $\nabla S = O(1)$ and $\partial_t S + v \cdot \nabla S = O(\varepsilon)$,

Non-adiabatic: $\nabla S = O(1)$ and $\partial_t S + v \cdot \nabla S = O(1)$.

We claim that the fluid is asymptotically incompressible only in the almost isentropic or adiabatic cases. The heuristic argument is the following. When ε goes to 0, the pressure fluctuations converge to 0. Consequently, the limit entropy and density are functions of the temperature alone. Therefore, the evolution equation for the entropy and the continuity equation furnish us with two values for the convective derivative of the temperature. By equating both values, it is found that the limit divergence constraint is of the form $\operatorname{div} v = \kappa C_p \operatorname{div}(k \nabla T)$. Hence, $\operatorname{div} v = 0$ implies that $\kappa \nabla T = 0$, which means that the limit flow is adiabatic.

The previous heuristic shows that the incompressible limit is a special case of the low Mach number limit. Most of the known results, which we now briefly review, are devoted to the incompressible limit. The mathematical analysis of the incompressible limit began in the early Eighties with works of Ebin, Kreiss, Klainerman & Majda, Schochet, Isozaki, Ukai and others. The reader who wants some good introductions to this subject is referred to the surveys papers of Danchin [18], Gallagher [22] and Schochet [33].

For the isentropic equations, the analysis is well-developed even for solutions which are not regular. Indeed, the incompressible limit of the isentropic Navier-Stokes equations has been rigorously justified for weak solutions by Desjardins and Grenier [19], Lions and Masmoudi [28] (see also [20]). For viscous gases, global well-posedness in critical spaces was established by Danchin in [15], and the limit $\varepsilon \to 0$ was justified in the periodic case in [16], with the whole space case earlier achieved in [17]. The study of the incompressible limit of the isentropic equations is a very vast subject, and we should also mention, among many others, the very interesting results of Hoff [27] and Dutrifoy and Hmidi [21].

For the non-isentropic Euler equations ($\mu = 0 = \kappa$) with general initial data, Métivier and Schochet have proved some theorems [31, 32] that supersede a number of earlier results (a part of their study is extended in [2] to the boundary case). In particular, they have proved the existence of classical solutions on a time interval independent of ε. The key point is to prove uniform estimates for the acoustic components. This is where the difference between almost isentropic and almost adiabatic enters. The reason is the following: the acoustics components are propagated by a wave equation whose coefficients are functions of the density, hence of the entropy. In the isentropic case, these coefficients are almost constant (the derivatives are of order of $O(\varepsilon)$). By contrast, in the non isentropic case, these coefficients are variable. This changes the nature of the linearized equations. The main obstacle is precisely that the linearized equations are not uniformly well-posed in Sobolev spaces. Hence, it is notable that one can prove that the solutions exist and are uniformly bounded for a time independent of ε.

In [3] we start a rigorous analysis of the corresponding problem for the general case in which the combined effects of large temperature variations and thermal conduction are taken into account. As first anticipated in [29], this case yields some new problems concerning the nonlinear coupling of the equations. To see this, it is convenient to rewrite System (1.1) in terms of (P, v, T). For the sake of simplicity, consider perfect gases such that $P = R\rho T$ and $e = C_V T$ for two positive constants R and C_V. By setting $\gamma = 1 + R/C_V$, we find that

$$
\begin{cases}
\partial_t P + v \cdot \nabla P + \gamma P \operatorname{div} v = (\gamma - 1)\kappa \operatorname{div}(k\nabla T) + (\gamma - 1)\varepsilon \mathcal{Q}, \\[2mm]
\rho(\partial_t v + v \cdot \nabla v) + \dfrac{\nabla P}{\varepsilon^2} = \mu \operatorname{div} \sigma, \\[2mm]
\rho C_V(\partial_t T + v \cdot \nabla T) + P \operatorname{div} v = \kappa \operatorname{div}(k\nabla T) + \varepsilon \mathcal{Q},
\end{cases}
\tag{1.2}
$$

where $\mathcal{Q} := \varepsilon \mu \sigma \cdot Dv$.

Equations (1.2) are supplemented with initial data:

$$
P_{|t=0} = P_0, \quad v_{|t=0} = v_0 \quad \text{and} \quad T_{|t=0} = T_0.
\tag{1.3}
$$

Furthermore, to balance the acoustic components, we consider as usual the case where $P = \text{Cte} + O(\varepsilon)$ so that $\partial_t v = O(\varepsilon^{-1})$.

Theorem 1.1 (From [3]). *Let $d \geqslant 1$ and $\mathbb{D}$ denote either the whole space $\mathbb{R}^d$ or the torus $\mathbb{T}^d$. Consider an integer $s > 1 + d/2$. For all positive $\underline{P}$, $\underline{T}$ and M_0, there is a positive time T and a positive M such that for all $a := (\varepsilon, \mu, \kappa) \in (0, 1] \times [0, 1] \times [0, 1]$ and all initial data (P_0^a, v_0^a, T_0^a) such that P_0^a and T_0^a take positive values and such that*

$$
\varepsilon^{-1} \left\| P_0^a - \underline{P} \right\|_{H^{s+1}(\mathbb{D})} + \left\| v_0^a \right\|_{H^{s+1}(\mathbb{D})} + \left\| T_0^a - \underline{T} \right\|_{H^{s+1}(\mathbb{D})} \leqslant M_0,
$$

the Cauchy problem for (1.2)–(1.3) has a unique solution (P^a, v^a, T^a) such that $(P^a - \underline{P}, v^a, T^a - \underline{T}) \in C^0([0, T]; H^{s+1}(\mathbb{D}))$ and

$$
\sup_{a \in A} \sup_{t \in [0, T]} \left\{ \varepsilon^{-1} \left\| P^a(t) - \underline{P} \right\|_{H^s(\mathbb{D})} + \left\| v^a(t) \right\|_{H^s(\mathbb{D})} + \left\| T^a(t) - \underline{T} \right\|_{H^s(\mathbb{D})} \right\} \leqslant M.
$$

With Theorem 1.1 in hands, by using a Theorem of [31] about the decay to zero of the local energy, we can prove that the penalized terms converge strongly in $L^2(0, T; H^{s'}_{\text{loc}}(\mathbb{R}^d))$ ($s' < s$) to zero. Strong compactness in time for the slow variables is clear from the Ascoli's Theorem, so that one can rigorously justify, at least in the whole space case, the convergence towards the limit system. We mention that the analysis extend to general gases and the combustion equations setting (see [4]). Let us also mention that the research of numerical algorithms valid for all flow speeds is a very active field, see the paper by H. Guillard in this volume.

The previous problems have been discussed in many other places (see [1, 18, 22, 33]), so we will not enter into the details. Instead, let us mention some related results for the lifespan of the classical solutions of the non-isentropic Euler equations

$$\begin{cases} \partial_t \rho + v \cdot \nabla \rho + \rho \operatorname{div} v = 0, \\ \rho(\partial_t v + v \cdot \nabla v) + \nabla P = 0, \\ \partial_t S + v \cdot \nabla S = 0, \end{cases} \tag{1.4}$$

where ρ, P and S are related by an equation of state (say, $\rho^\gamma = P e^{-S}$).

The link between the study of the lifespan of smooth solutions and the low Mach number limit is via the above mentioned result of Métivier and Schochet:

Theorem 1.2 (From [31]). *With notations as above, there exists a positive time T such that for all $\varepsilon \in (0,1]$, all $k \in [0,1]$ and all initial data*

$$P_0^\varepsilon(x) := \underline{P} e^{\varepsilon p(x/\varepsilon^k)}, \quad v_0^\varepsilon := \varepsilon v(x/\varepsilon^k), \quad S_0^\varepsilon := S_0(x/\varepsilon^k),$$

with $(p_0, v_0, S_0) \in H^s$, the Cauchy problem for (1.4) has a unique solution $(P^\varepsilon, v^\varepsilon, S^\varepsilon)$ in $C^0([0, T/\varepsilon^{1-k}]; H^s(\mathbb{D}))$ with norm uniformly bounded.

Note that this result follows from Theorem 1.1 (applied with $\mu = \kappa = 0$) by scaling space and time variables. The case $k = 1$ corresponds to high-frequency solutions with large-amplitude polarized on the entropy (see [12, 13] for recent results concerning large-amplitude high-frequency solutions). The case $k = 0$ corresponds to small pressure and initial velocity. If in addition we require $S_0^\varepsilon = O(\varepsilon)$ then it is elementary proved that the solutions exists at least for a time of order of $1/\varepsilon$ (one of the main concerns in [31, 3] was precisely to prove that this remains true for large entropy variations). However, when $S_0^\varepsilon = O(\varepsilon)$, one has much more precise results in some specific situations. For 2D isentropic axymetric flows, as established by Alinhac in [5], one can give an exact asymptotic result for the lifespan of smooth solutions. Recently, Godin has obtained a similar sharp result for 3D spherically symmetric flows with entropy variations (of order of ε) and initial data which are small perturbation of a constant state. More precisely, consider a perfect gas ($P = \rho^\gamma e^S$, $\gamma > 1$) and initial data of the form

$$\rho(0, x) = \underline{\rho} + \varepsilon \rho^0(r) + \varepsilon^2 \rho^1(r, \varepsilon), \quad r := |x|,$$
$$v(0, x) = \varepsilon v^0(r)\omega + \varepsilon^2 v^1(r, \varepsilon)\omega, \quad \omega := x/|x|, \tag{1.5}$$
$$S(0, x) = \underline{S} + \varepsilon S^0(r) + \varepsilon^2 S^1(r, \varepsilon).$$

Theorem 1.3 (Godin [25]). *Assume that $\rho(x, 0) > 0$ and that ρ^j, $v^j\omega$ and S^j are C^∞ functions of x and supported in a fixed compact set $|x| \leq M$. Assume in addition that, for all $x \in \mathbb{R}^3$, $\alpha \in \mathbb{N}^3$ and $\varepsilon \in (0,1]$ small enough,*

$$\left|\partial_x^\alpha \rho^1\right| + \left|\partial_x^\alpha(v^1\omega)\right| + \left|\partial_x^\alpha S^1\right| \leq C_\alpha.$$

Then, the lifespan T_ε (the supremum of all positive times of existence of classical solutions of (1.4) with initial data given by (1.5)) satisfies

$$\lim_{\varepsilon \to 0} \varepsilon \ln T_\varepsilon = \tau^*,$$

where $\tau^* \in (0, +\infty]$ is finite if and only if $|v^0| + |\rho^0/\underline{\rho} + S^0/\gamma| \not\equiv 0$.

2. Rotating fluids equations

2.1. Strong magnetic field of variable amplitude

We first present a result of Gallagher and Saint-Raymond whose aim is to study the behavior of a fluid submitted to a strong external magnetic field. This problem has been widely studied when the magnetic field is constant. In this case, explicit computations allow us to use the filtering techniques. For variable magnetic field, the analysis is more involved. To begin with, the first task is to understand the coupling between the penalization term and the convective derivative. Hence, it is natural to begin by considering a system where the mean velocity can be decoupled from the rest of the system, such is Euler system of pressureless gas dynamics which has the form of a Burger-type system. Consider the equations

$$\begin{cases} \partial_t \varphi + v \cdot \nabla \varphi + \operatorname{div} v = 0, & \varphi(0, x) = \rho_0(x), \\ \partial_t v + v \cdot \nabla v = \varepsilon^{-1} L v, & v(0, x) = v_0(x), \end{cases} \tag{2.1}$$

where the unknown $(\varphi, v) = (\varphi, v_1, v_2)$ is a function of the variable $(t, x) \in \mathbb{R} \times \mathbb{R}^2$ with values in $\mathbb{R} \times \mathbb{R}^2$ and the penalization operator is given by:

$$L v := b(x) v^\perp = \begin{pmatrix} -b(x) v_2 \\ b(x) v_1 \end{pmatrix},$$

where $b = b(x)$ is a given C^∞ real-valued function which is bounded together with all its derivatives.

If $\nabla b = 0$ then L is skew-symmetric so that the perturbations terms do not appear in the linear and nonlinear energy estimates. In this case, if $(\varphi_0, v_0) \in H^s(\mathbb{R}^2)$ with $s > 2$, the general theory of quasi-linear hyperbolic equations implies that there exists $T > 0$ such that for all $\varepsilon > 0$ the Cauchy problem for (2.1) has a unique solution in $C^0([0, T]; H^{s-1} \times H^s(\mathbb{R}^2))$.

When $\nabla b \neq 0$, L is no longer skew-symmetric and there are no uniform estimates in Sobolev spaces. To make this precise, introduce the group of oscillations $e^{\tau L}$ defined by

$$(e^{\tau L} v)(t, x) = \cos(b(x)\tau)v(t, x) - \sin(b(x)\tau)v^\perp(t, x).$$

The filtering technique consists of computing the equation satisfies by the unknown $e^{(-t/\varepsilon)L} v$. It does not work here – even for the linear equation $\partial_t w_\varepsilon = \varepsilon^{-1} L w_\varepsilon$ – since for all $s > 0$ the family $\left\| e^{-(t/\varepsilon)L} \right\|_{H^s \to H^s}$ is unbounded. Moreover, it does not work also because the interaction with the convective derivative implies that one cannot expect $\partial_t e^{(-t/\varepsilon)L} v$ to be bounded in any space of distributions. Yet, in [23], it is proved that the classical solutions of (2.1) exist for a time independent of ε.

Theorem 2.1 (Gallagher & Saint-Raymond [23]**).** *Assume* $b(x) \geqslant \underline{b} > 0$ *and that* (φ_0, v_0) *belongs to* $X^s := W^{s-1,\infty} \times W^{s,\infty}$ *for some* $s > 2$. *Then there exists* $T > 0$ *such that for all* $\varepsilon \in (0, 1]$, *there exists a unique* $(\varphi_\varepsilon, v_\varepsilon) \in L^\infty(0, T; X^s)$ *solution of* (2.1).

As already mentioned, energy estimates are useless for inhomogeneous functions b. The alternative consists of rewriting system (2.1) by means of characteristics. It is then possible to give a precise statement to the idea that the rotation dictates the evolution in that it has a drastic influence over the transport by the velocity (for pressureless gases). More precisely, by using the non-stationary phase theorem, one can prove that, for all $x \in \mathbb{R}^2$ and $\varepsilon > 0$, the fluid particle trajectory starting from x at time 0, denoted by $X^\varepsilon(t, x)$ stays in a ball of size $O(\varepsilon)$ around x. Furthermore, it is possible to give an explicit approximation of $X^\varepsilon(t, x) - x$ at any order with respect to ε. Not only does this allows to prove Theorem 2.1, but also to determine the global asymptotic (for v_ε and φ_ε) of the Euler system of pressureless gases. We refer the reader to the original paper [23] (see also [24]) for precised statements.

2.2. Inhomogeneous rotation

The general governing equations of an incompressible rotating fluid read

$$\partial_t v + v \cdot \nabla v + \nabla p + \frac{1}{\varepsilon} B(t, x) \times v - \Delta_h v - \nu \partial_3^2 v = 0,$$

$$\operatorname{div} v = 0, \tag{2.2}$$

where the space variable is $x = (x_h, x_3) = (x_1, x_2, x_3) \in \mathbb{R}^3$ and the unknown is the velocity v with values in $\mathbb{R}^3$. The subscript h indicates derivation in the horizontal variables, so that $\Delta_h = \partial_1^2 + \partial_2^2$. The constant coefficient of viscosity $\nu \geqslant 0$ can be zero.

The given function B is assumed to be C^∞ and bounded, together with all its derivatives. The analysis is well-developed in the case where $B(t, x)$ is a constant. As in the previous problems, in the general variable coefficients case, we cannot obtain the non-linear energy estimates by differentiating the equations nor by localizing in the frequency space by means of Littlewood-Paley operators. Yet, if B depends only upon a group of space variables then one has still uniform a priori estimates for some derivatives (∂_2 and ∂_3 if $B = B(x^1)$). Indeed, as recently shown by Majdoub and Paicu in [30], based on the incompressible condition and on the particular form of the non-linear term, one can hope to find a closed set of estimates which, in the end, allows to prove a uniform existence result. This strategy requires working in spaces of anisotropic regularity $H^{\sigma,s}$ (σ and s are given real numbers), which are equipped with the norm

$$\|u\|_{H^{\sigma,s}}^2 := \int_{\mathbb{R}^3} \langle \xi_h \rangle^\sigma \langle \xi_3 \rangle^s |\widehat{u}(\xi)|^2 \, d\xi, \qquad \xi_h = (\xi_1, \xi_2).$$

Theorem 2.2 (Majdoub & Paicu [30]**).** *Suppose that* $B = B(t, x_1)$. *Given* $s > 1/2$ *and an incompressible initial data* $v_0 \in H^{0,s}(\mathbb{R}^3)$, *there exists* $T > 0$ *such that* for all $\varepsilon > 0$ the Cauchy problem for (2.2) has a unique solution v_ε such that $v_\varepsilon \in C^0([0,T]; H^{0,s}(\mathbb{R}^3)) \cap L^2(0,T; H^{1,s}(\mathbb{R}^3))$.

As already mentioned, the spaces $H^{\sigma,s}$ of co-normal regularity are well adapted to the problem. In [30], the previous result is complemented by a couple of Theorems showing that the full H^s regularity is propagated by the flow, one of which is the following:

Theorem 2.3 (From [30]). *Suppose that $B = B(t, x_1)$. Given $s > 1/2$ and an incompressible initial data $v_0 \in H^s(\mathbb{R}^3)$, there exists $T > 0$ such that for all $\varepsilon > 0$ the Cauchy problem for (2.2) has a unique solution v_ε such that $v_\varepsilon \in C^0([0,T]; H^s(\mathbb{R}^3))$ and $\nabla_h v_\varepsilon \in L^2(0,T; H^s(\mathbb{R}^3))$.*

To deduce this result from Theorem 2.2, the main technical ingredient consists of controlling the lifespan of the solutions by a norm that does not involve more than one spatial derivative with respect to x_h.

Theorem 2.4 (From [30]). *Suppose that $B = 0$. Given $s > 1/2$ and an incompressible initial data $v_0 \in H^s(\mathbb{R}^3)$, there exists $T > 0$ such that the Cauchy problem for (2.2) has a unique solution such that $v \in C^0([0,T]; H^s)$ and $\nabla_h v \in L^2(0,T; H^s(\mathbb{R}^3))$. Moreover, either the lifespan T_* is $+\infty$ or*

$$\lim_{t \to T_*} \int_0^t \|\nabla_h v(t')\|_{L^\infty_{x_3}(L^2_{x_h})}^2 \left(1 + \|v(t')\|_{L^\infty_{x_3}(L^2_{x_h})}^2\right) dt' = +\infty,$$

where $\|f\|_{L^\infty_{x_3}(L^2_{x_h})} := \sup_{x_3 \in \mathbb{R}} \left(\int_{\mathbb{R}^2} |f(x_1, x_2, x_3)|^2 \, dx_1 \, dx_2\right)^{1/2}.$

3. Planetary geostrophic equations

The primitive equations, written in a non-dimensional way, are:

$$\begin{cases} \partial_t v + u \cdot \nabla v + f \dfrac{v^\perp}{\varepsilon} + \dfrac{\nabla_x p}{\varepsilon} - \nu_x \Delta_x v - \nu_z \partial_z^2 v = 0, \\[2mm] \dfrac{\partial_z p}{\varepsilon} = \dfrac{T}{\mathrm{Fr}^2}, \\[2mm] \mathrm{div}_x v + \partial_z w = 0, \\[2mm] \partial_t T + u \cdot \nabla T - \nu'_x \Delta_x T - \nu'_z \partial_z^2 T + \dfrac{\mathrm{Bu}}{\varepsilon} w = Q, \end{cases} \tag{3.1}$$

where the unknowns are the velocity $u = (v, w) = (v_1, v_2, w) \in \mathbb{R}^3$, the pressure p and the temperature T.

The space variables $x \in \mathbb{R}^2$ and $z \in \mathbb{R}$ denote the horizontal and vertical coordinates. The Coriolis parameter f, Rosby number ε, Froude number Fr and Burger number Bu are assumed positive and independent of the space variables. So are the viscous parameters ν and ν'. Finally, $Q = Q(t, x)$ is a given source term.

The relative importance of competing physical process dictates the nature of the limit system. For instance, the convergence towards the quasi-geostrophic equations corresponds to the case where

$$\varepsilon \ll 1, \quad \mathrm{Fr} = \varepsilon^{1/2}, \quad \mathrm{Bu} = 1.$$

This problem has developed since the pioneering works of Lions, Temam and Wang (for some recent advances see the results of Charve [8, 9] as well as [7] and the book of Chemin, Desjardins, Gallagher and Grenier [10]). Below, we concentrate on the thermohaline circulation problem, which corresponds to

$$\varepsilon \ll 1, \quad \mathrm{Fr} = \varepsilon^{1/2}, \quad \mathrm{Bu} = \varepsilon, \tag{3.2}$$

so that, formally, the limit system reads

$$\begin{cases} fv^{\perp} + \nabla_x p = 0, \quad \partial_z p = T, \quad \mathrm{div}_x\, v + \partial_z w = 0, \\ \partial_t T + u \cdot \nabla T - \nu'_x \Delta_x - \nu'_z \partial_z^2 T + w = Q, \end{cases} \tag{3.3}$$

which is known as the planetary geostrophic equations and widely used to model large scale ocean dynamics.

The derivation of the planetary geostrophic equations has been just justified by Bresch, Gérard-Varet and Grenier. They proved that, for small initial data, the solutions exist on a time independent of ε and converge to the solution of the limit system. Before giving a precise statement, let us explain the main new phenomena. When $\mathrm{Bu} = O(\varepsilon)$, the singular operator is not skew-symmetric and hence the large terms do no disappear from the energy estimate. To give this statement a precise result, assume that $f = 1$, $\nu = \nu' = 1$ and that the domain is the 3-dimensional torus. With these assumptions, the energy identity reads

$$\frac{1}{2}\frac{d}{dt}\int_{\mathbb{T}^3}|\lambda v|^2 + |T|^2 + \int_{\mathbb{T}^3}|\lambda\nabla_{x,z}v|^2 + |\nabla_{x,z}T|^2 = \int_{\mathbb{T}^3}QT,$$

where

$$\lambda := \sqrt{\frac{\mathrm{Fr}^2\,\mathrm{Bu}}{\varepsilon}}.$$

The problem presents itself: if ε, $\mathcal{F}$ and Bu are as in (3.2), then $\lambda = \sqrt{\varepsilon}$. Hence, the energy estimate provides no information concerning the speed v. Yet, v appears in the nonlinear term at order $O(1)$ (in the convection term $v \cdot \nabla T$). To get around this difficulty is rather tricky and is partly based upon a reduction process introduced by Cheverry in [11]: one derives a normal form of the system, in which the viscous part ensures the control of the non-symmetric terms.

Let us give a precise statement for fluid initially at rest

$$v_{|t=0} = 0, \quad T_{|t=0} = 0, \tag{3.4}$$

driven by a large thermohaline source term $Q \in C^\infty([0,T] \times \mathbb{T}^3)$. Introduce the precised limit system

$$\begin{cases} v^\perp + \nabla_x p = 0, \quad \partial_z p = T, \quad \text{div}_x\, v + \partial_z w = 0, \\ \partial_t T + u \cdot \nabla T - \nu \Delta_{x,z} T + w = Q, \\ (\partial_t - \Delta_x) \Delta_x \bar{p} + \text{curl}_x\, \overline{(v \cdot \nabla_x v)} = 0, \end{cases} \tag{3.5}$$

where, given $g = g(x,z)$, $\bar{g}$ is the average $\bar{g}(x) := \int_\mathbb{T} g(x,z)\, dz$.

Theorem 3.1 (Bresch, Gérard-Varet & Grenier [6]). *For all $\tau > 0$ and $Q \in C^\infty([0,\tau] \times \mathbb{T}^3)$ there exists $\tau_* > 0$ and a unique solution $(u^0, T^0) \in C^\infty([0,\tau_*] \times \mathbb{T}^3)$ of (3.4)–(3.5).*

Assume in addition that $Q \in C_0^\infty((0,\tau) \times \mathbb{T}^3)$. Then, for all $s \in \mathbb{N}$, there exists $\varepsilon_0 >$ and $\delta > 0$ such that if $\varepsilon \leq \varepsilon_0$ and $\sup_{t,x,z} |\partial_z T^0(t,x,z)| \leq \delta$, then there exists a unique solution $(u^\varepsilon, T^\varepsilon) \in C^\infty([0,\tau_*] \times \mathbb{T}^3)$ of (3.4)–(3.1). Moreover, $(u^\varepsilon, T^\varepsilon)$ converges strongly to (u^0, T^0) in $L^\infty(0,\tau_*; H^s(\mathbb{T}^3))$.

4. Relaxation limits

Consider the isothermal Euler equations with a large relaxation term:

$$\begin{cases} \partial_t \rho + v \cdot \nabla \rho + \rho \, \text{div}\, v = 0, \\ \rho(\partial_t v + v \cdot \nabla v) + \nabla \rho + \lambda \rho v = 0. \end{cases} \tag{4.1}$$

The study of the relaxation limit concerns the asymptotic limit where the large parameter λ goes to $+\infty$. The analysis contains at least two parts: a uniform boundedness result for appropriate λ-dependent weighted norms, and a convergence result on a large time scale. Moreover, with a damping term in hands, a classical issue is to prove that (as for the Navier-Stokes equations), the classical solutions with small enough initial data are global in time. The following result, obtained by Coulombel and Goudon, solves both questions and complements previous works of Natalini and Hanouzet [26], Sideris, Thomases and Wang [34] and Yong [35].

Theorem 4.1 (Coulombel & Goudon [14]). *Use the notation $m := \rho v$ and let $\underline{\rho}$ be a positive constant. For all $d \geq 1$ and $\mathbb{N} \ni s > d/2 + 1$ there exist two constants δ and C such that for all $\lambda \in [1, +\infty)$ and all initial data (ρ_0, m_0) satisfying*

$$c_0 := \left\| \rho_0 - \underline{\rho} \right\|_{H^s(\mathbb{R}^d)}^2 + \left\| m_0 \right\|_{H^s(\mathbb{R}^d)}^2 \leq \delta,$$

there exists a unique global in time solution $(\rho_\lambda, v_\lambda)$ to (4.1) such that

$$\sup_{t \geq 0} \left\{ \left\| \rho_\lambda(t) - \underline{\rho} \right\|_{H^s(\mathbb{R}^d)}^2 + \left\| m_\lambda(t) \right\|_{H^s(\mathbb{R}^d)}^2 \right\} + \lambda \int_0^{+\infty} \left\| m_\lambda(t) \right\|_{H^s(\mathbb{R}^d)}^2 \, dt \leq C c_0.$$

To prove that the solutions are global in time, the idea is that, for small enough solutions, the third order non-linear terms appearing in the energy estimates are bounded by the positive quadratic term coming from the damping. Yet, the matrix multiplying the damping term is degenerate in that one merely controls the velocity, which is not enough to close the estimates. To be precise, it is convenient to use the entropic variables. Set

$$U := (\sigma, v) \quad \text{with} \quad \sigma := \log(\rho/\underline{\rho}) - \frac{1}{2} |v|^2.$$

Then, one can rewrite the system (4.1) under the form

$$A_0(v) \partial_t U + \sum_{1 \leqslant j \leqslant d} A_j(v) \partial_j U + \lambda \begin{pmatrix} 0 \\ v \end{pmatrix} = 0, \tag{4.2}$$

where each A_α is a matrix-valued function smooth in its arguments and furthermore symmetric; moreover A_0 is definite positive. An important feature of (4.2) is that the matrices A_α depend only on v. After a first round of estimates this yields

$$n_T^2 := \sup_{t \in [0,T]} \|(\sigma, v)(t)\|_{H^s(\mathbb{R}^d)}^2 + \lambda \int_0^T \|v(t)\|_{H^s(\mathbb{R}^d)}^2 \, dt \leqslant C n_0^2 + C N_T^3, \tag{4.3}$$

where C depends only on $\|(U, \nabla U)\|_{L^\infty((0,T) \times \mathbb{R}^d)}$ and

$$N_T^2 := n_T^2 + \frac{1}{\lambda} \int_0^T \|\nabla \sigma(t)\|_{H^{s-1}(\mathbb{R}^d)}^2 \, dt.$$

To establish (4.3), one has to distinguish two different steps. First, by differentiating System (4.2) and using the special structure of the equations, one can express all the source terms as products of three terms estimated by N_T, which yields the desired estimate for the H^{s-1}-norm of ∇U. The estimate of the L^2 estimate also requires a careful analysis. In [14] this is done by a kinetic-type analysis.

To obtain a closed set of estimates it next remains to estimate

$$(\sqrt{\lambda})^{-1} \|\nabla \sigma\|_{L^2(0,T;H^{s-1})}.$$

To do so, the desired stability condition used in [14, 35] comes from what is known as the Kawashima–Shizuta condition. This key structural assumption means that the matrices A_α satisfy the following property: introduce the matrix-valued symbol

$$K(\xi) := |\xi|^{-1} \begin{pmatrix} 0 & \xi^T \\ -\xi & 0 \end{pmatrix},$$

then $K(\xi) A_0(0)$ is skew-symmetric and

$$K(\xi) \sum_{1 \leqslant j \leqslant d} \xi_j A_j(0) = |\xi|^{-1} \begin{pmatrix} |\xi|^2 & 0 \\ 0 & -\xi \otimes \xi \end{pmatrix}.$$

We now consider the limit of solutions of (4.1) as λ goes to $+\infty$. Since m_λ converges to zero, the equation $\partial_t \rho_\lambda + \operatorname{div} m_\lambda = 0$ implies that the limit equation for the density is only $\partial_t \rho = 0$. By contrast, the following result shows that the asymptotic limit is nontrivial in a large time scale. It is found that, after the appropriate rescaling, the momentum goes to zero and the density converges towards the solution of the heat equation.

Theorem 4.2. *Using the same notations as in Theorem* 4.1, *set*

$$\varrho_\lambda(t,x) := \rho_\lambda(\lambda t, x), \qquad V_\lambda := v_\lambda(\lambda t, x).$$

Then, $\varrho_\lambda - \rho$ *is uniformly bounded in* $C^0(\mathbb{R}_+; H^s)$ *and* $\lambda\varrho_\lambda V_\lambda$ *is uniformly bounded in* $L^2(\mathbb{R}_+; \dot{H}^s)$. *Moreover, for all* $T > 0$ *and all* $s' < s$, *the family* $\{\varrho_\lambda\}$ *converges in* $C^0([0,T]; H^{s'}_{loc}(\mathbb{R}^d))$ *to the unique solution of the heat equation* $\partial_t\varrho - \Delta\varrho = 0$ *with initial data* $\varrho(0) = \rho_0$.

Note that one can observe the asymptotic behavior on a large time scale since the previous uniform stability result was devoted to global in time solutions.

References

[1] T. ALAZARD – "Alentours de la limite incompressible", in *Séminaire: Équations aux Dérivées Partielles*, 2004–2005, École Polytech., 2005, Exp. No. XXIV.

[2] T. ALAZARD – "Incompressible limit of the nonisentropic Euler equations with solid wall boundary conditions", *Adv. in Differential Equations* **10** (2005), p. 19–44.

[3] T. ALAZARD – "Low Mach number of the full Navier–Stokes equations", *Arch. Ration. Mech. Anal.* (2005), in press.

[4] T. ALAZARD – "Low Mach number flows, and combustion", preprint.

[5] S. ALINHAC – "Temps de vie des solutions régulières des équations d'Euler compressibles axisymétriques en dimension deux", *Invent. Math.* **111** (1993), no. 3, p. 627–670.

[6] D. BRESCH, D. GÉRARD-VARET & E. GRENIER – "Derivation of the planetary geostrophic equations", preprint ENS, 2005.

[7] F. CHARVE – "Etude de phénomènes dispersifs en mécanique des fluides géophysiques", Thèse de doctorat de l'école Polytechnique, 2004.

[8] F. CHARVE – "Global well-posedness and asymptotics for a geophysical fluid system", *Comm. Partial Differential Equations* **29** (2004), no. 11-12, p. 1919–1940.

[9] F. CHARVE – "Convergence of weak solutions for the primitive system of the quasigeostrophic equations", *Asymptot. Anal.* **42** (2005), no. 3-4, p. 173–209.

[10] J.-Y. CHEMIN, B. DESJARDINS, I. GALLAGHER & E. GRENIER – "Basis of mathematical geophysics", to appear, 2005.

[11] C. CHEVERRY – "Propagation of oscillations in real vanishing viscosity limit", *Comm. Math. Phys.* **247** (2004), no. 3, p. 655–695.

[12] C. CHEVERRY, O. GUÈS & G. MÉTIVIER – "Oscillations fortes sur un champ linéairement dégénéré", *Ann. Sci. École Norm. Sup.* (4) **36** (2003), no. 5, p. 691–745.

[13] C. CHEVERRY, O. GUÈS & G. MÉTIVIER – "Large-amplitude high-frequency waves for quasilinear hyperbolic systems", *Adv. Differential Equations* **9** (2004), no. 7-8, p. 829–890.

[14] J.-F. COULOMBEL & T. GOUDON – "The strong relaxation limit of multidimensional isothermal euler equations", to appear in Trans. Am. Soc., 2005.

[15] R. DANCHIN – "Global existence in critical spaces for flows of compressible viscous and heat-conductive gases", *Arch. Ration. Mech. Anal.* **160** (2001), no. 1, p. 1–39.

[16] R. DANCHIN – "Zero Mach number limit for compressible flows with periodic boundary conditions", *Amer. J. Math.* **124** (2002), no. 6, p. 1153–1219.

[17] R. DANCHIN – "Zero Mach number limit in critical spaces for compressible Navier-Stokes equations", *Ann. Sci. École Norm. Sup.* (4) **35** (2002), no. 1, p. 27–75.

[18] R. DANCHIN – "Low mach number limit for viscous compressible flows", in *special issue vol. 39 No. 3 (May-June 2005)*, M2AN Math. Model. Numer. Anal., 2005.

[19] B. DESJARDINS & E. GRENIER – "Low Mach number limit of viscous compressible flows in the whole space", *R. Soc. Lond. Proc. Ser. A Math. Phys. Eng. Sci.* **455** (1999), no. 1986, p. 2271–2279.

[20] B. DESJARDINS, E. GRENIER, P.-L. LIONS & N. MASMOUDI – "Incompressible limit for solutions of the isentropic Navier-Stokes equations with Dirichlet boundary conditions", *J. Math. Pures Appl.* (9) **78** (1999), no. 5, p. 461–471.

[21] A. DUTRIFOY & T. HMIDI – "The incompressible limit of solutions of the two-dimensional compressible Euler system with degenerating initial data", *Comm. Pure Appl. Math.* **57** (2004), no. 9, p. 1159–1177.

[22] I. GALLAGHER – "Résultats récents sur la limite incompressible", *Séminaire Bourbaki 2003–2004* Exp. No. **926**.

[23] I. GALLAGHER & L. SAINT-RAYMOND – "On pressureless gases driven by a strong inhomogeneous magnetic field", *SIAM J. Math. Anal.* **36** (2005), no. 4, p. 1159–1176.

[24] I. GALLAGHER & L. SAINT-RAYMOND – "Résultats asymptotiques pour des fluides en rotation inhomogène, in *Séminaire: Équations aux Dérivées Partielles, 2003–2004,* École Polytech., 2004, Exp. No. III.

[25] P. GODIN – "The lifespan of a class of smooth spherically symmetric solutions of the compressible euler equations with variable entropy in three space dimensions several spaces variables", *Arch. Ration. Mech. Anal.* **177** (2005), no. 3, p. 479–511.

[26] B. HANOUZET & R. NATALINI – "Global existence of smooth solutions for partially dissipative hyperbolic systems with a convex entropy", *Arch. Ration. Mech. Anal.* **169** (2003), no. 2, p. 89–117.

[27] D. HOFF – "The zero-Mach limit of compressible flows", *Comm. Math. Phys.* **192** (1998), no. 3, p. 543–554.

[28] P.-L. LIONS & N. MASMOUDI – "Incompressible limit for a viscous compressible fluid", *J. Math. Pures Appl.* (9) **77** (1998), no. 6, p. 585–627.

[29] A. MAJDA – *Compressible fluid flow and systems of conservation laws in several space variables*, Applied Mathematical Sciences, vol. 53, Springer-Verlag, New York, 1984.

[30] M. MAJDOUB & M. PAICU – "Uniform local existence for inhomogeneous rotating fluid equations", in preparation, 2005.

[31] G. MÉTIVIER & S. SCHOCHET – "The incompressible limit of the non-isentropic Euler equations", *Arch. Ration. Mech. Anal.* **158** (2001), no. 1, p. 61–90.

[32] G. MÉTIVIER & S. SCHOCHET – "Averaging theorems for conservative systems and the weakly compressible Euler equations", *J. Differential Equations* **187** (2003), no. 1, p. 106–183.

[33] S. SCHOCHET – "The mathematical theory of low mach numbers flows", in *special issue vol. 39 No. 3 (May-June* 2005), M2AN Math. Model. Numer. Anal., 2005.

[34] T.C. SIDERIS, B. THOMASES & D. WANG – "Long time behavior of solutions to the 3D compressible Euler equations with damping", *Comm. Partial Differential Equations* **28** (2003), no. 3-4, p. 795–816.

[35] W.-A. YONG – "Entropy and global existence for hyperbolic balance laws", *Arch. Ration. Mech. Anal.* **172** (2004), no. 2, p. 247–266.

Thomas Alazard
MAB Université de Bordeaux I
F-33405 Talence Cedex, France
e-mail: `thomas.alazard@math.u-bordeaux1.fr`

Analysis and Simulation of Fluid Dynamics

Advances in Mathematical Fluid Mechanics, 15–31

Recent Mathematical Results and Open Problems about Shallow Water Equations

Didier Bresch, Benoît Desjardins and Guy Métivier

Abstract. The purpose of this work is to present recent mathematical results about the shallow water model. We will also mention related open problems of high mathematical interest.

Mathematics Subject Classification (2000). 35Q30.

Keywords. Viscous and inviscid flow, shallow-water model, lake equations, quasi-geostrophic equations, weak and strong solutions, degenerate viscosities.

1. Introduction

The shallow water equations are the simplest form of equation of motion that can be used to describe the horizontal structure of the atmosphere. They describe the evolution of an incompressible fluid in response to gravitational and rotational accelerations. The solutions of the shallow water equations represent many types of motion, including Rossby waves and inertia-gravity waves. The aim of this paper is to discuss several problems related to the general form of shallow water equations written as follows, on a two-dimensional domain Ω:

$$\text{St } \partial_t h + \text{div}(hu) = 0, \tag{1}$$

$$\text{St } \partial_t(hu) + \text{div}(hu \otimes u) = -h \frac{\nabla(h-b)}{\text{Fr}^2} - h \frac{fu^\perp}{\text{Ro}}$$

$$+ \frac{2}{\text{Re}} \text{div}(hD(u)) + \frac{2}{\text{Re}} \nabla(h\text{div}u) - \frac{1}{\text{We}} h\nabla\Delta h + \mathcal{D}, \tag{2}$$

where $h \in \mathbf{R}$ denotes the height of the free surface, u is the mean horizontal velocity and f a function depending on the latitude y in order to describe the variability of the Coriolis force. In addition, b is a given function and describes the topography of the bottom level of the fluid, $b > 0$ on Ω and b can vanish on $\partial\Omega$; moreover, $b \leq h$. In the second equation, $\perp$ denotes the direct rotation of angle $\frac{\pi}{2}$, namely $G^\perp = (-G_2, G_1)$ when $G = (G_1, G_2)$. The numbers St, Ro,

Re, Fr and We respectively denote the Strouhal number, the Rossby number, the Reynolds number, the Froude number and the Weber number. Some damping terms $\mathcal{D}$ coming from friction may be added or not. We will discuss this point in Section 3. Remark that in 1871, Saint–Venant wrote in a note[1] a system which describes the flow of a river and corresponds to the inviscid shallow water model written here.

Equations (1) and (2), respectively express the conservation of height, momentum energy. System (1)–(2) is supplemented with initial conditions

$$h|_{t=0} = h_0, \qquad (hu)|_{t=0} = q_0, \tag{3}$$

The functions h_0, q_0, are assumed to satisfy

$$h_0 \geq 0 \quad \text{a.e. on } \Omega, \quad \text{and} \quad \frac{|q_0|^2}{h_0} = 0 \quad \text{a.e. on} \quad \{x \in \Omega \,/\, h_0(x) = 0\}. \tag{4}$$

The formal derivation of such system from the Navier–Stokes equations with free boundary may be found in [37]. Validity of such approximation will be discussed in the last section.

2. Conservation of potential vorticity

The inviscid case. For the two-dimensional inviscid shallow water equations, the vorticity ω is a scalar, defined as

$$\omega = \partial_x u_2 - \partial_y u_1$$

To derive an evolution equation for ω, we take the curl of the momentum equation divided by h, and use the identity

$$\text{curl}\left(\text{St }\partial_t u + u \cdot \nabla u + f \frac{u^\perp}{\text{Ro}}\right) = \text{St }\partial_t \omega + u \cdot \nabla \omega + \left(\omega + \frac{f}{\text{Ro}}\right)\text{div}u + u_2 \frac{\partial_y f}{\text{Ro}}.$$

We can eliminate the term involving div$u$, multiplying the vorticity equation by h and the conservation of mass equation by $\omega + f/\text{Ro}$ and write the difference of the two obtained identities. This yields

$$h\left(\text{St }\partial_t + u \cdot \nabla\right)\left(\omega + \frac{f}{\text{Ro}}\right) - \left(\omega + \frac{f}{\text{Ro}}\right)\left(\text{St }\partial_t h + u \cdot \nabla h\right) = 0.$$

Therefore :

$$\left(\text{St }\partial_t + u \cdot \nabla\right)\frac{\omega_R}{h} = 0, \quad \omega_R := \omega + \frac{f}{\text{Ro}}. \tag{5}$$

The quantity ω_R is called the *relative vorticity*. The equation means that the ratio of ω_R and the effective depth h is conserved along the particle trajectories of the flow. This constraint is called the *potential vorticity*. It provides a powerful constraint in large scale motions of the atmosphere. If $\omega + f/\text{Ro}$ is constant initially, the only way that $\omega + f/\text{Ro}$ remains constant at a latter time is if h itself is constant. In the general, the conservation of potential vorticity tells us that if h increases

---

[1]de Saint Venant, *Théorie du mouvement non permanent des eaux, avec application aux crues des rivières et è l'introduction des marées dans leur lit*, C.R.Ac. Sc. Paris, LXXIII, 1871, 147–154

then $\omega + f/\mathrm{Ro}$ must increase, and conversely, if h decreases, then $\omega + f/\mathrm{Ro}$ must decrease.

The viscous case. If h is assumed to be constant equal to 1, $\mathcal{D} = 0$ and $f(y) = \beta y$ with β a constant, a system on the relative vorticity $\omega + f$ and stream function is easily written. It reads

$$\mathrm{St}\, \partial_t(\omega + f) + u \cdot \nabla(\omega + f) - \frac{1}{\mathrm{Re}}\Delta(\omega + f) = 0, \quad -\Delta\Psi = \omega, \quad u = \nabla^\perp\Psi.$$

When h is not constant, even when it does not depend on time, there are no simple equation for ω_R nor ω_P, analogous to (5). The term

$$\mathrm{curl}\Big(\frac{1}{h}(2\nu\mathrm{div}(hD(u)) + \nabla(2h\mathrm{div}u))\Big)$$

generates cross terms between derivatives of h and u. As a consequence, even in the case where $h = b(x)$ is a given function depending only on x, we are not able to get global existence and uniqueness of a strong solution for this viscous model if we allow b to vanish on the shore. The lack of equation for the vorticity also induces problems in the proof of convergence from the viscous case to the inviscid case when the Reynolds number tends to infinity. This is as an interesting open problem. Other interesting questions are to analyze the domain of validity of viscous shallow-water equations in bounded domain, and to determine the relevant boundary conditions when h vanishes on the shore.

3. LERAY solutions

3.1. A new mathematical entropy

We introduce in this section a new mathematical entropy that has been recently discovered in [11] for more general compressible flows namely Korteweg system. The domain Ω which is considered is either the periodic domain or the whole space. In [11] the capillarity coefficient σ is taken equal to 0, more precisely the Weber number is equal to ∞, and the considered system may be written in the following form

$$\partial_t\rho + \mathrm{div}(\rho u) = 0, \tag{6}$$

$$\partial_t(\rho u) + \mathrm{div}(\rho u \otimes u) = -\nabla p(\rho) + \mathrm{div}(2\mu(\rho)D(u)) + \nabla(\lambda(\rho)\mathrm{div}u). \tag{7}$$

The energy identity for such system reads

$$\frac{1}{2}\frac{d}{dt}\int_\Omega (\rho|u|^2 + 2\pi(\rho)) + \int_\Omega \Big(2\mu(\rho)|D(u)|^2 + \lambda(\rho)|\mathrm{div}u|^2\Big) = 0,$$

where π denotes the internal energy per unit volume given by

$$\pi(\rho) = \rho\int_{\bar{\rho}}^{\rho} \frac{p(s)}{s^2}ds$$

for some constant reference density $\bar{\rho}$.

In [11], a new mathematical entropy has been discovered that helps to get a great variety of mathematical results about compressible flows with density dependent viscosities. Namely if $\lambda(\rho) = 2(\mu'(\rho)\rho - \mu(\rho))$, then the following equality holds

$$\frac{1}{2}\frac{d}{dt}\int_\Omega \left(\rho|u + 2\nabla\varphi(\rho)|^2 + 2\pi(\rho)\right) + \int_\Omega \frac{p'(\rho)\mu'(\rho)}{\rho}|\nabla\rho|^2 + \int_\Omega 2\mu(\rho)|A(u)|^2 = 0$$

where $A(u) = (\nabla u - {}^t\nabla u)/2$ and $\rho\varphi'(\rho) = \mu'(\rho)$. For the reader's convenience, we recall here the alternate proof given in [9].

Proof of the new mathematical entropy identity. Using the mass equation we know that for all smooth function $\xi(\cdot)$

$$\partial_t \nabla\xi(\rho) + (u \cdot \nabla)\nabla\xi(\rho) + \sum_i \nabla u_i \partial_i \xi(\rho) + \nabla(\rho\xi'(\rho)\mathrm{div} u) = 0.$$

Thus, using once more the mass equation, we see that $v = \nabla\xi(\rho)$ satisfies:

$$\partial_t(\rho v) + \mathrm{div}(\rho u \otimes v) + \rho\sum_i \nabla u_i \partial_i \xi(\rho) + \rho\nabla(\rho\xi'(\rho)\mathrm{div} u) = 0$$

which gives, using the momentum equation on u,

$$\partial_t(\rho(u + v)) + \mathrm{div}(\rho u \otimes (u + v)) - \mathrm{div}(2\mu(\rho)D(u)) - \nabla(\lambda(\rho)\mathrm{div} u)$$
$$+\nabla p(\rho) + \rho\sum_i \nabla u_i \partial_i \xi(\rho) + \rho\nabla(\rho\xi'(\rho)\mathrm{div} u) = 0.$$

Next, we write the diffusion term as follows:

$$-\mathrm{div}(2\mu(\rho)D(u)) = -\mathrm{div}(2\mu A(u)) - 2\sum_i \nabla u_i \partial_i \mu - 2\nabla(\mu\mathrm{div} u) + 2\mathrm{div} u\nabla\mu$$

where $A(u) = (\nabla u - {}^t\nabla u)/2$. Therefore, the equation for $u + v$ reads

$$\partial_t(\rho(u + v)) + \mathrm{div}(\rho u \otimes (u + v)) - 2\mathrm{div}(\mu(\rho)A(u)) - 2\mu'(\rho)\sum_i \nabla u_i \partial_i \rho$$

$$-2\nabla(\mu(\rho)\mathrm{div} u) + 2\mu'(\rho)\nabla\rho\mathrm{div} u + \nabla p(\rho) - \nabla(\lambda(\rho)\mathrm{div} u) + \rho\xi'(\rho)\sum_i \nabla u_i \partial_i \rho$$

$$+\nabla(\rho^2\xi'(\rho)\mathrm{div} u) - \rho\xi'(\rho)\nabla\rho\mathrm{div} u = 0.$$

This equation can be simplified under the form

$$\partial_t(\rho(u + v)) + \mathrm{div}(\rho u \otimes (u + v)) - 2\mathrm{div}(\mu(\rho)A(u)) + \nabla p(\rho)$$
$$+\nabla((\rho^2\xi'(\rho) - 2\mu - \lambda)\mathrm{div} u) + (\rho\xi'(\rho) - 2\mu'(\rho))\sum_i \nabla u_i \partial_i \rho$$

$$+(2\mu'(\rho) - \rho\xi'(\rho))\nabla\rho\,\mathrm{div} u = 0 \qquad (8)$$

If we choose ξ such that $2\mu'(\rho) = \xi'(\rho)\rho$, then $\lambda = \xi'(\rho)\rho^2 - 2\mu$ and the last three terms cancel, which implies:

$$\partial_t(\rho(u + v)) + div(\rho u \otimes (u + v)) - 2\mathrm{div}(\mu(\rho)A(u)) + \nabla p(\rho) = 0.$$

Multiplying this equation by $(u+v)$ and the mass equation by $|u+v|^2/2$ and adding we easily get the new mathematical entropy equality. We just have to observe that

$$\int_\Omega \text{div}(\mu(\rho)A(u)) \cdot v = 0,$$

since v is a gradient. The term involving $\nabla p(\rho)$ gives

$$\int_\Omega \nabla p(\rho) \cdot (u+v) = \int_\Omega \rho \nabla \pi'(\rho) \cdot u + \int_\Omega \frac{p'(\rho)\mu'(\rho)}{\rho} |\nabla \rho|^2$$

$$= \frac{d}{dt} \int_\Omega \pi(\rho) + \int_\Omega \frac{p'(\rho)\mu'(\rho)}{\rho} |\nabla \rho|^2 \qquad (9)$$

where $\pi(\rho) = \rho \int_{\overline{\rho}}^\rho p(s)/s^2 ds$ with $\overline{\rho}$ a constant reference density.

We remark that the mathematical entropy estimate gives an extra information on ρ, namely

$$\mu'(\rho)\nabla \rho/\sqrt{\rho} \in L^\infty(0,T;L^2(\Omega))$$

assuming $\mu'(\rho_0)\nabla \rho_0/\sqrt{\rho_0} \in L^2(\Omega)$ initially.

Such an information is crucial in the analysis of viscous compressible flows with density dependent viscosities and it should help in various cases. Recent applications have been given. For instance, in [38], A. MELLET and A. VASSEUR study the stability of isentropic compressible Navier–Stokes equations with barotropic pressure law $p(\rho) = a\rho^\gamma$ with $\gamma > 1$ in dimension $d = 1, 2$ and 3. The diffusive term is assumed under the form $-\text{div}(2\mu(\rho)\nabla u) - \nabla(\lambda(\rho)\text{div}u)$.

Another interesting result concerns the full compressible Navier–Stokes equations. Existence of global weak solutions has been obtained in [10] assuming perfect gas law except close to vacuum where cold pressure is used to get compactness on the temperature. This completes the result recently obtained by E. FEIREISL in [22] where the temperature satisfies an inequality instead of an equality in the sense of distributions and where the assumptions on the equation of state prevent from considering perfect polytropic gas laws even away from vacuum. Note that the two above results do not involve the same kind of viscosity laws. Something has therefore to be understood.

We remark that the relation between λ and μ

$$\lambda(\rho) = 2(\mu'(\rho)\rho - \mu(\rho)) \qquad (10)$$

and the conditions

$$\mu(\rho) \geq c, \qquad \lambda(\rho) + 2\mu(\rho)/d \geq 0$$

cannot be fulfilled simultaneously. Indeed the viscosity μ has to vanish in vacuum. In [27], D. HOFF and D. SERRE prove that only viscosities vanishing with the density may prevent failure of continuous dependence on initial data for the Navier–Stokes equations of compressible flow. Our relation (10) between λ and

μ push to consider such degenerate viscosities. We refer the reader to [38] for interesting mathematical comments on the relation imposed between λ and μ.

Remark on the viscous shallow water equations. We stress that, in the equation written at the beginning of the paper, the viscous term does not satisfy the conditions imposed above. Indeed, we have $\mu = \rho$ but $\lambda \not\equiv 0$. As a consequence, the usual viscous shallow water equations are far from being solved for weak solutions except in 1D where $-2\text{div}(hD(u)) - 2\nabla(h\text{div}u) = -4\partial_x(h\partial_x u)$

Equations with capillarity. Capillarity can be taken into account, as it is observed in [11], introducing a term of the form $-\rho\nabla(G'(\rho)\Delta G(\rho))/$We at the right hand side of the momentum equation. It turns out that it suffices to choose a viscosity equal to $\mu(\rho) = G(\rho)$. This gives the mathematical entropy

$$\frac{1}{2}\frac{d}{dt}\int_\Omega \left(\rho|u + 2\nabla\varphi(\rho)|^2 + 2\pi(\rho) + \frac{1}{\text{We}}|\nabla\mu(\rho)|^2\right) + \int_\Omega \frac{p'(\rho)\mu'(\rho)}{\rho}|\nabla\rho|^2$$
$$+\frac{1}{\text{We}}\sigma\int_\Omega \mu'(\rho)|\Delta\mu(\rho)|^2 + \int_\Omega 2\mu(\rho)|A(u)|^2 = 0 \qquad (11)$$

where $A(u) = (\nabla u - {}^t\nabla u)/2$ and $\rho\varphi'(\rho) = \mu'(\rho)$. Applications of such mathematical entropy will be provided in [14], to approximations of hydrodynamics. It may also be used to construct suitably smooth sequences of approximate solutions to non capillary models corresponding to the limit of infinite Weber number.

The temperature dependent case. For the full compressible Navier–Stokes equation, the mathematical energy equality reads

$$\frac{d}{dt}\int_\Omega(\frac{1}{2}\rho|u|^2 + \pi(\rho)) + \int_\Omega 2\mu(\rho)|D(u)|^2 + \int_\Omega \lambda(\rho)|\text{div}u|^2 = -\int_\Omega \nabla p \cdot u,$$

and the new mathematical entropy equality reads

$$\frac{d}{dt}\int_\Omega(\frac{1}{2}\rho|u + v|^2 + \pi(\rho)) + \int_\Omega 2\mu(\rho)|A(u)|^2 = -\int_\Omega \nabla p \cdot (u + v)$$

where $A(u) = (\nabla u - {}^t\nabla u)/2$. To close the estimate, it is sufficient to prove that the extra terms $\int_\Omega \nabla p \cdot u$ and $\int_\Omega \nabla p \cdot (u + v)$ can be controlled by the left-hand side. This has been done in [10], in which the global existence of weak solutions is proved for the full compressible Navier–Stokes equations with perfect polytropic pressure laws modified by a cold pressure component (at zero temperature) to control density close to vacuum. The proof uses the new mathematical entropy to control the density far from vacuum.

3.2. Weak solutions with drag terms

When drag terms such as $\mathcal{D} = r_0 u + r_1 h|u|u$ are present in the Saint–Venant model, with diffusion term equal to $-2\text{div}(hD(u))$, the existence of global weak solution is proved in [12] without capillarity term ($r_1 > 0$ and $r_0 > 0$). The authors mention that in 1D, r_1 may be taken equal to 0.

Of course such drag terms is helpful from a mathematical view point since it gives the extra information we need on u to prove stability results. Anyway, we will see in the last section that the derivation of the shallow water equations is far from being understood. We will give an example where drag forces appear from the underlying 3D incompressible free boundary Navier–Stokes equations with Dirichlet conditions at the bottom and an example where no drag terms appear when other boundary conditions are considered, namely no slip conditions.

3.3. Dropping drag terms

A. MELLET and A. VASSEUR, in [38], show how to ignore the drag terms for the shallow water equations without capillary terms assuming the diffusive term under the form $-\mathrm{div}(h\nabla u)$. This is useful for the mathematical analysis. Indeed, controlling $\nabla\sqrt{\rho}$ implies further information on u assuming this regularity initially. More precisely, multiplying the momentum equation by $|u|^{\delta}u$ with δ small enough and assuming that $\sqrt{\rho_0}u_0 \in L^{2+\delta}(\Omega)$, one proves that $\sqrt{\rho}u \in L^{\infty}(0,T;L^{2+\delta}(\Omega))$ using the estimates given by the new mathematical entropy. This new piece of information is sufficient to pass to the limit in the nonlinear term without the help of drag terms. Note that such technic hold for general μ and λ satisfying the relation needed for the new mathematical entropy.

Important remark. Note that the new mathematical entropy has been used to get stability results for approximate solutions for various models: compressible barotropic Navier–Stokes equations, compressible Navier-Stokes equations with thermal conductivity, Korteweg equations... In [12], [10] for instance, the authors claim that such approximate solutions may be built to get global existence of weak solutions, i.e., to build actual sequences of suitably smooth approximate solutions. The details will be given in [8] for readers convenience.

4. Strong solutions

There are several results about the local existence of strong solutions for the shallow water equations written as follows

$$\partial_t h + \mathrm{div}(hu) = 0, \tag{12}$$

$$\partial_t(hu) + \mathrm{div}(hu \otimes u) - \nu\mathrm{div}(h\nabla u) + h\nabla h = 0. \tag{13}$$

Following the energy method of A. MATSUMURA and T. NISHIDA, it is natural to show the global (in time) existence of classical solutions to the dissipative shallow water equations on a different domain. The external force field and the initial data being assumed to be small in a suitable space. Such result has been proved in [30], [17]. In [45], a global existence and uniqueness theorem of strong solutions for the initial-value problem for the viscous shallow water equations is established for small initial data and no forcing. Polynomial L^2 and decay rates are established and the solution is shown to be classical for $t > 0$.

More recently, in [47], is studied the Cauchy problem for viscous shallow water equations. The authors work in the Sobolev spaces of index $s > 2$ to obtain local solutions for any initial data, and global solutions for small initial data. The proof is based on Littlewood-Paley decomposition of solutions. The result reads

Theorem 4.1. *Let $s > 0$, u_0, $h_0 - \overline{h_0} \in H^{s+2}(R^2)$, $\|h_0 - \overline{h_0}\|_{H^{s+2}} \ll \overline{h_0}$. Then there exists a positive time T, a unique solution (u, h) of Cauchy problem (12)–(13) such that*

$$u, h - \overline{h_0} \in L^\infty(0, T; H^{2+s}(R^2)), \qquad \nabla u \in L^2(0, T; H^{2+s}(R^2)).$$

Furthermore, there exists a constant c such that if

$$\|h_0 - \overline{h_0}\|_{H^{s+2}(R^2)} + \|u_0\|_{H^{s+2}(R^2)} \le c$$

then we can choose $T = +\infty$.

Nothing has been done so far, using for instance the new mathematical entropy, to obtain better results such as local existence of strong solution with initial data including vacuum. Remark that such a situation is important from a physical point of view: the dam break situation.

5. Other viscous terms in the literature

In [34], different diffusive terms are proposed, namely $-2\nu \mathrm{div}(hD(u))$, or $-\nu h\Delta u$ or else $-\nu\Delta(hu)$. The reader is referred to [3] for the study of such diffusive terms in the low Reynolds approximation. Let us give some comments around the last two propositions.

Diffusive term equal to $-\nu h\Delta$. It is shown by P. GENT in [23], that this form, which is frequently used for the viscous adiabatic shallow-water equations, is energetically inconsistent compared to the primitive equations. An energetically consistent form of the shallow-water equations is then given and justified in terms of isopycnal coordinates. This energetical form is exactly the form considered in the present paper. Examples are given of the energetically inconsistent shallow-water equations used in low-order dynamical systems and simplified coupled models of tropical air-sea interaction and the El Nino-Southern Oscillation phenomena.

From a mathematical point of view, this inconsistency can be easily identified. Let us multiply, formally, the momentum equation by u/h. We get

$$\frac{d}{dt}\int_\Omega |u|^2 + \nu \int_\Omega |\nabla u|^2 + \frac{1}{2}\int_\Omega u \cdot \nabla |u|^2 + \int_\Omega u \cdot \nabla h = 0.$$

The last term may be written

$$\int_\Omega u \cdot \nabla h = -\int_\Omega \log h \, \mathrm{div}(hu) = \int_\Omega \log h \, \partial_t h = \frac{d}{dt}\int_\Omega (h \log h - h).$$

The third term may be estimated as follows

$$\left|\int_\Omega |u|^2 \mathrm{div} u\right| \le \|u\|_{L^4(\Omega)}^2 \|u\|_{H^1(\Omega)} \le c\|u\|_{H^1(\Omega)}^2 \|u\|_{L^2(\Omega)}.$$

Thus if we want to get dissipation, we have look at solutions such that

$$\|u\|_{L^2(\Omega)} < \nu/c,$$

that means sufficiently small velocity solutions. Such an analysis has been performed in [43], looking at solutions of the above system such that u bounded in $L^2(0,T;H^1(\Omega))$, $h \in L^\infty(0,T;L^1(\Omega))$, $h \log h \in L^\infty(0,T;L^1(\Omega))$. The mathematical difficulty is to prove the convergence of sequences of approximate solutions $(\rho_n u_n)_{n \in N}$ in the stability proof. For this, the author uses strongly the fact that h_n and $h_n \log h_n$ are uniformly bounded in $L^\infty(0,T;L^1(\Omega))$. Using Dunford-Pettis theorem and Trudinger-Moser inequality, he can conclude by a compactness argument.

Remark that using this diffusive term, several papers have been devoted to the low Reynolds approximation namely the following system

$$\partial_t h + \operatorname{div}(hu) = 0, \quad \partial_t u - \Delta u + \nabla h^\alpha = f,$$

with $\alpha \geq 1$. The most recent one, see [33], deals with blow up phenomena, if the initial density contains vacuum, using uniform bounds with respect to time of the L^∞ norm on the density. It would be interesting to understand what happens without simplification by h in the momentum equation allowing the height to vanish. Remark that such simplification has been also done in [36] to study high rotating and low Froude number limit of inviscid shallow water equations.

Diffusive term equal to $-\nu\Delta(hu)$. Using such diffusion term also gives energetical inconsistency. Only results about the existence of global weak solutions for small data have been obtained.

6. The quasi-geostrophic model

The well-known quasigeostrophic system for zero Rossby and Froude number flows has been used extensively in oceanography and meteorology for modelling and forecasting mid-latitude oceanic and atmospheric circulation. Deriving this system requires a (singular) perturbation expansion. The quasigeostrophic equation expresses conservation of the zero-order potential vorticity of the flow. In 2D, that means neglecting the stratification, such a model can be derived from the shallow water equations. We just have to choose $Fr = Ro = O(\varepsilon)$ and $b = 1 + \varepsilon\eta_b$ with $\eta_b = O(1)$ and let ε go to 0. It yields the following two-dimensional system

$$St\,\partial_t u + \operatorname{div}(u \otimes (u + \eta_b)) = -\mathcal{D} - \nabla p + \frac{1}{Re}\Delta u - St\partial_t\Delta^{-1}u, \quad (14)$$

$$\operatorname{div} u = 0, \quad (15)$$

We also note the presence of the new term $\partial_t\Delta^{-1}u$ coming from the free surface, which cannot be derived from the standard rotating Navier–Stokes equations in a fixed domain. To the knowledge of the authors, there exists only one mathematical paper concerning the derivation of such models from the viscous Shallow water equations. It concerns global weak solutions, see [12].

7. The lake equations

The so-called lake equations arise as the shallow water limit of the rigid lid equations – three-dimensional Euler or Navier–Stokes equations with a rigid lid upper boundary condition – in a horizontal basin with bottom topography. It could also be seen as a low Froude approximation of the shallow water equation assuming $b = O(1)$. Neglecting the Coriolis force, it gives

$$\text{St}\,\partial_t(bu) + \text{div}(bu \otimes u)$$
$$= -b\nabla p + \frac{2}{\text{Re}}\text{div}(bD(u)) + \frac{2}{\text{Re}}\nabla(b\text{div}u), \qquad (16)$$
$$\text{div}(bu) = 0, \qquad (17)$$

7.1. The viscous case

This model has been formally derived and studied by D. LEVERMORE and B. SAMARTINO, see [32]. In this paper, assuming that the depth b is positive and smooth up to the boundary of Ω, they prove that the system is globally well posed. Note that such model has been used to simulate the currents in Lake Erie. Concerning the boundary conditions, they consider the no-slip boundary conditions

$$u \cdot n = 0, \quad \tau \cdot ((\nabla u + {}^t\nabla u)n) = -\beta u \cdot \tau.$$

where n and τ are the outward unit normal and unit tangent to the boundary. They assume $\beta \geq \kappa$ where κ is the curvature of the boundary.

7.2. The inviscid case

Let us assume formally that $\text{Re} \to \infty$ to model an inviscid flow and consider a two-dimensional bounded domain Ω. We get the following system

$$\text{St}\,\partial_t(bu) + \text{div}(bu \otimes u) + b\nabla p = 0, \qquad (18)$$
$$\text{div}(bu) = 0, \quad u \cdot n|_{\partial\Omega} = 0. \qquad (19)$$

That means a generalization of the standard two-dimensional incompressible Euler equation obtained if $b \equiv 1$.

A strictly positive bottom function. YUDOVICH's method may be applied using the fact that the relative vorticity ω/b is transported by the flow. More precisely, the inviscid lake equation may be written using a stream-relative vorticity formulation under the following form

$$\text{St}\,\partial_t\left(\frac{\omega}{b}\right) + u \cdot \nabla\left(\frac{\omega}{b}\right) = 0, \qquad (20)$$
$$-\text{div}(\nabla\Psi/b) = \omega, \quad \omega = \text{curl}, \quad \Psi|_{\partial\Omega} = 0. \qquad (21)$$

Assuming $b \geq c > 0$ smooth enough, L^p regularity on the stream function Ψ remains true, that is

$$\|\Psi\|_{W^{2,p}(\Omega)} \leq Cp\|\omega\|_{L^p(\Omega)}$$

where C does not depend on p. Such elliptic estimates with non-degenerate coefficients come from [1]-[2]. This result allows D. LEVERMORE, M. OLIVER, E. TITI, in [31], to conclude to global existence and uniqueness of strong solutions.

A degenerate bottom function. This is the case when b vanishes on the boundary (the shore). Suppose that φ is a function equivalent to the distance to the boundary, that is $\varphi \in C^\infty(\overline{\Omega})$, $\Omega = \{\varphi > 0\}$ and $\nabla\varphi \neq 0$ on $\partial\Omega$. Assuming that $b = \varphi^a$ where $a > 0$ and the problem on the stream function may be written under the form

$$-\varphi\Delta\Psi + a\nabla\varphi \cdot \nabla\Psi = \varphi^{a+1}\omega \text{ in } \Omega, \qquad \Psi|_{\partial\Omega} = 0. \tag{22}$$

The case $a = 1$, that is $b = \varphi$, is physically the most natural. This equation belongs to a well-known class of degenerate elliptic equations (see [25, 5]).

In [15], the authors prove that, for such degenerate equation, the L^p regularity estimate remains true. The analysis is based on Schauder's estimates solutions of (22) and on a careful analysis of the associated Green function which depends on the degenerate function b.

Using such estimate, the authors are able to follow the lines of the proof given by YUDOVITCH to get the existence and uniqueness of a global strong solution. Moreover as a corollary, they prove that the boundary condition $u \cdot n|_{\partial\Omega} = 0$ holds on the velocity.

Remark. To the authors' knowledge, there exists only one paper dealing with the derivation of the lake equations from the Euler equations, locally in time, see [42].

Remark. It would be very interesting to investigate the influence of b on properties which are known for the two-dimensional Euler equations, see [4].

An interesting open problem: Open sea boundary conditions. Let us mention here an open problem which has received a lot of attention from applied mathematicians, especially A. KAZHIKHOV. Consider the Euler equations, or more generally the inviscid lake equations, formulated in a two-dimensional bounded domain. When the boundary is of inflow type, all the velocity components are prescribed. Along an impervious boundary, flux vanishes everywhere. This is known as a slip condition. Along a boundary of outflow type, the flux normal to the boundary surface is prescribed for all points of the boundary.

A local existence and uniqueness theorem is proved in a class of smooth solutions by A. KAZHIKHOV in [28]. A global existence theorem is also proved under the assumption that the flow is almost uniform and initial data are small. But the question of global existence and uniqueness in the spirit of V. YUDOVITCH's results remains open, see [48]. In this paper, the author shows how the boundary conditions can be augmented in this more general case to obtain a properly posed problem. Under the additional condition $\text{curl } v|_{S^-} = \pi(x,t)$, where $\pi(x,t)$ is –

modulo some necessary restrictions – arbitrary, the author shows the existence, in the two-dimensional case, of a unique solution for all time. The method is constructive, being based on successive approximations, and it brings out clearly the physical basis for the additional condition. To understand it in a better way, open sea boundary conditions could be helpful for shallow-water equations for instance with an application to strait of Gibraltar modelling.

8. Multi-level models

We now give an example where it could be important to write multi-level shallow water equations: it concerns the modelling of the dynamics of water in the Alboran sea and the strait of Gibraltar. In this sea, two layers of water can be distinguished: the surface Atlantic water penetrating into the Mediterranean through the strait of Gibraltar and the deeper, denser Mediterranean water flowing into the Atlantic. This observation shows that, if we want to use two-dimensional models to simulate such phenomena, we have to consider at least two layers models. The model which is usually used to study this phenomena considers sea water as composed of two immiscible layers of different densities. In this model, waves appear not only at the surface but also at the interface. It is assumed that for the phenomena under consideration, the wavelength is sufficiently large to make accurate the shallow water approximation in each layer. Therefore the resulting equations form a coupled system of shallow water equations. Concerning viscous bi-layer shallow water equations, to the authors knowledge, only papers with the diffusion $-h\Delta u$ in each layer has been studied, that is the energetically inconsistent one, see [40]. Nothing has been done with the structure studied in the mono-layer case in [12]–[38]. We also note that there are very few mathematical studies concerning the propagation of waves in multi-levels geophysical models.

Shallow water flow with non-constant density. Shallow water equations taking into account non-constant density of the material are subject to investigation. There exists only few results in this direction. They may be seen as perturbations of the known results for the standard shallow water equations. Let us for instance comment on the model studied recently in [26]. This model reads

$$\partial_t(\rho h) + \partial_x(\rho h u) - 0, \tag{23}$$
$$\partial_t(\rho h u) + \partial_x(\rho h u^2 + \tfrac{1}{2}\beta(x)\rho h^2) = \rho h g. \tag{24}$$

where $\rho = h^\alpha$, $\alpha \geq 0$ being a constant, $\beta = \beta(x)$ and $g = g(v,x)$ are given functions. Here h stands for height, ρ for density and v for velocity, and the whole models an avalanche down an inclined slope.

Moreover, the author considers the following particular choices for β and g: $\beta(x) = k\cos(\gamma(x))$, $g(v,x) = \sin(\gamma(x)) - \text{sign}(v)\cos(\gamma(x))\tan(\delta_F(x))$, where δ_F and γ are given functions. Here sign is the sign function, with $\text{sign}(0) = [-1,1]$.

After a change of unknown, this system can be rewritten in the form

$$\frac{\partial}{\partial t}u + \frac{\partial}{\partial x}F(u) \in \tilde{G}(u,x),$$

with essentially the same structure as the system of isentropic gas dynamics in one dimension of space (see for instance [35]), except for the fact that there is an inclusion instead of an equality. A precise (natural) definition of entropy solutions of such systems and a long-time existence theorem of such solutions, under the assumption that the initial height is bounded below by a positive constant, which corresponds to avoiding vacuum may be performed using well-known tools, without additional difficulties.

To propose and study better shallow water models taking into account the density variability of the material could be an interesting research area.

Remark. Using the mathematical entropy estimate, perhaps it could be possible to find a physical viscous and capillary approximation for the 1D Euler equation replacing the viscous mathematical approximation used in [35], namely

$$\partial_t\rho + \partial_x(\rho u) = \varepsilon\partial_x^2\rho, \tag{25}$$

$$\partial_t(\rho u) + \partial_x(\rho u^2) + \nabla_x p(\rho) = \varepsilon\partial_x^2(\rho u). \tag{26}$$

9. Derivation of shallow water equations

Recently D. COUTAND, S. SHKOLLER, see [20]–[21], have written two papers dedicated to the free surface incompressible Euler and Navier–Stokes equations with or without surface tension. We also mention two recent papers dedicated to the formal derivation of viscous shallow water equations from the Navier–Stokes equations with free surface, see [24] for 1D shallow water equations and see [37] for 2D shallow water equations.

It would be interesting to prove mathematically such formal derivations. We make here a few remarks concerning the hypothesis which have been used to derive formally these viscous shallow water equations with damping terms.

First hypothesis. The viscosity is of order ε, meaning that the viscosity is of the same order than the depth, and the asymptotic analysis is performed at order 1.

Second hypothesis. The boundary condition at the bottom for the Navier–Stokes equations is taken using wall laws. Namely the boundary conditions are of the form $(\sigma n)_{\text{tang}} = r_0 u$ on the bottom with $u \cdot n = 0$. These boundary conditions can lead to a drag term. (there is no such drag term if $r_0 = 0$).

Dirichlet boundary condition. Suppose that one starts with standard Dirichlet boundary conditions instead of a wall law which by itself is a modelled view of boundary layers near to the bottom. Then we get a linear drag term due to the parabolic profile of the velocity, see [46] and the quadratic term $\partial_x(hu^2)$ is replaced by $6\partial_x(hu^2)/5$.

To conclude, we stress that the rigorous derivation of applicable shallow water equations is far from being understood and a deep mathematical analysis is needed in this direction. Let us mention for the reader convenience some recent works made in that direction by [6], [7] in order to try to propose some generalization of shallow water equations which take into account order one variation in the slope of the bottom. We also mention recent works by J.-P. VILA, see [46], where a precise asymptotic is performed in the description of the velocity profile. Looking at the asymptotic with adherence condition on the bottom and free surface conditions on the surface, he proves that depending on the Ansatz for the horizontal velocity and for the viscosity coefficient, we can formally get various asymptotic inviscid models at the main order.

Acknowledgments

The first author would like to thank Professor ZhouPing XIN for his kind invitation to visit Hong-Kong in November 2005 under the financial support of the Institute of Mathematical Sciences, the Chinese University of Hong-Kong. Thanks also to the staff for their efficiency namely: Lily, Caris and Jason and to members or visitors in IMS for interesting discussions: Jing, Hai-Liang, Dong-Juan, and "tea for two" for instance. He is also supported by a Rhône-Alpes fellowship obtained in 2004 on problems dedicated to viscous shallow water equations.

References

[1] S. AGMON, A. DOUGLIS, L. NIRENBERG. Estimates near the boundary for solutions of elliptic partial differential equations I, *Comm. Pures and Appl. Math.*, 12, (1959), 623–727.

[2] S. AGMON, A. DOUGLIS, L. NIRENBERG. Estimates near the boundary for solutions of elliptic partial differential equations II, *Comm. Pures and Appl. Math.*, 17, (1964), 35–92.

[3] C. BERNARDI, O. PIRONNEAU. On the shallow water equations at low Reynolds number. *Commun. Partial Diff. Eqs.* 16, 59–104 (1991).

[4] A.L. BERTOZZI, A.J. MAJDA. *Vorticity and incompressible flows*. Cambridge University Press, (2001).

[5] P. BOLLEY, J. CAMUS, G. MÉTIVIER, Estimations de Schauder et régularité Hölderienne pour une classe de problèmes aux limites singuliers, *Comm. Partial Diff. Equ.* 11 (1986), 1135–1203.

[6] F. BOUCHUT, A. MANGENEY-CASTELNAU, B. PERTHAME, J.P. VILOTTE. A new model of Saint-Venant and Savage-Hutter type for gravity driven shallow water flows. *C.R. Acad. Sci. Paris*, série I, 336(6):531–536, (2003).

[7] F. BOUCHUT, M. WESTDICKENBERG. Gravity driven shallow water models for arbitrary topography. *Comm. in Math. Sci.*, 2(3):359–389, (2004).

[8] D. BRESCH, B. DESJARDINS. Numerical approximation of compressible fluid models with density dependent viscosity. In preparation (2005).

[9] D. BRESCH, B. DESJARDINS, J.-M. GHIDAGLIA. On bi-fluid compressible models. In preparation (2005).

[10] D. BRESCH, B. DESJARDINS. Existence globale de solutions pour les équations de Navier–Stokes compressibles complètes avec conduction thermique. *C. R. Acad. Sci.,* Paris, Section mathématiques. Submitted (2005).

[11] D. BRESCH, B. DESJARDINS. Some diffusive capillary models of Korteweg type. *C.R. Acad. Sciences,* Paris, Section Mécanique. Vol. **332** no. 11 (2004), pp. 881–886.

[12] D. BRESCH, B. DESJARDINS. Existence of global weak solutions for a 2D viscous shallow water equations and convergence to the quasi-geostrophic model. *Commun. Math. Phys.,* **238**, 1-2, (2003), pp. 211–223.

[13] D. BRESCH, M. GISCLON, C.K. LIN. An example of low Mach (Froude) number effects for compressible flows with nonconstant density (height) limit. *M2AN,* Vol. 39, N° 3, pp. 477–486, (2005).

[14] D. BRESCH, A. JUENGEL, H.-L. LI, Z.P. XIN. Effective viscosity and dispersion (capillarity) approximations to hydrodynamics. Forthcoming paper, (2005).

[15] D. BRESCH, G. MÉTIVIER. Global existence and uniqueness for the lake equations with vanishing topography: elliptic estimates for degenerate equations. To appear in *Nonlinearity,* (2005).

[16] D. BRESCH, B. DESJARDINS, C.K. LIN. On some compressible fluid models: Korteweg, lubrication and shallow water systems. *Comm. Partial Differential Equations,* **28**, 3–4, (2003), p. 1009–1037.

[17] A.T. BUI. Existence and uniqueness of a classical solution of an initial boundary value problem of the theory of shallow waters. *SIAM J. Math. Anal.* 12 (1981) 229–241.

[18] J.F. CHATELON, P. ORENGA. Some smoothness and uniqueness for a shallow water problem. *Adv. Diff. Eqs.,* 3, 1 (1998), 155–176.

[19] C. CHEVERRY. Propagation of oscillations in Real Vanishing Viscosity Limit, *Commun. Math. Phys.* 247, 655–695 (2004).

[20] D. COUTAND, S. SHKOLLER. Unique solvability of the free-boundary Navier-Stokes equations with surface tension, *Arch. Rat. Mech. Anal.* (2005).

[21] D. COUTAND, S. SHKOLLER. Well-posedness of the free-surface incompressible Euler equations with or without surface tension, Submitted (2005).

[22] E. FEIREISL. *Dynamics of viscous compressible fluids.* Oxford Science Publication, Oxford, (2004).

[23] P.R. GENT. The energetically consistent shallow water equations. *J. Atmos. Sci.,* 50, 1323–1325, (1993).

[24] J.F. GERBEAU, B. PERTHAME. Derivation of Viscous Saint-Venant System for Laminar Shallow Water; Numerical Validation, Discrete and Continuous Dynamical Systems, Ser. B, Vol. 1, Num. 1, 89–102, (2001).

[25] C. GOULAOUIC, N. SHIMAKURA, Régularité Hölderienne de certains problèmes aux limites dégénérés, *Ann. Scuola Norm. Sup. Pisa,* 10, (1983), 79–108.

[26] P. GWIAZDA. An existence result for a model of granular material with non-constant density. *Asymptotic Analysis,* **30**,

[27] D. HOFF, D. SERRE. The failure of continuous dependence on initial data for the Navier-Stokes equations of compressible flow. *SIAM J. Appl. Math.,* 51(4):887–898, (1991).

[28] A.V. KAZHIKHOV, Initial-boundary value problems for the Euler equations of an ideal incompressible fluid. *Moscow Univ. Math. Bull.* 46 (1991), no. 5, 10–14.

[29] A.V. KAZHIKHOV, A. VEIGANT. Global solutions of equations of potential fluids for small Reynolds number. *Diff. Eqs.*, 30, (1994), 935–947.

[30] P.E. KLOEDEN, Global existence of classical solutions in the dissipative shallow water equations. *SIAM J. Math. Anal.* 16 (1985), 301–315.

[31] D. LEVERMORE, M. OLIVER, E.S. TITI, Global well-posedness for models of shallow water in a basin with a varying bottom, *Indiana Univ. Math. J.* 45 (1996), 479–510.

[32] D. LEVERMORE, B. SAMMARTINO. A shallow water model in a basin with varying bottom topography and eddy viscosity, *Nonlinearity,* Vol. 14, n. 6, 1493–1515 (2001).

[33] J. LI, Z.P. XIN. Some Uniform Estimates and Blowup Behavior of Global Strong Solutions to the Stokes Approximation Equations for Two-Dimensional Compressible Flows. To appear in *J. Diff. Eqs.* (2005)

[34] P.-L. LIONS. *Mathematical topics in fluid dynamics, Vol. 2, Compressible models.* Oxford Science Publication, Oxford, (1998).

[35] P.-L. LIONS, B. PERTHAME, P. E. SOUGANIDIS, Existence of entropy solutions to isentropic gas dynamics System. *Comm. Pure Appl. Math.* 49 (1996), no. 6, 599–638.

[36] A. MAJDA. *Introduction to PDEs and waves for the atmosphere and ocean.* Courant lecture notes in Mathematics, (2003).

[37] F. MARCHE. Derivation of a new two-dimensional shallow water model with varying topography, bottom friction and capillary effects. Submitted (2005).

[38] A. MELLET, A. VASSEUR. On the isentropic compressible Navier-Stokes equation. Submitted (2005).

[39] L. MIN, A. KAZHIKHOV, S. UKAI. Global solutions to the Cauchy problem of the Stokes approximation equations for two-dimensional compressible flows. *Comm. Partial Diff. Eqs.*, 23, 5-6, (1998), 985–1006.

[40] M.L. MUOZ-RUIZ, F.-J. CHATELON, P. ORENGA. On a bi-layer shallow-water problem. *Nonlinear Anal. Real World Appl.* 4 (2003), no. 1, 139–171.

[41] A. NOVOTNY, I. STRASKRABA. *Introduction to the mathematical theory of compressible flow.* Oxford lecture series in Mathematics and its applications, (2004).

[42] M. OLIVER, Justification of the shallow water limit for a rigid lid flow with bottom topography, *Theoretical and Computational Fluid Dynamics* 9 (1997), 311–324.

[43] P. ORENGA. Un théorème d'existence de solutions d'un problème de shallow water, *Arch. Rational Mech. Anal.* 130 (1995) 183–204.

[44] L. SUNDBYE. Existence for the Cauchy Problem for the Viscous Shallow Water Equations. *Rocky Mountain Journal of Mathematics,* 1998, 28 (3), 1135–1152.

[45] L. SUNDBYE. Global existence for Dirichlet problem for the viscous shallow water equations. *J. Math. Anal. Appl.* 202 (1996), 236–258.

[46] J.-P. VILA. Shallow water equations for laminar flows of newtonian fluids. Paper in preparation and private communication, (2005).

[47] W. WANG, C.-J. XU. The Cauchy problem for viscous shallow water equations *Rev. Mat. Iberoamericana* 21, no. 1 (2005), 1–24.

[48] V.I. YUDOVICH. The flow of a perfect, incompressible liquid through a given region. *Soviet Physics Dokl.* 7 (1962) 789–791.

Didier Bresch
LMC–IMAG, (CNRS–INPG–UJF)
F-38051 Grenoble cedex, France

and

Institute of Mathematical Sciences
The Chinese University of Hong-Kong
Shatin, NT Hong-Kong
e-mail: `didier.bresch@imag.fr`

Benoît Desjardins
CEA/DIF, B.P. 12
F-91680 Bruyères
le Châtel, France

and

E.N.S. Ulm, D.M.A.
45 rue d'Ulm
F-75230 Paris cedex 05, France
e-mail: `Benoit.Desjardins@cea.fr`

Guy Métivier
MAB, Université Bordeaux 1
351, cours de la Libération
F-33405 Talence cedex, France
e-mail: `metivier@math.u-bordeaux.fr`

Analysis and Simulation of Fluid Dynamics
Advances in Mathematical Fluid Mechanics, 33–44
© 2006 Birkhäuser Verlag Basel/Switzerland

Direct Numerical Simulation and Analysis of 2D Turbulent Flows

Charles-Henri Bruneau

Abstract. Efficient methods are used to approximate incompressible Navier-Stokes equations. 2D turbulent flows are simulated in the cavity and behind arrays of cylinders in a channel. They confirm on one hand the presence of an attractor and on the other hand the coexistence of both direct enstrophy and inverse energy cascades. The use of a threshold directly on the vorticity intensity or on the wavelets packets coefficients separate the flow into two parts, each part corresponding to one cascade.

1. Introduction

The aim of this work is to study the behavior of incompressible flows at high Reynolds numbers. The first step is to develop a method robust enough to compute flows in various geometries and at high Reynolds numbers. That means that the method must be able to reproduce accurately the results available in the literature but must also be robust enough to compute transient and turbulent flows. In addition it is necessary to well represent obstacles and to reach a good efficiency to get long time simulations. The method has been tested on the lid-driven cavity problem and compared successfully to the results available in the literature up to Re = 10, 000 [7]; in particular the first Hopf bifurcation has been found close to Re = 8, 000 for this problem like in [2], [9] and [15] where the estimation was obtained via completely different methods including the computation of the first eigenvalues.

Here we give the outlines of the modelling, the approximation and the resolution. Then we show some results at Re = 10, 000 for the lid-driven cavity problem for which there are much less results available and bigger discrepancies between the various simulations. We then move to Re = 100, 000 to study the qualitative behavior of the solution. To reach this goal we perform several computations with different initial data and show that the solutions are driven to the same stage, confirming the presence of an attractor in 2D.

In addition to this academic test case, we compute turbulent flows in a channel behind arrays of cylinders. This numerical simulation corresponds to soap film experiments and the results are in good agreement with the one observed ([5] and [11]). In particular, we show the coexistence of the direct enstrophy cascade with a decay in k^{-3} and an inverse energy cascade which slope is very close to $k^{-5/3}$. Moreover the analysis reveals that the core of vortices have the dynamics of the enstrophy cascade and the filaments of vorticity and spirals have the dynamics of the inverse energy cascade.

2. Numerical simulation

Let Ω be the computational domain, the 2D unsteady Navier-Stokes equations read:

$$\begin{cases} \partial_t U - \dfrac{1}{\mathrm{Re}} \Delta U + (U.\nabla)U + \nabla p &= 0 \quad \text{in }]0, T[\times \Omega \\ \nabla.U &= 0 \quad \text{in }]0, T[\times \Omega \end{cases} \tag{1}$$

where $U = (u, v)$ is the velocity, p is the pressure and T is the simulation time. These equations are coupled to an initial datum $U(0, x, y) = U_0(x, y)$ in Ω and relevant boundary conditions. Namely no-slip boundary conditions on the walls, non homogeneous Dirichlet boundary conditions for the constant lid in a cavity or the incoming Poiseuille flow in a channel or the constant upstream flow in an open domain and non-reflecting boundary conditions on the artificial frontiers of the domain if any [4]. These last boundary conditions read:

$$\sigma(U, p)\, n + \tfrac{1}{2}(U \cdot n)^- (U - U_{\mathrm{reff}}) = \sigma(U_{\mathrm{reff}}, p_{\mathrm{reff}})\, n \tag{2}$$

where $(U_{\mathrm{reff}}, p_{\mathrm{reff}})$ is a reference flow chosen to convey properly the vortices out of the computational domain without any reflections. With these boundary conditions we have a well-posed problem and we do not need to add a sponge domain to get realistic results. In numerous applications it is also necessary to handle obstacles with a no-slip boundary condition imposed on their boundary. Instead of using a body fitting mesh, we prefer here to use the penalisation method that allows to keep a Cartesian mesh on a rectangular box. This method consists in adding a penalisation term U/K in the momentum equation to yield:

$$\begin{cases} \partial_t U - \dfrac{1}{\mathrm{Re}} \Delta U + (U.\nabla)U + \dfrac{U}{K} + \nabla p &= 0 \quad \text{in }]0, T[\times \Omega \\ \nabla.U &= 0 \quad \text{in }]0, T[\times \Omega \end{cases} \tag{3}$$

where K is the adimensional coefficient of permeability of the medium for the obstacles. It was shown that for a small coefficient in the body this corresponds to solve Darcy law [1]. If an obstacle is a solid body a very small coefficient is necessary to get a velocity close to zero in the body. In practise we take $K = 10^{-8}$ in such cases. To have the same approximation on the whole domain we also add this term in the fluid domain and take $K = 10^{16}$ in order this term vanishes and thus only the Navier-Stokes equations are solved in the fluid. This set of penalised

Navier-Stokes equations is also called Brinkman-Navier-Stokes equations and we can see in [6] how they are derived from Brinkman equation.

Now we have fixed the modelling, we can think of the approximation and decide to use a Cartesian mesh on a rectangular box. The first point is to compute the velocity and the pressure directly from the Navier-Stokes equations written in primitive variables. To reduce the numerical dissipation, we choose to treat the nonlinear convection term explicitly in time. So we can write the semi-discretized system for a Gear second order scheme:

$$
\begin{cases}
\dfrac{3U^n}{2\delta t} - \dfrac{1}{\text{Re}}\Delta U^n + \dfrac{U^n}{K} + \nabla p^n = \dfrac{2U^{n-1}}{\delta t} - 2(U^{n-1}\cdot\nabla)U^{n-1} \\
\qquad\qquad\qquad\qquad\qquad\qquad\quad -\dfrac{U^{n-2}}{2\delta t} + (U^{n-2}\cdot\nabla)U^{n-2} \quad \text{in } \Omega \\
\nabla\cdot U^n = 0 \qquad\qquad\qquad\qquad\qquad\qquad\qquad \text{in } \Omega
\end{cases}
\tag{4}
$$

that gives the required accuracy for long time behavior of the flow at high Reynolds numbers. Then the space discretization is achieved by finite differences on staggered cells as shown on Figure 1. The second order centred scheme is used for

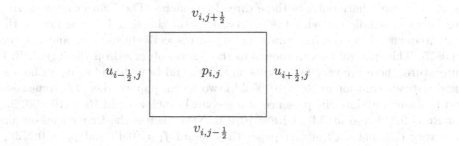

FIGURE 1. A staggered cell

the discretization of every linear terms whereas a third order upwind scheme is used to discretize convection terms (see [7] for the construction of this scheme). It appears that the accuracy of the method is directly linked to the accuracy of the approximation of these terms as well as the stability of the simulation. As the goal is to simulate high Reynolds flows, the approximation must not introduce too much dissipation that could change the behavior of the flow but must be stable enough to avoid oscillations or blow up. To reach a good efficiency and minimise the storage (in particular in 3D), we use a multigrid algorithm to solve the problem. It is well known that the relaxation smoothers capture very fast the high frequencies on a given grid. So here we take a set of grid for instance on a square cavity from 4×4 cells on the coarsest grid up to 1024×1024 cells on the finest grid. So we use a V-cycle starting on the finest grid with only two iterations of smoother going down to the coarsest grid and only one when going up with the correction on the low frequencies including the mean value. The smoother is a cell by cell Gauss-Seidel procedure that strongly couples the five unknowns of a staggered cell [7].

3. Flows in the square lid-driven cavity

For such a classical problem it is still very difficult to obtain reliable results and
to have a complete description of the bifurcations. After the first Hopf bifurcation
that occurs around Re = 8000 (see [7] for comparisons with other results of the
literature), the stable solution remains purely periodic for a wide range of Reynolds
numbers but at Re = 10,000 it seems that there are some tiny fluctuations in the
signal obtained at a monitoring point on a very fine grid 2048 × 2048. Indeed,
if we observe carefully such a signal on Figure 2 we see that, in addition to the
main periodic signal of period approximately equal to 1.65, there is a big pattern
corresponding to these fluctuations which is reproduced for instance from times
$t \approx 40$ to $t \approx 51$ and from times $t \approx 51$ to $t \approx 62$. This pattern corresponds a priori
to the addition of another periodic phenomenon with a period around 11. But, if
we select an entire number of these patterns and perform a Fourier analysis, we
see that the spectrum reveals three frequencies. The main frequency $f_1 = 0.61$
gives a peak of huge amplitude but there are two more peaks of lower frequencies
$f_2 = 0.175$ and $f_3 = 0.4375$ and small amplitude. In addition we can see in the
spectrum some harmonics of these three frequencies. The Fourier analysis seems
to be in contradiction with the observations on the signal but is not as these
two frequencies are commensurate and equal respectively to twice and five times
0.0875. This last value corresponds to the period observed on the signal. In the
literature, there are very few results on the second branch and some authors still
find a steady solution at Re = 10,000! In two recent papers, there are some results
on this that validate the presence of a second branch around Re = 10,000, in [2]
at Re = 9,765 and in [14] at Re = 10,300. Nevertheless the frequencies obtained
are very different, in the first paper they found $f_1 = 0.45$ and $f_2 = 0.2736$, in
the second paper they found $f_1 = 0.0017$ and $f_2 = 0.069$. Thus it is difficult
to conclude. In [2], they found that the frequency of the first branch $f_1 = 0.45$ is
conserved in the second branch obtained when a second incommensurate frequency
becomes active. In the present work, we find that the main frequency has changed
and there are two additional frequencies. Our results are confirmed by a third
paper giving results at Re = 10,000 [13] as it is found $f_1 = 0.59$, $f_2 = 0.137$ and
$f_3 = 0.43$. These last results are very close to ours but the numerical method is very
close either. The question is still open and we can imagine many scenarios, namely
that there are two branches very close to each other, the second one corresponding
to the description in [2] and the third one corresponding to the description above.

This first test and the available results of the literature show that it is not obvious whatever the approximation method is to describe the exact behavior of the
solution and to give accurate values of the frequencies for instance. Nevertheless
we think that it is possible to give some indications on the behavior of the solutions
at high Reynolds numbers even if we can not assert that the values are quantita-
tively correct. We perform computations at high Reynolds number Re = 100,000
to study the behavior of the turbulent solution with respect to the initial datum.
One simulation starts from rest and another simulation is initialised by the sta-

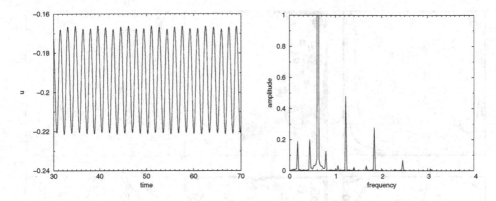

FIGURE 2. Horizontal velocity history (left) and power spectrum (right) at monitoring point (14/16,1/16) for Re = 10,000 on grid 2048 × 2048.

ble steady solution obtained at Re = 5,000. The stream-function, vorticity and pressure fields are represented on Figures 3 and 4 for both cases at times $t = 10$ and $t = 200$. We observe that the solutions are very different at the beginning but reach about the same stage after a significant simulation time. To reinforce this statement are shown on Figure 5 the evolution of the global quantities energy and enstrophy defined by

$$E = \frac{1}{2} \int_\Omega \|U\|^2 dx, \quad Z = \frac{1}{2} \int_\Omega \|\omega\|^2 dx$$

where $\omega = \partial_x v - \partial_y u$ is the vorticity. At time $t = 0$ the energy is $E = 0$ and increases for the first simulation while it starts from $E = 4.73 \times 10^{-2}$ and decreases for the second. Both curves seem to go to the same asymptotic state. In the meantime, the enstrophy histories which are very different at the beginning converge to each other. This confirms the presence of an attractor for Navier-Stokes equations in 2D as we can see that the final stages are very similar. Indeed, if we analyse the signals at the end of the simulations at a monitoring point, we get about the same spectra (Figure 6).

4. Simulation of soap film flows

In this section we simulate the flow at Re = 50,000 behind arrays of cylinders in a channel of width W (Figure 7). This corresponds to a vertical soap film flow in which the driving is given by an array of horizontal cylinders and two arrays of vertical cylinders along the wall to reinforce the injection scale (see [5] and [11]). On Figure 7 is plotted the vorticity field for both the numerical simulation and the experiment. We see clearly that the same patterns are present in the flow although the small eddies are not easy to visualise on the picture of the experiment. To qualify these flows, we want to analyse more clearly the fully developed

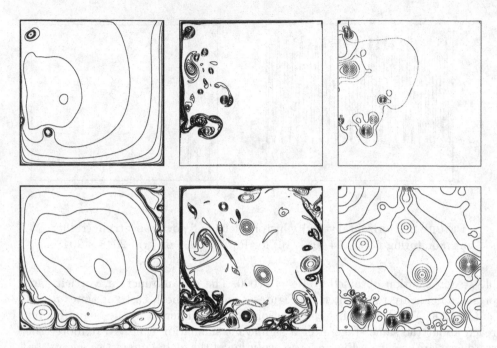

FIGURE 3. Evolution of the solution at Re = 100,000 computed from
initialisation at rest. From left to right stream-function, vorticity and
pressure fields on grid 1024×1024 at times $t = 10$ (top) and $t = 200$
(bottom).

turbulent flow. To do so we store the temporal evolution of both components of
the velocity at monitoring points inside the flow behind the horizontal cylinders.
Fourier analysis of this temporal signals gives the power spectrum of the longitu-
dinal and transverse components of the velocity. If the turbulence is isotropic, and
if these one-dimensional spectra show power law scaling, the scaling exponents
can be identified with the exponents of the energy density spectrum. Recall that
in two-dimensional turbulence, the energy density spectrum scales as $E(k) \sim k^{-3}$
in the enstrophy cascade range $(k > k_{\text{injection}})$ while it scales as $E(k) \sim k^{-5/3}$
in the inverse energy range $(k < k_{\text{injection}})$. The results for the power spectra of
the two velocity components are displayed in Figure 8, where the two components
have roughly equal amplitudes for most of the range of scales examined (i.e., from
$W/1000$ to W) showing that the turbulence created is isotropic. Second, one can
identify two different scaling of the spectra: at small scales, a scaling consistent
with -3 is seen, while at larger scales, a scaling consistent with $-5/3$ is observed.
These two ranges are located above and below the injection scale that corresponds
to the diameter of the small cylinders. To better understand the dynamics of the
flow we perform a 2D analysis. We decide to separate the flow patterns into two
parts, namely the vortices and the remaining as in [3]. For an empirical value of the

FIGURE 4. Evolution of the solution at Re = 100,000 computed from initialisation by the steady solution at Re = 5,000. From left to right stream-function, vorticity and pressure fields on grid 1024 × 1024 at times $t = 10$ (top) and $t = 200$ (bottom).

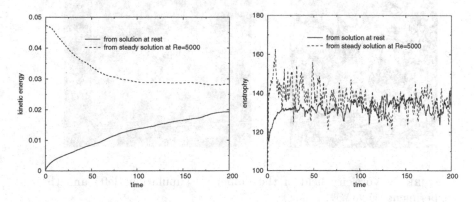

FIGURE 5. Comparison of the energy (left) and the enstrophy (right) histories at Re = 100,000 with the two initialisations.

cut we can isolate the vortices directly on the vorticity field. Then the remaining is essentially made of the vorticity filaments in between. Now we can analyse each part separately as it is shown on Figure 9 computing the 2D Fourier spectrum. On

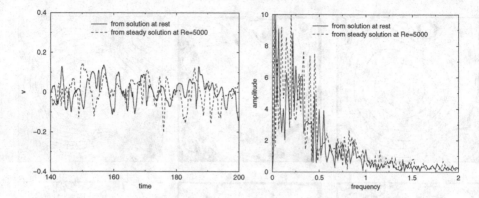

FIGURE 6. Zoom of vertical velocity component history (left) and corresponding power spectrum (right) at monitoring point $(14/16, 1/16)$ for $Re = 100,000$.

FIGURE 7. Vorticity field of the numerical simulation (left) and the experiment (right).

the whole field we find again the coexistence of the two cascades on both sides of the injection scale whereas the two separate parts have only one dynamics. The direct enstrophy cascade is located from the injection scale to the small scales and the inverse energy cascade is located from the injection scale to the large scales. The

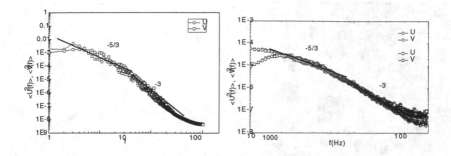

FIGURE 8. Velocity power spectra of the numerical simulation (left) and the experiment (right).

vortices and the filaments have a different dynamics but it is difficult to analyse their respective role as the spectra show a dissipative queue as pointed out in [10].

Another way to separate the different structures of the flow is to use a threshold in a wavelets packets decomposition of the flow (see [8]). We select as in the previous case the vortices and the remaining made of both the filaments of vorticity and the spirals inside the vortices as shown in an academic context in [12]. So the flow is split into two coherent parts, namely the vortices and the vorticity filaments and spirals that have their own dynamics (Figure 10).

5. Conclusions

Although in 2D we can afford to perform accurate direct numerical simulations of complex incompressible flows, it is still a challenge to describe the whole behavior of the flow with respect to the Reynolds number. In this paper, we try to shed some light on some issues that are theoretically derived for a long time, namely the presence of an attractor and the coexistence of the two turbulent cascades. It appears that with an accurate simulation tool it is possible to describe at least qualitatively the behavior of the solutions. In addition a wavelet packets analysis reveals, on a real flow with obstacles subject to no-slip and artificial boundary conditions in a bounded box, that the whole flow can be divided into two sub flows with the vortices on one hand and the filaments of vorticity and spirals on the other hand that have their own dynamics.

6. Acknowledgements

This paper gathers results of joint works with Mazen Saad for the lid-driven cavity problem; Hamid Kellay for the comparison of 2D turbulent flows numerical simulations and soap film experiments; Patrick Fischer, Zsolt Peter and Alain Yger for the analysis of 2D images of turbulent flows. The author would like to thank warmly each of them.

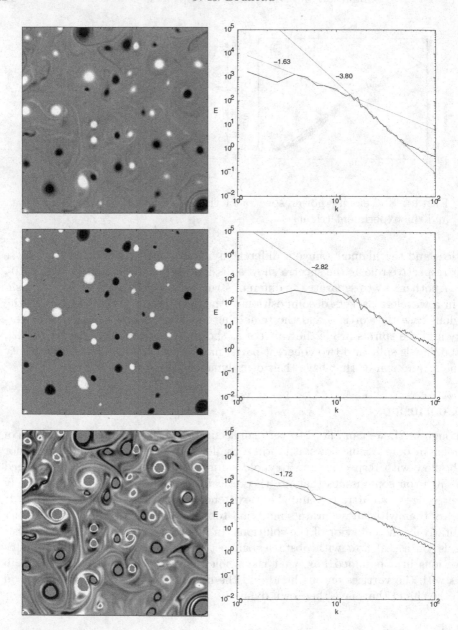

FIGURE 9. The whole vorticity field (top), the vortex cores (middle) and the vorticity filaments (bottom) with their corresponding energy spectrum. The threshold is applied directly on the vorticity field.

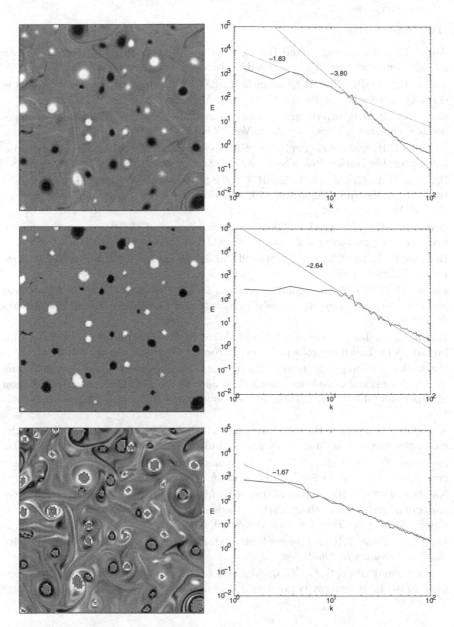

FIGURE 10. The whole vorticity field (top), the vortex cores (middle) and the spirals and vorticity filaments (bottom) with their corresponding energy spectrum. The threshold is applied on the wavelets packets coefficients.

References

[1] Angot Ph., Bruneau C.H. and Fabrie, P. (1999): "A penalization method to take into account obstacles in incompressible viscous flows", *Numer. Math.*, Vol. 81.

[2] Auteri F., Parolini N. and Quartapelle L. (2002): "Numerical investigation on the stability of singular driven cavity flow", *J. Comput. Phys.*, Vol. 183 n° 1.

[3] Borue V. (1994): "Inverse energy cascade in stationary two-dimensional homogeneous turbulence", *Phys. Rev. Lett.*, Vol. 72, 1475.

[4] Bruneau C.H. and Fabrie P. (1994): "Effective downstream boundary conditions for incompressible Navier-Stokes equations", *Int. J. Num. Meth. Fluids*, Vol. 19 n° 8.

[5] Bruneau C.H., Greffier O., Kellay H. (1999): "Numerical study of grid turbulence in two dimensions and comparison with experiments on turbulent soap films", *Phys. Rev. E*, Vol. 60, R1162.

[6] Bruneau C.H., Mortazavi I. (2004): "Passive control of the flow around a square cylinder using porous media", *Int. J. Num. Meth. Fluids*, Vol. 46, 415.

[7] Bruneau C.H., Saad M. (2005): "The 2D lid-driven cavity problem revisited", *Computers & Fluids*, to appear.

[8] Fischer P. (2005): "Multiresolution analysis for 2D turbulence. Part 1: Wavelets vs cosine packets, a comparative study", *Discrete and Continuous Dynamical Systems B*, Vol. 5, 659.

[9] Fortin A., Jardak M., Gervais J.J. and Pierre R. (1997): "Localization of Hopf bifurcations in fluid flow problems", *Int. J. Numer. Methods Fluids*, Vol. 24 n° 11.

[10] Gkioulekas E., Tung K.K. (2005): "On the double cascades of energy and enstrophy in two dimensional turbulence. Part 2. Approach to the KLB limit and interpretation of experimental evidence", *Discrete and Continuous Dynamical Systems B*, Vol. 5, 103.

[11] Kellay H., Bruneau C.H., Wu X.L. (2000): "Probability density functions of the enstrophy flux in two dimensional grid turbulence", *Phys. Rev. Lett.* Vol. 84, 5134.

[12] Kevlahan N., Farge M. (1997): "Vorticity filaments in two-dimensional turbulence: creation, stability and effect", *J. Fluid Mech.* Vol. 346, 49.

[13] Nobile E. (1996): "Simulation of time-dependent flow in cavities with the additive-correction multigrid method. Part 1: Mathematical formulation, Part 2: Applications", *Num. Heat Transfer, Part B* Vol. 30, 341.

[14] Peng Y.F., Shiau Y.H. and Hwang R.R. (2002): "Transition in a 2-D lid-driven cavity flow", *Computers & Fluids*, Vol. 32, 337.

[15] Sahin M. and Owens R.G.(2003): "A novel fully-implicit finite volume method applied to the lid-driven cavity problem. Part I and II", *Int. J. Numer. Methods Fluids* Vol. 42 n° 1.

Charles-Henri Bruneau
CNRS UMR 5466 – INRIA FUTURS Equipe MC2
MAB, Université Bordeaux 1
F-33405 Talence, France
e-mail: bruneau@math.u-bordeaux1.fr

Analysis and Simulation of Fluid Dynamics
Advances in Mathematical Fluid Mechanics, 45–68
© 2006 Birkhäuser Verlag Basel/Switzerland

Numerical Capture of Shock Solutions of Nonconservative Hyperbolic Systems via Kinetic Functions

Christophe Chalons and Frédéric Coquel

Abstract. This paper reviews recent contributions to the numerical approximation of solutions of nonconservative hyperbolic systems with singular viscous perturbations. Various PDE models for complex compressible materials enter the proposed framework. Due to lack of a conservative form in the limit systems, associated weak solutions are known to heavily depend on the underlying viscous regularization. This small scales sensitiveness drives the classical approximate Riemann solvers to grossly fail in the capture of shock solutions. Here, small scales sensitiveness is encoded thanks to the notion of kinetic functions so as to consider a set of generalized jump conditions. To enforce for validity these jump conditions at the discrete level, we describe a systematic and effective correction procedure. Numerical experiments assess the relevance of the proposed method.

1. Introduction

We survey some of the recent numerical methods for approximating the solutions of nonlinear hyperbolic systems with viscous perturbations, in the form:

$$\mathcal{A}_0(\mathbf{v}_\epsilon)\partial_t\mathbf{v}_\epsilon + \mathcal{A}_1(\mathbf{v}_\epsilon)\partial_x\mathbf{v}_\epsilon = \epsilon\partial_x(\mathcal{D}(\mathbf{v}_\epsilon)\partial_x\mathbf{v}_\epsilon), \quad x \in \mathbb{R}, \, t > 0. \qquad (1)$$

Here, the central issue stems from that neither $\mathcal{A}_0$ nor $\mathcal{A}_1$ coincide with Jacobian matrices so that the nonlinear PDE model (1) does not take the standard form of systems in conservation form. Prominent models from the Physics of complex compressible materials actually enter the present framework. The nonconservative terms in (1) are generically the by-product of simplifying modelling assumptions. In most instances, these assumptions intend to bypass the need for dealing with intricate mechanisms taking place at a too fine scale. Averaged multiphase flows [26], [27] or averaged turbulent flows [1], [6] provide typical major examples. Also do plasmas models when scales smaller than the Debye length are neglected [17], [14].

At last, a recent multifluid model [15] falls into the present setting. After [14], [1], [6], a surprising property met by most (if not all) of these models stays in the existence of an admissible change of variable such that (1) recasts in the form:

$$\partial_t \mathbf{u}_\epsilon + \partial_x \mathcal{F}(\mathbf{u}_\epsilon) = \epsilon \mathcal{R}(\mathbf{u}_\epsilon, \partial_x \mathbf{u}_\epsilon, \partial_{xx} \mathbf{u}_\epsilon), \quad x \in \mathbb{R}, \ t > 0, \tag{2}$$

where in contrast with (3), the regularization term now stands in (genuine) non-conservative form while the left-hand side is conservative. Let us stress from now on that this property will play a central role hereafter in view of the difficulties we now enter.

In realistic applications, all the reported models have to be tackled in the regime of a large Reynolds number: *i.e.*, the rescaling parameter $\epsilon > 0$ in (1) is small, with typical order of magnitude 10^{-6}. In view of the genuine nonlinearities in the underlying hyperbolic operator in (1), solutions under consideration involve in general propagating viscous shock layers which differ from their end states $\mathbf{u}_-$ and $\mathbf{u}_+$ only in a $\mathcal{O}(\epsilon)$-interval of stiff transition. Obviously, mesh refinements of practical interest cannot afford for a proper resolution of such small scales. Hence and away from solid boundaries, we are urged to consider the singular limit $\epsilon \to 0^+$ in (1). Due to the lack of conservative form, discontinuous limit solutions cannot be understood in the classical sense of distributions. However for these limit solutions, the nonconservative terms in the limit system of (1) exhibit products of discontinuous functions with measures and these must be given some suitable meaning. This difficulty has received several significant contributions over the past decade after the works by LeFloch [21], Dal Maso, LeFloch and Murat [18] to tackle the singular limit in (1) and by Berthon, Coquel and LeFloch [5] to handle the limit system in the distinct but equivalent form (2). We also refer to Colombeau [11], Colombeau, Leroux [12] for a theory in a distinct functional framework.

These suitable theories inevitably come with the property that shock solutions in the limit system are inherently regularization dependent: two distinct viscous mechanisms in (1) or equivalently in (2) generically give birth to two different families of shock solutions in the limit system. The small-scales sensitiveness in the discontinuous solutions is in complete opposition with the independence property met in the usual conservative setting. After [21], [18], we hereafter shade light on the roots of such a sensitivity. In [21], [18], sensitiveness is encoded in terms of a fixed family of paths connecting possible end states in the viscous shock profiles in (1) while in [5], it is encoded in the so-called kinetic functions. Roughly speaking, these kinetic-functions can be regarded as the mass of bounded Borel measures concentrated on the shock solutions to (2) so as to give rise to generalized Rankine-Hugoniot jump conditions. More precisely, the mass of these bounded measures actually coincide with entropy dissipation rates coming with the viscous shock profiles of(2). Both approaches complement each other and are introduced in this review since they are involved in the numerical procedures to be discussed.

These theories are exemplified, in this paper, on an important class of models encompassing several turbulence and multifluid as well modelings. We refer the reader to [1] for the so-called (k, ε) model and variants of it, to [6] for the multi-scales turbulence approach and to [15] for multifluid descriptions. All these models naturally take the form (2) for extended Navier-Stokes equations when considering several independent specific entropies for governing independent pressure laws. The limit system is seen to yield an natural extension of the classical Euler equations involving bounded Borel measures concentrated along the discontinuity curves of the solutions and vanishing everywhere else.

The inherent small scale sensitiveness of weak solutions for the limit system in (1) or (2) makes their numerical approximation a particularly challenging issue. The core of the difficulty indeed stems from the property of shock solutions to be regularization dependent: the artificial dissipation terms induced by numerical methods tend to corrupt the discrete shocks. Large failures in the celebrated Godunov method in the proper capture of shock solutions to (58) are well exemplified in Chalons [6], approximate Riemann solvers grossly fail as well as illustrated in Berthon [1] and Chalons [6]. We refer to Hou, LeFloch [20] for an analysis in the scalar setting. By contrast, the Glimm method stays free from artificial diffusion and has been shown to converge to the correct solutions [23]. Our main purpose in this paper is to illustrate, after [3], [9] and the related works we quote here-after, how to enforce roughly speaking the artificial diffusion in classical numerical methods to mimic the exact dissipation mechanism. More precisely, the procedure we describe intends to keep all the independent discrete rate of entropy dissipations in the exact balance prescribed by the kinetic functions coming with (2). We refer the reader to Tadmor [28] for a precise link between numerical viscosity and discrete entropy rates. As a consequence of preserved balances, a far much better agreement is achieved between exact and discrete solutions: errors are virtually negligible for shocks of moderate amplitude in the models investigated in [1], [6].

The format of the present paper is as follows. The second section highlights the roots in the small scales sensitiveness of shock solutions of (1) and (2) as well to then introduce both theories of family of paths and kinetic functions in the class of piecewise Lipshitz continuous functions. The third section describes the main properties of the extended Navier-Stokes equations and their limit model, the extended Euler equations, with a special emphasis put on the description of full sets of extended generalized Rankine-Hugoniot jump relations respectively derived from the two theories. The last section then explains the origin in the failure of classical Riemann solvers so as to naturally introduce a systematic and effective correction procedure.

2. Shock solutions for nonconservative hyperbolic systems

Given a smooth matrix-valued function $\mathcal{D} \geq 0$, this section introduces some of the mathematical tools, developed over the past decade, to handle first order systems with singular viscous perturbation built from $\mathcal{D}$:

$$\mathcal{A}_0(\mathbf{v}_\epsilon)\partial_t \mathbf{v}_\epsilon + \mathcal{A}_1(\mathbf{v}_\epsilon)\partial_x \mathbf{v}_\epsilon = \epsilon \partial_x (\mathcal{D}(\mathbf{v}_\epsilon)\partial_x \mathbf{v}_\epsilon). \quad x \in \mathbb{R}, \ t > 0, \quad (3)$$

By singular, it is classically meant that (3) is addressed in the regime of a vanishing rescaling parameter $\epsilon \to 0^+$. Here, the unknown $\mathbf{v}_\epsilon$ belongs to some open convex subset $\Omega_\mathbf{v} \subset \mathbb{R}^n$. The matrix-valued functions $\mathcal{A}_i : \Omega_\mathbf{v} \to \mathcal{M}_n(\mathbb{R})$, $i = 0, 1$, are supposed to be smooth with $\mathcal{A}_0(\mathbf{v})$ invertible and $\mathcal{A}_0^{-1}(\mathbf{v})\mathcal{A}_1(\mathbf{v})$ $\mathbb{R}$-diagonalizable for all states $\mathbf{v} \in \Omega_\mathbf{v}$. In other words, with fixed $\epsilon > 0$, (3) is nothing but a nonlinear hyperbolic system with viscous perturbation. For simplicity in the discussion, all the fields in the underlying hyperbolic model (*i.e.* obtained formally when setting $\epsilon = 0$ in (3)) are supposed to be genuinely nonlinear. Here, the central assumption is that neither $\mathcal{A}_0$ nor $\mathcal{A}_1$ coincide with Jacobian matrices: namely the nonlinear PDE model (3) does not write in conservation form. Motivated by several models from the Physics, we shall assume the existence of a smooth change of variable $\mathbf{v} \in \Omega_\mathbf{v} \to \mathbf{u}(\mathbf{v}) \in \Omega_\mathbf{u}$ so that the smooth solutions of (3) obey the following equivalent form:

$$\partial_t \mathbf{u}(\mathbf{v}_\epsilon) + \partial_x \mathcal{F}(\mathbf{u}(\mathbf{v}_\epsilon)) = \epsilon \mathcal{R}(\mathbf{u}(\mathbf{v}_\epsilon), \partial_x \mathbf{u}(\mathbf{v}_\epsilon), \partial_{xx}\mathbf{u}(\mathbf{v}_\epsilon)), \quad x \in \mathbb{R},\ t > 0, \qquad (4)$$

which we shall write for short:

$$\partial_t \mathbf{u}_\epsilon + \partial_x \mathcal{F}(\mathbf{u}_\epsilon) = \epsilon \mathcal{R}(\mathbf{u}_\epsilon, \partial_x \mathbf{u}_\epsilon, \partial_{xx}\mathbf{u}_\epsilon), \quad x \in \mathbb{R},\ t > 0. \qquad (5)$$

Notice of course that in contrast with (3), the regularization term in (5):

$$\mathcal{R}(\mathbf{u}(\mathbf{v}_\epsilon), \partial_x \mathbf{u}(\mathbf{v}_\epsilon), \partial_{xx}\mathbf{u}(\mathbf{v}_\epsilon)) \equiv \nabla_\mathbf{v} \mathbf{u}(\mathbf{v}_\epsilon)\mathcal{A}_0^{-1}(\mathbf{v}_\epsilon)\partial_x (\mathcal{D}(\mathbf{v}_\epsilon)\partial_x \mathbf{v}_\epsilon), \qquad (6)$$

is in general nonconservative while the underlying first order operator in (5) now stands in conservation form. To shade further light in such a change of variable, let us consider the underlying hyperbolic system in (3) (*i.e.*, again setting formally $\epsilon = 0$) so as to evaluate the following scalar product:

$$\nabla_\mathbf{v} u_i(\mathbf{v}) \cdot \partial_t \mathbf{v} + \nabla_\mathbf{v} u_i(\mathbf{v}) \cdot \mathcal{A}_0^{-1}(\mathbf{v})\mathcal{A}_1(\mathbf{v})\partial_x \mathbf{v} = 0, \qquad (7)$$

where for any given $i \in \{1, \dots, n\}$, the smooth function $u_i : \Omega_\mathbf{v} \to \mathbb{R}$ denotes the ith component of the vector-valued function $\mathbf{v} \to \mathbf{u}(\mathbf{v})$. Due to (5), the above scalar equation necessarily recasts in conservation form:

$$\partial_t u_i(\mathbf{v}) + \partial_x \mathcal{F}_i(\mathbf{u}(\mathbf{v})) = 0, \qquad (8)$$

so that, by definition, the scalar functions u_i, $\mathcal{F}_i$ play the role of an entropy pair (trivial or non trivial) for smooth solutions of the underlying hyperbolic system in (3). Hence, the ith component of the regularization term (6) is nothing but the associated entropy rate of production. At the present stage, we do not assume convexity in the mapping $\mathbf{v} \to u_i(\mathbf{v})$ nor suppose that each of the possible (second order) nonconservative products in (6) keeps a constant sign (say negative). Rephrasing the above observations in (6)–(8), the existence of the change of variable $\mathbf{v} \to \mathbf{u}(\mathbf{v})$ in (5) thus requires the existence of as many additional entropy pairs with independent gradients for (3) than there exist scalar equations involving genuine nonconservative products in (3). Such a requirement might sound rather restrictive but surprisingly, most of the important nonconservative models for complex compressible materials actually achieve it. The interest in the equivalent form (5) over (3) stays in that it allows for a mathematical framework to handle the singular limit $\epsilon \to 0^+$ which turns to be really tractable from the numerical standpoint. The next section introduces such a framework due to Berthon, Coquel, LeFloch [5].

2.1. Definition of weak solutions

In this paragraph, we first address formal issues concerning the singular limit in (3) to then motivate precise definitions that are needed in the forthcoming sections devoted to applications. Since by assumption neither $\mathcal{A}_0$ nor $\mathcal{A}_1$ do coincide with Jacobian matrices, the underlying hyperbolic system in (3), obtained setting $\epsilon = 0$:

$$\mathcal{A}_0(\mathbf{v})\partial_t \mathbf{v} + \mathcal{A}_1(\mathbf{v})\partial_x \mathbf{v} = 0, \quad x \in \mathbb{R}, \ t > 0, \tag{9}$$

writes in nonconservative form just like its viscous form (3). Therefore, one cannot formally pass to the limit $\epsilon \to 0^+$ in (3) in the usual sense of distributions, so as to recover (9) in the classical weak sense. But a weak sense is needed since in general, the nonlinear hyperbolic system (9) does not admit smooth solutions: propagating shock waves appear in finite time in smooth initial data. Nevertheless and provided that some suitable estimates on the sequence $\mathbf{v}_\epsilon$ and its derivatives $\partial_t \mathbf{v}_\epsilon$, $\partial_x \mathbf{v}_\epsilon$ are satisfied, one reasonably expects to get $\epsilon \partial_x \mathcal{D}(\mathbf{v}_\epsilon)\partial_x \mathbf{v}_\epsilon \to 0$, $\epsilon \to 0$, together with:

$$\mathcal{A}_0(\mathbf{v}_\epsilon)\partial_t \mathbf{v}_\epsilon \rightharpoonup \mathcal{A}_0(\mathbf{v})\partial_t \mathbf{v}, \quad \mathcal{A}_1(\mathbf{v}_\epsilon)\partial_x \mathbf{v}_\epsilon \rightharpoonup \mathcal{A}_1(\mathbf{v})\partial_x \mathbf{v}, \tag{10}$$

vaguely in the sense of measures so that (9) could be reached in this rather vague sense. The central difficulty stems from the non conservative products in (10): they involve products of discontinuous functions with measures and thus, they are generally not stable with respect to weak convergence (see [18] and [24] for a definition and counterexamples). At present, several successful and stable definitions exist in the BV framework: we refer the reader to LeFloch [21], [22], Dal Maso, LeFloch, Murat [18] and LeFloch, Tzavaras [24]. These definitions lead to a solution of the Riemann problem in the class of hyperbolic systems with genuine nonlinearity for initial data sufficiently flat [18]. Existence of weak solutions of (9) has been then established on the ground of the random choice method [23]. Weak solutions to the nonconservative system (9) can be thus defined in the class of BV functions. After the classical results by Volpert [29] and Federer [16], such functions can be manipulated as if they were piecewise Lipschitz continuous functions. For the simplicity in this brief review and without significant loss of generality, we restrict from now on attention to piecewise Lipschitz continuous functions.

After [21], [18], and [24], here is now the key issue to be put forward. These suitable theories necessarily come with the property that the singular limits entering (10) intrinsically depends on the sequence $\mathbf{v}_\epsilon$ via the choice of the viscous regularization in (3). The core of this sensitiveness basically finds its root in the non conservative products $\mathcal{A}_0(\mathbf{v}) \times \partial_t \mathbf{v}$ and $\mathcal{A}_1(\mathbf{v}) \times \partial_x \mathbf{v}$ in (9). Indeed and without reference to the singular limit in (3), such products are ambiguous: already the simplest formal product $H \times \delta$ can be found to be equal to $\alpha\delta$ with α an arbitrary positive constant. Hence, the measures $\mathcal{A}_0(\mathbf{v}) \times \partial_t \mathbf{v}$ and $\mathcal{A}_1(\mathbf{v}) \times \partial_x \mathbf{v}$ cannot be uniquely defined: this nonuniqueness precisely gives room for the shock solutions to (9) to be regularization dependent. As underlined first by LeFloch [21], uniqueness in the definition of the nonconservative products can be restored with explicit reference to the precise shape of $\mathcal{D}$ in (3).

In sharp contrast is the independence property met by shock solutions with respect to small scale effects in the setting of conservative (genuinely) nonlinear systems. Indeed assuming $\mathcal{A}_0$ and $\mathcal{A}_1$ to coincide with the jacobian matrices of some flux functions $\mathcal{F}_0$ and $\mathcal{F}_1$, then from suitable estimates on the derivatives of $\mathbf{v}_\epsilon$, $\partial_t \mathcal{F}_0(\mathbf{v}_\epsilon) \rightharpoonup \partial_t \mathcal{F}_0(\mathbf{v})$ and $\partial_x \mathcal{F}_1(\mathbf{v}_\epsilon) \rightharpoonup \partial_x \mathcal{F}_1(\mathbf{v})$ in the usual sense of distributions with $\epsilon \mathcal{D}(\mathbf{v}_\epsilon)\partial_x \mathbf{v}_\epsilon \rightharpoonup 0$, so that we get at points of jump in the limit function $\mathbf{v}$:

$$[-\sigma(\mathcal{F}_0(\mathbf{v}_+) - \mathcal{F}_0(\mathbf{v}_-)) + (\mathcal{F}_1(\mathbf{v}_+) - \mathcal{F}_1(\mathbf{v}_-))]\delta_{x-\sigma t} = 0, \qquad (11)$$

for some speed of propagation σ. These so-called Rankine-Hugoniot jump conditions stay completely free from the particular shape of the viscous matrix $\mathcal{D}$ and allow to define $\mathbf{v}_+ = \mathbf{v}_+(\mathbf{v}_-, \sigma)$ (at least locally) independently of $\mathcal{D}$. As already claimed, such a property cannot hold for nonconservative hyperbolic systems and at points of jump, possible exit states $\mathbf{v}_+$ do depend on $\mathbf{v}_-$ and σ but also deeply on the shape of $\mathcal{D}$: $\mathbf{v}_+ = \mathbf{v}_+(\mathcal{D}; \mathbf{v}_-, \sigma)$.

Let us now turn considering the equivalent form (5). By contrast to (3), the nonconservative regularization term $\epsilon \mathcal{R}(\mathbf{u}_\epsilon, \partial_x \mathbf{u}_\epsilon, \partial_{xx} \mathbf{u}_\epsilon)$ cannot be expected to converge to zero in the sense of measures as ϵ goes to zero but instead to a (vector-valued) bounded Borel measure $\mu_{\mathbf{u}}\{\mathcal{D}\}$ concentrated on the discontinuities of the limit function $\mathbf{u}$. Such a measure vanishes in the region of continuity of $\mathbf{u}$ and has a non trivial mass, we denote $\mathcal{K}_{\mathcal{D}}(\mathbf{u}_-, \sigma)$, along any curve of discontinuity of $\mathbf{u}$ (see hereafter for the notations). From the very definition (6) of the regularization term $\epsilon \mathcal{R}(\mathbf{u}_\epsilon, \partial_x \mathbf{u}_\epsilon, \partial_{xx} \mathbf{u}_\epsilon)$, it is clear that the mass of $\mu_{\mathbf{u}}\{\mathcal{D}\}$ generically depends on the precise shape of $\mathcal{D}$. The viscosity matrix $\mathcal{D}$ being prescribed in (6), the exit state $\mathbf{u}_+$ at points of jump must then solves the following set of generalized jump relations:

$$-\sigma(\mathbf{u}_+ - \mathbf{u}_-) + (\mathcal{F}(\mathbf{u}_+) - \mathcal{F}(\mathbf{u}_-)) = \mathcal{K}_{\mathcal{D}}(\mathbf{u}_-, \sigma), \qquad (12)$$

hence, another illustration of the inherent small-scale sensitiveness of shock solutions to (9).

Let us now address precise definitions for weak solutions to (9) in the class of piecewise Lipschitz continuous functions. These definitions, needed in the forthcoming sections, heavily rely on the properties of travelling solutions to (3). For fixed $\epsilon > 0$, these are smooth solutions to (3), and thus equivalently to the companion system (5), of the form:

$$\begin{cases} \mathbf{v}_\epsilon(x, t) = \mathbf{w}_\epsilon(x - \sigma t) = \mathbf{w}_\epsilon(\xi), \\ \lim_{\xi \to \pm\infty} \mathbf{w}_\epsilon(\xi) = \mathbf{v}_\pm, \quad \lim_{\xi \to \pm\infty} \frac{d}{d\xi}\mathbf{w}_\epsilon(\xi) = 0, \end{cases} \qquad (13)$$

where σ denotes the speed of the wave and $\mathbf{v}_-$, $\mathbf{v}_+$ are two states in $\Omega_{\mathbf{v}}$. A solution to (3) of the form (13) must thus solve the following system of ordinary differential equations:

$$(\mathcal{A}_1(\mathbf{w}_\epsilon) - \sigma\mathcal{A}_0(\mathbf{w}_\epsilon))\mathbf{w}_\epsilon' = \epsilon(\mathcal{D}(\mathbf{w}_\epsilon)\mathbf{w}_\epsilon')', \quad \mathbf{w}_\epsilon' = \frac{d}{d\xi}\mathbf{w}_\epsilon(\xi). \tag{14}$$

Now considering the rescaled function $\mathbf{w} : \mathbb{R} \to \Omega_\mathbf{v}$ defined by:

$$\mathbf{w}\left(\frac{\xi}{\epsilon}\right) = \mathbf{w}_\epsilon(\xi), \tag{15}$$

then $\mathbf{w}$ must solve the next ODE problem free from the parameter ϵ:

$$(\mathcal{A}_1(\mathbf{w}) - \sigma\mathcal{A}_0(\mathbf{w}))\mathbf{w}' = (\mathcal{D}(\mathbf{w})\mathbf{w}')', \tag{16}$$

while achieving the same asymptotic conditions as those stated independently from ϵ in (13). Notice that in the present nonconservative case, (16) cannot be integrated once to give rise to a first order system like in the conservative framework. Assuming $\mathcal{D}(\mathbf{w})$ invertible for all $\mathbf{w} \in \Omega_\mathbf{v}$, one merely has to consider the extended dynamical system:

$$\begin{cases} \mathbf{r}' = (\mathcal{A}_1(\mathbf{w}) - \sigma\mathcal{A}_0(\mathbf{w}))\mathcal{D}^{-1}(\mathbf{w})\mathbf{r}, \\ \mathbf{w}' = \mathcal{D}^{-1}(\mathbf{w})\mathbf{r}, \end{cases} \tag{17}$$

for which the set of critical points, *i.e.*, $(\mathbf{r}, \mathbf{w}) = (0, \mathbf{w})$, is *a priori* unknown.

Remark 1. Several authors have established sufficient conditions on the viscosity matrix $\mathcal{D}$ ensuring the existence of small-amplitude travelling wave solutions to hyperbolic systems with viscous regularization: we refer to the work by Majda, Pego [25] and the references therein. Besides, let us stress that the left state $\mathbf{v}_-$ being fixed, the speed σ has to be properly prescribed so as to meet the Lax compression condition $\lambda_k(\mathbf{v}_-) > \sigma$ in order to give rise to a small-amplitude travelling wave solution (see [25] for the details). We tacitly assume from now on that all the reported conditions on $\mathcal{D}$ and σ are met without further reference.

Let a state $\mathbf{v}_-$ be given and some velocity σ be prescribed so that there exists a critical point $(0, \mathbf{v}_+) = (0, \mathbf{v}_+(\mathbf{v}_-, \sigma))$ that can be reached exponentially fast in the future by a smooth solution $(\mathbf{r}, \mathbf{w})$ of (17) and connecting exponentially fast $(0, \mathbf{v}_-)$ in the past. Notice that generally speaking, $\mathbf{v}_+$ does depend on the fixed viscosity matrix $\mathcal{D}$: namely $\mathbf{v}_+ = \mathbf{v}_+(\mathcal{D}; \mathbf{v}_-, \sigma)$. From (15), we are now in a position to define a sequence of solutions $\{\mathbf{w}_\epsilon\}_{\epsilon>0}$ to (14) with the required asymptotic conditions (13). This sequence obeys $\|\mathbf{w}_\epsilon'\|_{L^1} = \|\mathbf{w}'\|_{L^1} < \infty$ and thus converges strongly in L^1_{loc} to the step function:

$$\mathbf{v}(x,t) = \mathbf{v}_- + (\mathbf{v}_+(\mathcal{D}; \mathbf{v}_-, \sigma) - \mathbf{v}_-)H(x - \sigma t), \quad x \in \mathbb{R}, \ t > 0, \tag{18}$$

where H denotes the usual Heaviside function. These considerations have led LeFloch [21] to state:

Definition 2.1. *The limit function (18) is a shock solution to (9), compatible with the viscosity matrix $\mathcal{D}$ in (3).*

To go one step further in the characterization of shock solutions to (9) in the sense of Definition 2.1, notice the next identities, valid for all $\epsilon > 0$:

$$\int_{\mathbb{R}_\xi} \mathcal{A}_i(\mathbf{w}_\epsilon(\xi))\mathbf{w}'_\epsilon(\xi)d\xi = \int_{\mathbb{R}_\xi} \mathcal{A}_i(\mathbf{w}(\xi))\mathbf{w}'(\xi)d\xi, \quad i = 0,1, \tag{19}$$

while in view of the asymptotic conditions expressed in (13):

$$\int_{\mathbb{R}_\xi} (\mathcal{D}(\mathbf{w}_\epsilon(\xi))\mathbf{w}'_\epsilon(\xi))' d\xi = 0. \tag{20}$$

Following [21], let us consider an increasing one to one function $\psi : (0,1) \to \mathbb{R}$ so as to introduce the following path connecting $\mathbf{v}_-$ to $\mathbf{v}_+$ in the phase space $\Omega_\mathbf{v}$:

$$\phi_{\mathcal{D}}(s; \mathbf{v}_-, \mathbf{v}_+) = \mathbf{w}(\psi(s)), \quad s \in (0,1). \tag{21}$$

Equipped with these notations, we observe that integrating once the ODE system (14) yields the following set of extended Rankine-Hugoniot relations:

$$\begin{aligned} -\sigma \int_0^1 \mathcal{A}_0(\phi_{\mathcal{D}}(s; \mathbf{v}_-, \mathbf{v}_+))\tfrac{\partial \phi_{\mathcal{D}}}{\partial s}(s; \mathbf{v}_-, \mathbf{v}_+)ds \\ + \int_0^1 \mathcal{A}_1(\phi_{\mathcal{D}}(s; \mathbf{v}_-, \mathbf{v}_+))\tfrac{\partial \phi_{\mathcal{D}}}{\partial s}(s; \mathbf{v}_-, \mathbf{v}_+)ds = 0, \end{aligned} \tag{22}$$

to be solved by (18). It can be shown that the identity (22) stays invariant by change of parametrization of the path (21). Next, $\mathcal{D}$ being fixed, letting the left state $\mathbf{v}_-$ run in $\Omega_\mathbf{v}$ and the speed σ (suitably) in $\mathbb{R}$ give rise (at least locally) to a complete family of travelling waves to (3) and thus to a whole family of shock solutions according to Definition 2.1. In turn, (21) allows to define a family of paths $\phi_{\mathcal{D}}$ so as to connect left and right states in the shock solutions of (9). This construction provides a particular example of the general theory of family of paths introduced by Dal Maso, LeFloch, Murat [18] so as to propose a weakly stable definition of nonconservative products in the BV framework. We shall not enter the details in this review and we refer the reader to [18] for the required material. After [18], a fixed family of path being fixed, the limit system (9) may be written:

$$\left[\mathcal{A}_0(\mathbf{v})\partial_t \mathbf{v} \right]_{\phi_{\mathcal{D}}} + \left[\mathcal{A}_1(\mathbf{v})\partial_x \mathbf{v} \right]_{\phi_{\mathcal{D}}} = 0, \tag{23}$$

in the sense of the next definition (see [18] for the BV framework):

Definition 2.2. *A piecewise Lipschitz continuous solution* $\mathbf{v} = \mathbf{v}(x,t)$ *is called a weak solution to* (23) *iff it satisfies in the strong sense* (9) *in each region of continuity while at points of jump it obeys* (22).

Some of the forthcoming developments will make use of this definition. Let us next turn considering another relevant definition for weak solutions when addressing the equivalent formulation (5) in the setting of the travelling solutions (13) to (3). Since such solutions are smooth, the sequence of functions $\mathbf{u}_\epsilon = \mathbf{u}(\mathbf{w}_\epsilon)$ equally solve (5) with the asymptotics $\lim_{\epsilon \to \pm\infty} \mathbf{u}_\epsilon(\xi) = \mathbf{u}_\pm = \mathbf{u}(\mathbf{v}_\pm)$, for all $\epsilon > 0$. Hence and with little abuse in the notations, the next identities hold true for all $\epsilon > 0$:

$$-\sigma\mathbf{u}'_\epsilon + (\mathcal{F}(\mathbf{u}_\epsilon))' = \epsilon\mathcal{R}(\mathbf{u}_\epsilon, \mathbf{u}'_\epsilon, \mathbf{u}''_\epsilon), \tag{24}$$

while the rescaled function $\tilde{u}(\xi) = u(w)$ defined from (15) satisfies:

$$-\sigma\tilde{u}' + (\mathcal{F}(\tilde{u}))' = \mathcal{R}(\tilde{u}, \tilde{u}', \tilde{u}''). \tag{25}$$

Again the sequence $\{u_\epsilon\}_{\epsilon>0}$ is seen to converge strongly in L^1_{loc} to the step function:

$$u(x,t) = u_- + (u_+(\mathcal{D}; u_-, \sigma) - u_-)H(x - \sigma t), \quad x \in \mathbb{R}, \, t > 0. \tag{26}$$

The very interest in the derivation of the limit function (26) stems from its characterization by the following set of generalized Rankine-Hugoniot jump conditions:

$$-\sigma(u_+ - u_-) + (\mathcal{F}(u_+) - \mathcal{F}(u_-)) = \mathcal{K}_{\mathcal{D}}(u_-, \sigma); \tag{27}$$

where the so-called kinetic function $\mathcal{K}_{\mathcal{D}}(v_-, \sigma) \in \mathbb{R}^n$ is defined thanks to the next
identity valid for all $\epsilon > 0$ and derived from (24)–(25):

$$\epsilon \int_{\mathbb{R}_\xi} \mathcal{R}(u_\epsilon, u_\epsilon', u_\epsilon'')d\xi = \int_{\mathbb{R}_\xi} \mathcal{R}(\tilde{u}, \tilde{u}', \tilde{u}'')d\xi \equiv \mathcal{K}_{\mathcal{D}}(u_-, \sigma). \tag{28}$$

Clearly, the vector-valued kinetic function defined in (28) solely depends on the
prescribed state v_- and velocity σ who gave birth to a travelling wave solution
to (5), the viscosity matrix $\mathcal{D}$ being fixed. Equipped with a kinetic-function built
from a prescribed viscosity matrix $\mathcal{D}$, Berthon, Coquel, LeFloch [5] introduce the
following notion of weak solutions:

Definition 2.3. *Let be given a smooth kinetic function $\mathcal{K}_{\mathcal{D}}$. A piecewise Lipschitz
solution $u = u(x,t)$ is called a weak solution of the nonconservative limit system
in (5) iff in each region of continuity, u solves in the classical sense:*

$$\partial_t u + \partial_x \mathcal{F}(u) = 0, \tag{29}$$

while at points of jump, it obeys the generalized Rankine-Hugoniot conditions (27).

In other words, defining $\mu_u\{\mathcal{D}\}$ the bounded Borel measure which vanishes
in the region of continuity of u and has the mass $\mathcal{K}_{\mathcal{D}}(u_-, \sigma)$ along any curve of
discontinuity of u, Definition 2.3 is equivalent to the requirement that u solves:

$$\partial_t u + \partial_x \mathcal{F}(u) = \mu_u\{\mathcal{D}\}, \quad x \in \mathbb{R}, \, t > 0. \tag{30}$$

3. The Euler equations with several independent entropies

This section describes some of the main properties of the following nonconservative
system with singular viscous perturbation:

$$\begin{cases} \partial_t \rho^\epsilon + \partial_x \rho u^\epsilon = 0, \\ \partial_t \rho u^\epsilon + \partial_x \left(\rho u^{\epsilon 2} + \sum_{i=1}^N p_i^\epsilon\right) = \epsilon \partial_x \left(\sum_{i=1}^N \mu_i \partial_x u^\epsilon\right) + \epsilon \partial_x \left(\sum_{i=1}^N \kappa_i \partial_x T_i^\epsilon\right), \\ \partial_t \rho \varepsilon_i^\epsilon + \partial_x \rho \varepsilon_i u^\epsilon + p_i^\epsilon \partial_x u^\epsilon = \epsilon \mu_i (\partial_x u^\epsilon)^2 + \epsilon \partial_x (\kappa_i \partial_x T_i^\epsilon), \quad i = 1, \ldots, N, \end{cases} \tag{31}$$

in the regime of an infinite Reynolds number $\mathcal{R}ey = 1/\epsilon \to +\infty$. Here and with
classical notations, $\rho, \rho u$ and $\{\rho\varepsilon_i\}_{i=1,\ldots,N}$ respectively stand for the density, the

momentum and N independent internal energies of a complex compressible material. Observe that the system (31) just reads as a natural extension of the usual Navier-Stokes equations when a single pressure is involved in the momentum equation. In (31), N pressure laws enter and are independently governed via N distinct internal energies $\rho\varepsilon_i$. Several models from the physics actually enter the present framework with $N > 1$ and we refer the reader to [1] ,[6] for detailed examples. The system (31) is given the following condensed form:

$$\partial_t\mathbf{v}_\epsilon + \mathcal{A}(\mathbf{v}_\epsilon)\partial_x\mathbf{v}_\epsilon = \epsilon\mathcal{B}(\mathbf{v}_\epsilon,\partial_x\mathbf{v}_\epsilon,\partial_{xx}\mathbf{v}_\epsilon), \quad x \in \mathbb{R},\ t > 0, \tag{32}$$

which natural phase space reads:

$$\Omega_\mathbf{v} = \{\mathbf{v} := (\rho, \rho u, \{\rho\epsilon_i\}_{1\leq i\leq N}) \in \mathbb{R}^{N+2}/\rho > 0,\ \rho u \in \mathbb{R},\ \rho\epsilon_i > 0,\ 1 \leq i \leq N\}.$$

At the present stage, (32) does not seem to fit with either the model (3) or (5) we have promoted in the last section. Actually, (32) will be seen hereafter to recast in some instances as (3) but always as (5).

3.1. Closure equations and basic properties

Let us first state the (general) closure equations we assume in (31). The internal energies are assumed to obey the second principle of the thermodynamics, i.e., for any given $i \in \{1,\ldots,N\}$, $\rho\varepsilon_i$ is associated with an entropy ρs_i solution of:

$$-T_i ds_i = d\varepsilon_i + p_i d\tau, \quad \tau = 1/\rho, \tag{33}$$

with the property that the mapping $(\tau, s_i) \to \varepsilon_i(\tau, s_i)$ is strictly convex. Thus we get the required thermodynamic closure equations from (33):

$$p_i(\tau, s_i) = -\partial_\tau\varepsilon_i(\tau, s_i), \quad T_i(\tau, s_i) = -\partial_{s_i}\varepsilon_i(\tau, s_i),$$

where the temperature $T_i(\mathbf{v})$ is classically assumed to stay positive on $\Omega_\mathbf{v}$. As a well known consequence, the well defined mapping $(\tau, \varepsilon_i) \to s_i(\tau, \varepsilon_i)$ is strictly convex and so is also, with little abuse in the notation, the mapping $(\rho, \rho\varepsilon_i) \to \{\rho s_i\}(\rho, \rho\varepsilon_i) := \rho s_i(\frac{1}{\rho}, \frac{\rho\varepsilon_i}{\rho})$. Each pressure law $p_i(\mathbf{v})$ is assumed in addition to obey the general Weyl's conditions for real gases (see [19] for the details). At last, the viscosity laws $\mu_i : \Omega_\mathbf{v} \to \mathbb{R}_+$ and the conductivity laws $\kappa_i : \Omega_\mathbf{v} \to \mathbb{R}_+$, $1 \leq i \leq N$, in (31) are assumed to be smooth non negative functions but with the requirement that for some fixed $\mu_0 > 0$:

$$\sum_{i=1}^{N} \mu_i(\mathbf{v}) > \mu_0, \quad \text{for all } \mathbf{v} \in \Omega_\mathbf{v}. \tag{34}$$

All the above assumptions are quite classical within the frame of the usual Navier-Stokes equations (i.e., when $N = 1$ in (31)). Owing to these assumptions, our first statement highlights the relationships with this usual setting:

Lemma 3.1. *The underlying first order system in* (31), *obtained formally setting* $\epsilon = 0$:

$$\begin{cases} \partial_t\rho + \partial_x\rho u = 0, \\ \partial_t\rho u + \partial_x(\rho u^2 + \sum_{i=1}^{N} p_i(\mathbf{v})) = 0, \\ \partial_t\rho\varepsilon_i + \partial_x\rho\varepsilon_i u + p_i(\mathbf{v})\partial_x u = 0, \quad i = 1,\ldots,N, \end{cases} \tag{35}$$

is hyperbolic over $\Omega_{\mathbf{v}}$, with the following increasingly arranged eigenvalues:

$$\lambda_1(\mathbf{v}) = u - c < \lambda_{j=2,\ldots,N+1}(\mathbf{v}) = u < \lambda_{N+2}(\mathbf{v}) = u + c, \;\; c^2(\mathbf{v}) = \sum_{i=1}^{N} c_i^2(\mathbf{v}),$$

where each of the partial sound speed follows from $c_i^2(\mathbf{v}) := (\partial_\rho p_i)_{s_i} > 0$. Under the Weyl's assumption on the pressure laws, the $1-$ and $(N+2)-$ fields are genuinely nonlinear. All the other intermediate fields are linearly degenerate.

The intermediate fields with $i \in \{2, \ldots, N+1\}$ coincide with a contact discontinuity across which the eigenvalue u stays continuous. In other words, discontinuities coming with these fields do not induce ambiguity in all the non conservative products $p_i(\mathbf{v}) \times \partial_x u$ involved in (35). By contrast, the two extreme fields are genuinely nonlinear and are thus responsible for the occurrence of shock waves where the velocity u and each of the partial pressures $p_i(\mathbf{v})$ can be seen to achieve non trivial jumps [1], [6]. This is already the case in the standard Euler equations with $N = 1$. Hence and for these extreme discontinuities, ambiguities arise in the nonconservative products entering (35).

3.2. Equivalent formulations

In order to study the singular limit $\epsilon \to 0$ in (31), let us implement the program sketched in Section 2 and thus exhibit all the nontrivial conservation laws (8) satisfied by the smooth solution of (35). The next statement provides such laws but when directly expressed in the presence of the viscous perturbations in (31):

Proposition 3.2. *Smooth solutions of (31) satisfy the following conservation law:*

$$\partial_t (\rho E)(\mathbf{v}^\epsilon) + \partial_x (\rho H u)(\mathbf{v}^\epsilon) = \epsilon \partial_x \left(\left(\sum_{i=1}^{N} \mu_i \right) u^\epsilon \partial_x u^\epsilon \right) + \sum_{i=1}^{N} \partial_x (\kappa_i \partial_x T_i^\epsilon), \quad (36)$$

where the total energy $\rho E : \Omega_{\mathbf{v}} \to \mathbb{R}_+$ and the total enthalpy $\rho H : \Omega_{\mathbf{v}} \to \mathbb{R}_+$ respectively read:

$$(\rho E)(\mathbf{v}) = \frac{(\rho u)^2}{2\rho} + \sum_{i=1}^{N} \rho \varepsilon_i, \quad (\rho H)(\mathbf{v}) = (\rho E)(\mathbf{v}) + \sum_{i=1}^{N} p_i(\mathbf{v}). \quad (37)$$

These solutions next obey the following N equations:

$$-T_i(\mathbf{v}^\epsilon) \times \{\partial_t (\rho s_i)(\mathbf{v}^\epsilon) + \partial_x (\rho s_i u)(\mathbf{v}^\epsilon)\} = \epsilon \mu_i (\partial_x u^\epsilon)^2 + \epsilon \partial_x (\kappa_i \partial_x T_i^\epsilon), \quad (38)$$

and thus also the N entropy balance equations:

$$\partial_t (\rho s_i)(\mathbf{v}^\epsilon) + \partial_x (\rho s_i u)(\mathbf{v}^\epsilon) = -\epsilon \frac{\mu_i}{T_i^\epsilon} (\partial_x u^\epsilon)^2 - \epsilon \kappa_i \left(\frac{\partial_x T_i^\epsilon}{T_i^\epsilon} \right)^2 - \epsilon \partial_x \left(\kappa_i \frac{\partial_x T_i^\epsilon}{T_i^\epsilon} \right). \quad (39)$$

Note from (39) that classical non linear transformations in the s_i yield further additional balance equations for governing $\varphi(s_1, \ldots, s_N)$ where $\varphi : \mathbb{R}^N \to \mathbb{R}$ denotes any given arbitrary smooth function. Nevertheless and without specific assumptions on the thermodynamic closure equations (see [1], [6]), none of these additional equations boils down to a non trivial additional conservation law. In the light of this result, we are led to introduce the well-defined change of variable $\mathbf{v} \to \mathbf{u}(\mathbf{v}) = \mathbf{u} = \{\rho, \rho u, \{\rho s_i\}_{1 \le i \le N}\}^T$ so as to recast (31) under the form (5) according to:

Proposition 3.3. *Smooth solutions of* (31) *obey equivalently the system:*

$$
\begin{cases}
\partial_t \rho^\epsilon + \partial_x \rho u^\epsilon = 0, \\
\partial_t \rho u^\epsilon + \partial_x \left(\rho u^{\epsilon 2} + \sum_{i=1}^N p_i^\epsilon \right) = \epsilon \partial_x \left(\sum_{i=1}^N \mu_i \partial_x u^\epsilon \right), \\
\partial_t \rho s_i^\epsilon + \partial_x \rho s_i u^\epsilon = -\epsilon \frac{\mu_i}{T_i^\epsilon} (\partial_x u^\epsilon)^2 - \epsilon \kappa_i (\partial_x ln T_i^\epsilon)^2 - \epsilon \partial_x \kappa_i \partial_x (ln T_i^\epsilon).
\end{cases}
\tag{40}
$$

Note that the conservation law for the total energy is recovered from (40) as an additional nontrivial law:

$$
\partial_t (\rho E)(\mathbf{u}^\epsilon) + \partial_x (\rho H u)(\mathbf{u}^\epsilon) = \epsilon \partial_x \left(\left(\sum_{i=1}^N \mu_i \right) u^\epsilon \partial_x u^\epsilon \right) + \sum_{i=1}^N \partial_x (\kappa_i \partial_x T_i^\epsilon).
\tag{41}
$$

The reason for promoting the equivalent system (40) stems from the important property:

Proposition 3.4. *The function* $\mathbf{u} \in \Omega_{\mathbf{u}} \to (\rho E)(\mathbf{u}) \in \mathbb{R}$ *is strictly convex.*

We refer to [6] for a proof and related convexity properties. Other relevant equivalent forms are actually available [1], [6] but will not be addressed here for shortness. Let us now illustrate that under specific modelling assumptions, the system (31) takes the form (3). This will help to shade light in the forthcoming numerical methods. Let assume the viscosity laws to be given by N non negative real numbers $\mu_i \in \mathbb{R}+$, $1 \le i \le N$, with up to some relabelling $\mu_N > 0$ so that the requirement (34) is met. Then, observe the following $(N-1)$ relations easily derived from (38):

$$
-T_N^\epsilon \frac{\mu_i}{\mu_N} \times \{\partial_t (\rho s_N)^\epsilon + \partial_x (\rho s_N u)^\epsilon\} = \epsilon \mu_i (\partial_x u^\epsilon)^2 + \epsilon \partial_x \left(\frac{\kappa_N \mu_i}{\mu_N} \partial_x T_N^\epsilon \right),
\tag{42}
$$

which subtracted from (38) yield the $(N-1)$ additional laws:

$$
T_i^\epsilon \times \{\partial_t (\rho s_i)^\epsilon + \partial_x (\rho s_i u)^\epsilon\} - T_N \frac{\mu_i}{\mu_N} \times \{\partial_t (\rho s_N)^\epsilon + \partial_x (\rho s_N u)^\epsilon\} =
$$
$$
\epsilon \partial_x \left(\frac{\kappa_N \mu_i}{\mu_N} \partial_x T_N^\epsilon - \kappa_i \partial_x T_i^\epsilon \right).
\tag{43}
$$

Equipped with (43), it can be shown (see [6] for a related proof):

Proposition 3.5. *Let be given* N *constant non negative viscosity coefficients* $\{\mu_i\}_{1 \le i \le N}$ *with* $\mu_N > 0$. *Then the smooth solutions of* (31) *obey equivalently in* *the **u**-variable the following system:*

$$\begin{cases} \partial_t \rho^\epsilon + \partial_x \rho u^\epsilon = 0, \\ \partial_t \rho u^\epsilon + \partial_x \left(\rho u^{\epsilon 2} + \sum_{i=1}^N p_i^\epsilon \right) = \epsilon \partial_x \left(\sum_{i=1}^N \mu_i \partial_x u^\epsilon \right) + \epsilon \partial_x \left(\sum_{i=1}^N \kappa_i \partial_x T_i^\epsilon \right), \\ T_i(\mathbf{u}^\epsilon) \times \{ \partial_t (\rho s_i)^\epsilon + \partial_x (\rho s_i u)^\epsilon \} - T_N(\mathbf{u}^\epsilon) \frac{\mu_i}{\mu_N} \times \{ \partial_t (\rho s_N)^\epsilon + \partial_x (\rho s_N u)^\epsilon \} = \\ \qquad\qquad \epsilon \partial_x \left(\frac{\kappa_N \mu_i}{\mu_N} \partial_x T_N^\epsilon - \kappa_i \partial_x T_i^\epsilon \right), \\ \partial_t (\rho E)(\mathbf{u}^\epsilon) + \partial_x (\rho H u) = \epsilon \partial_x \left(\left(\sum_{i=1}^N \mu_i \right) u^\epsilon \partial_x u^\epsilon \right) + \sum_{i=1}^N \partial_x (\kappa_i \partial_x T_i^\epsilon), \end{cases}$$

$$(44)$$

which condensed form clearly reads:

$$\mathcal{A}_0(\mathbf{u}_\epsilon) \partial_t \mathbf{u}_\epsilon + \mathcal{A}_1(\mathbf{u}_\epsilon) \partial_x \mathbf{u}_\epsilon = \epsilon \partial_x (\mathcal{D}(\mathbf{u}_\epsilon) \partial_x \mathbf{u}_\epsilon), \quad x \in \mathbb{R}, \ t > 0, \qquad (45)$$

with $\mathcal{A}_0$ invertible:

$$Det(\mathcal{A}_0(\mathbf{u})) = -\frac{1}{\mu_N} \left(\sum_{i=1}^N \mu_i \right) (\Pi_{i=1}^N T_i(\mathbf{u})) < 0, \quad \text{for all } \mathbf{u} \in \Omega_\mathbf{u}. \qquad (46)$$

Remark 2. Observe that for general viscosity laws, the above manipulations generally yield the rather cumbersome form:

$$\mathcal{A}_0(\mathbf{u}_\epsilon) \partial_t \mathbf{u}_\epsilon + \mathcal{A}_1(\mathbf{u}_\epsilon) \partial_x \mathbf{u}_\epsilon = \epsilon \mathcal{B}(\mathbf{u}_\epsilon, \partial_x \mathbf{u}_\epsilon, \partial_{xx} \mathbf{u}_\epsilon), \quad x \in \mathbb{R}, \ t > 0, \qquad (47)$$

*with a nonconservative regularization term. Precisely except when the conductivity laws are set to zero, in which case (44) is still valid but with ratios of viscosities depending on **u**.*

The equivalent form (40) with the additional law (41) stays free from modelling assumptions on both the viscosity and conductivity laws. (40)–(41) will play a central role in the numerical analysis of the system (31) in the singular limit $\epsilon \to 0$ we now address.

3.3. Singular limit

We discuss on the ground of the mathematical tools introduced in Section 2 the limit system obtained from (31) as the rescaling parameter ϵ goes to zero. For reasons put forward in Section 2, the $2N$ viscosity and conductivity laws entering the singular viscous perturbation in (31) are tacitly fixed from now on, except when otherwise specified. With little abuse in the notations, this set of $2N$ constitutive laws is referred hereafter as to the viscous closure $\mathcal{D}$.

Focusing ourselves on the notion of weak solutions in the class of piecewise Lipschitz continuous functions, we first report the main properties of the rescaled travelling wave solutions to (31) with $\epsilon = 1$ (see (15)–(16)):

$$\mathbf{u}(x - \sigma t) = \mathbf{u}(\xi), \quad \lim_{\xi \to \pm\infty} \mathbf{u}(\xi) = \mathbf{u}_\pm, \quad \lim_{\xi \to \pm\infty} \mathbf{u}'(\xi) = 0, \qquad (48)$$

for some speed $\sigma \in \mathbb{R}$ and states $\mathbf{u}_-$, $\mathbf{u}_+$ in $\Omega_{\mathbf{u}}$. Under the asymptotic conditions expressed in (48), the above function $\mathbf{u}$ has to solve (40) (again with $\epsilon = 1$) and thus the following $(N+2) \times (N+2)$ ODE system:

$$\begin{cases} -\sigma d_\xi \rho + d_\xi \rho u = 0, \\ -\sigma d_\xi \rho u + d_\xi \rho u^2 + \sum_{i=1}^N p_i(\mathbf{v})) = d_\xi \left(\sum_{i=1}^N \mu_i d_\xi u \right), \\ -\sigma d_\xi \rho s_i + d_\xi \rho s_i u = -\frac{\mu_i}{T_i}(d_\xi u)^2 - \kappa_i \left(\frac{d_\xi T_i}{T_i} \right)^2 - d_\xi \left(\kappa_i \frac{d_\xi T_i}{T_i} \right). \end{cases} \quad (49)$$

Classical considerations allow for studying travelling solutions coming solely with the first genuinely nonlinear field. Indeed, travelling solutions for the symmetrical extreme field are just recovered when reversing the sign of the velocity and ξ while exchanging the role of the endpoints $\mathbf{u}_-$ and $\mathbf{u}_+$. Berthon, Coquel [2], [4] have proved the existence in the large of solutions (48) to the dynamical system (49) with the closure equations discussed in Section 3.1 but under the simplifying assumption of zero heat conductivities. Such an assumption is made to allow for a Lasalle invariance principle for the large ODE system (49) which in connection with a suitable Lyapunov function yield the existence of an endpoint $\mathbf{u}_+ \in \Omega_{\mathbf{u}}$ as soon as the state $\mathbf{u}_-$ and the velocity σ are prescribed according to:

Proposition 3.6. *Let $\mathbf{u}_- \in \Omega_{\mathbf{u}}$ be given. Let us consider a velocity σ subject to the Lax compression condition:*

$$u_- - c(\mathbf{u}_-) > \sigma, \quad c^2(\mathbf{u}_-) = \sum_{i=1}^N \partial_\rho p_i(\mathbf{u}_-). \quad (50)$$

Then there exists a travelling wave solution (48) to (49), unique (up to some translation) and connecting some (unique) state $\mathbf{u}_+(\mathbf{u}_-, \sigma)$ at $+\infty$.

We conjecture that such a positive result persists in the case of general conductivity laws. Let us underline that due to the nonconservative nature of the dynamical system (49), barely little is known about the exit state $\mathbf{u}_+(\mathbf{u}_-, \sigma)$ except its existence and the property that its precise form heavily depends on the viscous closure $\mathcal{D}$ through the N ratios of the viscosity laws $\mu_i(\mathbf{u})/\sum_{j=1}^N \mu_j(\mathbf{u})$. The state $\mathbf{u}_-$ and the speed σ being fixed according to (50), Berthon, Coquel [2], [4] have proved the existence of a smooth manifold of codimension 2 in $\Omega_{\mathbf{u}}$ uniquely made of critical points in the future for the dynamical system (49) with the property that each of these critical points can be reached at $+\infty$ provided that N ratios of viscosities are suitably prescribed. In the setting of N constants viscosities for N given polytropic gases, Chalons, Coquel [8] have shown how to explicitly determine the exit state $\mathbf{u}_+(\mathcal{D}; \mathbf{u}_-, \sigma)$ when simply solving a (known) scalar nonlinear algebraic equation with coefficients depending on $\mathcal{D}$.

Remark 3. In the case of a general viscous closure $\mathcal{D}$, let us stress that the system (49) can be given a dimensionless form. Then all the possible dimensionless exit states $\tilde{\mathbf{u}}_+(\mathcal{D}; ., .)$ (inferred from Proposition 3.6 with $\mathbf{u}_-$ running in $\Omega_{\mathbf{u}}$ and σ in

$\mathbb{R}$ *according to* (50)) *entirely depend on a reduced set of dimensionless numbers. Typically, N numbers* : $\Omega_{\mathbf{u}} \times \mathbb{R} \rightarrow [0,1]^N$ *are in order:*

$$0 \le \mathcal{M} = \frac{c(\mathbf{u}_-)}{u_- - \sigma} \le 1, \quad 0 \le \mathcal{M}_i = \frac{\sqrt{\partial_\rho p_i(\mathbf{u}_-)}}{c(\mathbf{u}_-)} \le 1, \; i \in 1, \dots, N-1, \quad (51)$$

where $\mathcal{M}$ *denotes the inverse of the usual Mach number while the* $\mathcal{M}_i$ *can be referred as to thermodynamic Mach numbers (see* [8] *for related reduced numbers). This makes feasible the numerical tabulation of the dimensionless exit states* $\tilde{\mathbf{u}}_+(\mathcal{D}; \{\mathcal{M}_i\}(\mathbf{u}_-), \mathcal{M}(\mathbf{u}_-, \sigma))$ *on the compact domain* (51).

Next, the sequence $\mathbf{u}_\epsilon : \xi \in \mathbb{R} \rightarrow \mathbf{u}(\xi/\epsilon) \in \Omega_{\mathbf{u}}$ built from the travelling wave solution (48)–(49) is seen to converge strongly in L^1_{loc} when $\epsilon \rightarrow 0$ to a limit step function:

$$\mathbf{u}(x,t) = \mathbf{u}_+ + (\mathbf{u}_+(\mathcal{D}, \mathbf{u}_-, \sigma) - \mathbf{u}_-)H(x - \sigma t), \quad x \in \mathbb{R}, t > 0. \quad (52)$$

With the chainrule (24)–(27) proposed in Section 2, this limit function, referred as to a shock solution to the singular limit in (40), solves the generalized Rankine-Hugoniot jump conditions:

$$\begin{cases} -\sigma(\rho_+ - \rho_-) + ((\rho u)_+ - (\rho u)_-) = 0, \\ -\sigma((\rho u)_+ - (\rho u)_-) + ((\rho u^2 + \sum_{i=1}^N p_i(\mathbf{u}))_+ - (\rho u^2 + \sum_{i=1}^N p_i(\mathbf{u}))_-) = 0, \\ -\sigma((\rho s_i)_+ - (\rho s_i)_-) + ((\rho s_i u)_+ - (\rho s_i u)_-) = \mathcal{K}\{\mathcal{D}\}_i(\mathbf{u}_-, \sigma), \quad i = 1, \dots, N, \end{cases}$$

$$(53)$$

where for each $i \in \{1, \dots, N\}$, the kinetic functions are given by:

$$\mathcal{K}\{\mathcal{D}\}_i(\mathbf{u}_-, \sigma) = -\int_{\mathbb{R}_\xi} \left\{ \frac{\mu_i}{T_i}(d_\xi u)^2 + \kappa_i \left(\frac{d_\xi T_i}{T_i}\right)^2 \right\}(\mathbf{u}(\xi)) d\xi \le 0. \quad (54)$$

Remark 4. In practical issues, once the right state $\mathbf{u}_+(\mathcal{D}; \mathbf{u}_-, \sigma)$ *has been numerically solved (see Remark 3), the required kinetic functions are evaluated thanks to the identities (see* [8] *for instance):*

$$\mathcal{K}\{\mathcal{D}\}_i(\mathbf{u}_-, \sigma) = \rho_-(u_- - \sigma)((s_i)_+(\mathcal{D}; \mathbf{u}_-, \sigma) - (s_i)_-), \quad i = 1, \dots, N, \quad (55)$$

in place of the equivalent but cumbersome form (54). *Of course, dimensionless forms of* (55) *are again in order.*

Observe that the smooth travelling solution (48) also solves the additional conservation law (41) so as a by-product, we get the additional jump condition:

$$-\sigma((\rho E)(\mathbf{u}_+) - (\rho E)(\mathbf{u}_-)) + ((\rho H u)(\mathbf{u}_+) - (\rho H u)(\mathbf{u}_-)) = 0. \quad (56)$$

For forthcoming numerical reasons, it is then crucial to recognize that the next set of generalized Rankine-Hugoniot jump conditions can be built from (53)–(56) according to:

Proposition 3.7. *Assume that up to some relabelling* $\mu_N(\mathbf{u}) > 0$ *so that*

$$\mathcal{K}\{\mathcal{D}\}_N(\mathbf{u}_-, \sigma) < 0.$$

Then the shock solution (52) *solves in the* **u** *variable:*

$$
\begin{cases}
-\sigma(\rho_+ - \rho_-) + ((\rho u)_+ - (\rho u)_-) = 0, \\
-\sigma((\rho u)_+ - (\rho u)_-) + ((\rho u^2 + \sum_{i=1}^{N} p_i(\mathbf{u}))_+ - (\rho u^2 + \sum_{i=1}^{N} p_i(\mathbf{u}))_-) = 0, \\
\{((\rho s_i)_+ - (\rho s_i)_-) + ((\rho s_i u)_+ - (\rho s_i u)_-\} \\
\quad - \frac{\mathcal{K}\{\mathcal{D}\}_i(\mathbf{u}_-,\sigma)}{\mathcal{K}\{\mathcal{D}\}_N(\mathbf{u}_-,\sigma)} \{(\rho s_N)_+ - (\rho s_N)_-) + ((\rho s_N u)_+ - (\rho s_N u)_-\} = 0, \\
\hfill i = 1, \ldots, N-1, \\
-\sigma((\rho E)(\mathbf{u}_+) - (\rho E)(\mathbf{u}_-)) + ((\rho H u)(\mathbf{u}_+) - (\rho H u)(\mathbf{u}_-)) = 0.
\end{cases}
\tag{57}
$$

For any given $i \in \{1, \ldots, N\}$, let us define $\mu_{\mathbf{u}}\{\mathcal{D}\}_i$ the non positive bounded Borel measure which vanishes in the region of continuity of **u** and has the mass $\mathcal{K}\{\mathcal{D}\}_i(\mathbf{u}_-, \sigma)$ along any curve of discontinuity of **u**, we introduce:

Definition 3.8. *The singular limit system* (40), *in the class of piecewise Lipschitz continuous functions, takes the form of the following extended Euler equations:*

$$
\begin{cases}
\partial_t \rho + \partial_x \rho u = 0, \\
\partial_t \rho u + \partial_x (\rho u^2 + \sum_{i=1}^{N} p_i(\mathbf{u})) = 0, \\
\partial_t \rho s_i + \partial_x \rho s_i u = \mu_{\mathbf{u}}\{\mathcal{D}\}_i, \quad i = 1, \ldots, N.
\end{cases}
\tag{58}
$$

Weak solutions of (58) *obeys the additional non trivial conservation laws in the usual weak sense:*

$$
\partial_t (\rho E)(\mathbf{u}) + \partial_x (\rho H u)(\mathbf{u}) = 0.
\tag{59}
$$

Here, weak solutions of the extended Euler equations (58) are understood in the sense of Definition 2.3. For forthcoming numerical purposes, it is useful to consider the setting of viscous closures $\mathcal{D}$ with arbitrary conductivity laws but constant viscosities (or arbitrary viscosity laws with null conductivities). In view of the equivalent form (44), weak solutions of (58) equally solve in the sense of Definition 2.2:

$$
\begin{cases}
\partial_t \rho + \partial_x \rho u = 0, \\
\partial_t \rho u + \partial_x (\rho u^2 + \sum_{i=1}^{N} p_i(\mathbf{u})) = 0, \\
\left[T_i(\mathbf{u})\{\partial_t(\rho s_i) + \partial_x(\rho s_i u)\} \right]_{\phi_{\mathcal{D}}} \\
\quad - \left[T_N(\mathbf{u}) \frac{\mu_i}{\mu_N} \{\partial_t(\rho s_N) + \partial_x(\rho s_N u)\} \right]_{\phi_{\mathcal{D}}} = 0, \quad 1 \le i \le N-1, \\
\partial_t (\rho E)(\mathbf{u}) + \partial_x(\rho H u) = 0,
\end{cases}
\tag{60}
$$

where the nonconservative products take the form (22) at points of jump. To go one step further, it can be easily seen that at such points the following identities hold for any given $i \in \{1, \ldots, N-1\}$:

$$
\begin{aligned}
T_i(\mathbf{u}_-,\sigma)\{-\sigma((\rho s_i)_+ - (\rho s_i)_-) + ((\rho s_i u)_+ - (\rho s_i u)_-)\} \\
-T_N(\mathbf{u}_-,\sigma)\frac{\mu_i}{\mu_N}\{-\sigma((\rho s_N)_+ - (\rho s_N)_-) + ((\rho s_N u)_+ - (\rho s_N u)_-)\} = 0,
\end{aligned}
\tag{61}
$$

for some averaged temperatures $\{T_i(\mathbf{u}_-, \sigma)\}_{1 \leq i \leq N}$. These relations can be understood as a particular case of the generalized jump conditions entering (57). Let us conclude this section with the following (global) existence result of solutions to the Riemann problem for the extended Euler Equations (58)

Theorem 3.9. (*Chalons, Coquel* [8]). *Let be given N constant non-negative viscosity coefficients $\{\mu_i\}_{1 \leq i \leq N}$ with up to some relabelling $\mu_N > 0$ in the setting of N independent pressure laws for polytropic gases. Let be given two states $\mathbf{u}_L$ and $\mathbf{u}_R$ in $\Omega_\mathbf{u}$. Then the Cauchy problem (58) with initial data $\mathbf{u}_0(x) = \mathbf{u}_L, x < 0, \mathbf{u}_R, x > 0$ has an unique solution away from vacuum.*

We refer to [8] for a precise definition of vacuum. The proof of the above result requires a sharp characterization of the right states $\mathbf{u}_+(\mathcal{D}, \mathbf{u}_-, \sigma)$ coming with travelling wave solutions. Hence the setting under consideration.

4. Riemann solvers and kinetic functions

In this section, we address the numerical approximation of the weak solutions of the extended Euler equations (58). The small scale sensitiveness of shock solutions makes this issue particularly challenging. The core of the difficulty indeed stems from the property of shock solutions to be regularization dependent: the artificial dissipation terms induced by numerical methods tend to corrupt the discrete shocks. Our main purpose in this section is to illustrate how to enforce the artificial diffusion in classical numerical methods to mimic the exact dissipation mechanism. Kinetic functions play a central role in the correction procedure we describe hereafter. This procedure intends to keep all the independent discrete rate of entropy dissipation in the exact balance prescribed by the kinetic functions in the generalized jump conditions (57). A deeply related strategy has been first introduced by Berthon, Coquel [3] with $N = 2$ in terms of a local (cell by cell) nonlinear correction procedure, then extended to the general case $N \geq 2$ by Chalons, Coquel [7]. It has recently received several fully explicit versions in the case $N \geq 2$ in the works by Chalons, Coquel [9], [10] and has been successfully extended to problems with two space variables in [6]. For convenience, all these works address the numerical approximation of the equivalent form (60) under the assumption of general viscosity laws with null conductivity coefficients (or say constant viscosity coefficients for arbitrary conductivity laws, see Remark 2). In this review, we extend the correction procedure to the general case, thus tackling directly the formulation (58) of the extended Euler Equations. To avoid unnecessary technical details with approximate Riemann solvers, the extension is performed on the basis of the pure Godunov method. We first motivate the very need to correct this classical solver when pointing out the origin of its failure. Understanding its roots then dictates the procedure. We conclude when highlighting the deep relations in all the existing techniques.

4.1. Origin of the failure

For simplicity in the notations, we restrict ourselves to uniform cartesian discretization of $\mathbb{R}_t \times \mathbb{R}_x$ defined by a constant time step Δt and a constant space step Δx. Introducing $x_{j+1/2} = (j+1/2)\Delta x$ with $j \in \mathbb{Z}$ and $t^n = n\Delta t$ with $n \in \mathbb{N}$, the cartesian grids under consideration then read:

$$\cup_{j\in\mathbb{Z}, n\in\mathbb{N}} C_j^n, \quad C_j^n = [x_{j-1/2}, x_{j+1/2}) \times [t^n, t^{n+1}). \tag{62}$$

The approximate solution of the Cauchy problem (58) with $\mathbf{u}_0$ as initial data, we denote $\mathbf{u}_\lambda(x,t)$ with $\lambda = \Delta t/\Delta x$, is classically sought as a piecewise constant function at each time level t^n:

$$\mathbf{u}_\lambda(x, t^n) := \mathbf{u}_j^n, \quad \text{for all } x \in [x_{j-1/2}, x_{j+1/2}), \quad n > 0, \quad j \in \mathbb{Z}, \tag{63}$$

with when $n = 0$:

$$\mathbf{u}_j^0 = \frac{1}{\Delta x} \int_{x_{j-1/2}}^{x_{j+1/2}} \mathbf{u}_0(x)dx, \quad j \in \mathbb{Z}. \tag{64}$$

Assuming the approximate solution $\mathbf{u}_\lambda(x, t^n)$ to be known at a given time $t^n \geq 0$, this one is then defined for $t \in [t^n, t^{n+1})$ as the solution of the Cauchy problem (58) with $\mathbf{u}_\lambda(x, t^n)$ as initial data. Choosing Δt small enough, *i.e.* under the CFL restriction:

$$\lambda \max_{\mathbf{u}} \rho(\nabla_{\mathbf{u}}\mathcal{F}(\mathbf{u})) \leq \frac{1}{2}, \tag{65}$$

where the maximum is taken over all the $\mathbf{u}$ under consideration, $\mathbf{u}_\lambda(x,t)$ with $t \in [t^n, t^{n+1})$ is nothing but the juxtaposition of a sequence of non interacting adjacent Riemann solutions $\mathbf{w}((x - x_{j+1/2})/(t - t^n); \mathbf{u}_j^n, \mathbf{u}_{j+1}^n)$, centered at each cell interface $x_{j+1/2}$. Let us then classically consider the L^2-projection of this solution at time t^{n+1-} onto piecewise constant functions:

$$\mathbf{u}_j^{n+1-} = \frac{1}{\Delta x} \int_{x_{j-1/2}}^{x_{j+1/2}} \mathbf{u}_\lambda(x, t^{n+1-})dx, \quad j \in \mathbb{Z}. \tag{66}$$

Easy calculations based on the Green formula show that the averages (66) reexpress conveniently in the form:

$$\begin{cases} \rho_j^{n+1-} = \rho_j^n - \lambda\Delta\{\rho u\}(\mathbf{w}(0^+; \mathbf{u}_j^n, \mathbf{u}_{j+1}^n) \\ (\rho u)_j^{n+1-} = (\rho u)_j^n - \lambda\Delta\{\rho u^2 + \sum_{i=N}^N p_i\}(\mathbf{w}(0^+; \mathbf{u}_j^n, \mathbf{u}_{j+1}^n), \\ (\rho s_i)_j^{n+1-} = (\rho s_i)_j^n - \lambda\Delta\{\rho s_i u\}_{j+1/2}^n + \lambda \mu_{\mathbf{u}}\{\mathcal{D}\}_i(C_j^n), \end{cases} \tag{67}$$

where for any given $i \in \{1, \ldots, N\}$, $\mu_{\mathbf{u}}\{\mathcal{D}\}_i(C_j^n)$ denotes the (non positive) mass of the bounded Borel measure $\mu_{\mathbf{u}}\{\mathcal{D}\}_i$ taken over all the possible shock waves that propagate in the cell C_j^n.

According to the usual Godunov's procedure, one would update the approximate solution at time t^{n+1} when defining $\mathbf{u}_\lambda(x, t^{n+1}) = \mathbf{u}_j^{n+1-}$ for all $x \in [x_{j-1/2}, x_{j+1/2})$ and $j \in \mathbb{Z}$. However, such an updating formula would prevent the L^1 norm of the total energy to be preserved with time because of the next statement:

Lemma 4.1. *Under the CFL condition* (65), *the following inequality holds for all* $j \in \mathbb{Z}$:

$$
\begin{aligned}
\{\rho E\}(\mathbf{u}_j^{n+1-}) &\leq \frac{1}{\Delta x} \int_{x_{j-1/2}}^{x_{j+1/2}} \{\rho E\}(\mathbf{u}_\lambda(x, t^{n+1-})) dx \\
&= \{\rho E\}(\mathbf{u}_j^n) - \lambda\Delta\{\rho H u\}(\mathbf{w}(0^+; \mathbf{u}_j^n, \mathbf{u}_{j+1}^n)),
\end{aligned}
\tag{68}
$$

the first inequality being strict generally speaking.

The equality entering the above estimate simply follows from the property that the weak solution $\mathbf{u}_\lambda(x, t)$ with $t \in [t^n, t^{n+1})$ preserves the total energy in view of the additional conservation law (56) valid in $\mathcal{D}'$. Next, the inequality in (68) is just a consequence of the classical Jensen inequality when invoking the strict convexity property of the function $\mathbf{u} \in \Omega_{\mathbf{u}} \to \{\rho E\}(\mathbf{u}) \in \mathbb{R}$ stated in Proposition 3.4. It is well known that in general, the resulting inequality holds strictly. More precisely, it can be seen that (see Coquel, LeFloch [13] for instance):

$$
\begin{aligned}
\{\rho E\}(\mathbf{u}_j^{n+1-}) &= \frac{1}{\Delta x} \int_{x_{j-1/2}}^{x_{j+1/2}} \{\rho E\}(\mathbf{u}_\lambda(x, t^{n+1-})) dx \\
&\quad - \mathcal{O}(1)\left(\|\mathbf{u}_j^n - \mathbf{u}_{j-1}^n\|^2 + \|\mathbf{u}_{j+1}^n - \mathbf{u}_j^n\|^2\right),
\end{aligned}
\tag{69}
$$

for some positive $\mathcal{O}(1)$ depending on the convexity modulus of $\{\rho E\}(\mathbf{u})$. Rephrasing the inequality (68) and its precised form (69), the updating formulae (67) make the L^1-norm of the total energy to dramatically decrease with time as soon as non trivial shock solutions propagate in the discrete solution. Thus the classical Godunov method (67) cannot provide us with a relevant numerical method for approximating the discontinuous solutions under consideration. The estimate (69) is nothing but the origin of the reported failure in the proper capture of shock solutions to (58).

One would be tempted to promote the conservation of the total energy at the discrete level so as to understand (up to some relabelling) the governing equation for $\{\rho s_N\}(\rho, \rho u, \rho E, \{\rho s_i\}_{1 \leq i \leq N-1})$ as an additional nontrivial equation. Again and because of convexity properties not reported here, such a strategy can only grossly fail. Indeed, it can be easily shown that this time, the equivalent set of generalized jump conditions (57) cannot hold true at the discrete level (already in the case of a single propagating shock wave). We refer the reader to [7], [9] and [10] for closely related proofs. The correction procedure we now propose finds its root in the negative result we have just reported: we propose to enforce for validity at the discrete level the generalized Rankine-Hugoniot conditions (57).

4.2. Correction procedure

In order to restore the validity of the generalized jump conditions (57) at each time level, we consider a cell by cell procedure to take place as soon as $\mu_{\mathbf{u}}\{\mathcal{D}\}_N(\mathcal{C}_j^n) < 0$. The assumption of a non zero mass in the current cell $\mathcal{C}_j^n$ just expresses the fact that non trivial shock waves do propagate in, so that a correction is needed to counteract the negative effects of (69). We propose to keep unchanged the updated values of both the density and momentum in conservation form:

$$
\rho_j^{n+1} = \rho_j^{n+1-}, \quad (\rho u)_j^{n+1} = (\rho u)_j^{n+1-}, \quad \text{for all } j \in \mathbb{Z},
\tag{70}
$$

Next, the N entropies $(\rho s_i)_j^{n+1}$ are sought to be solutions of the following $(N-1)$ relations with $i \in \{1,\ldots,N-1\}$:

$$
\{(\rho s_i)_j^{n+1} - (\rho s_i)_j^n + \lambda\Delta\{\rho s_i u\}_{j+1/2}^n\}-
$$
$$
\frac{\mu_{\mathbf{u}}\{\mathcal{D}\}_i(\mathcal{C}_j^n)}{\mu_{\mathbf{u}}\{\mathcal{D}\}_N(\mathcal{C}_j^n)}\{(\rho s_N)_j^{n+1} - (\rho s_N)_j^n + \lambda\Delta\{\rho s_N u\}_{j+1/2}^n\} = 0,
\tag{71}
$$

supplemented by (see (68)):

$$
\{\rho E\}(\rho_j^{n+1-}, (\rho u)_j^{n+1-}, \{\rho s_i\}_j^{n+1}) \equiv \{\rho E\}_j^n - \lambda\Delta\{\rho H u\}(\mathbf{w}(0^+, \mathbf{u}_j^n, \mathbf{u}_{j+1}^n))
\tag{72}
$$

Let us underline that the identities (71)–(72) are discrete forms of the N last jump relations in (57). When focusing ourselves in the setting of general viscosity laws with zero conductivity coefficients (or say constant viscosity coefficients with arbitrary conductivity laws), it can be easily seen that for any given $i \in \{1,\ldots,N-1\}$, the ratio $\mu_{\mathbf{u}}\{\mathcal{D}\}_i(\mathcal{C}_j^n)/\mu_{\mathbf{u}}\{\mathcal{D}\}_N(\mathcal{C}_j^n)$ coincides with an averaged form of the ratio $\{\mu_i T_N/\mu_N T i\}(\mathbf{u})$ in the singular limit in (44) (see Proposition 3.5). In this sense, (71)–(72) just provide a consistent discrete approximation of the limit system in (44). Achieving such a consistency property stays at the basis of the prediction-correction discrete methods developed in [3], [9].

Closely related proofs in these works allow to prove the existence of a unique state $\mathbf{u}_j^{n+1} \in \Omega_{\mathbf{u}}$ solution of (70)–(71)–(72). The discrete method is thus well-defined. Let us stress that an essential ingredient in the proof stays in the property that the total energy has been over-dissipated in the prediction step! (see [3], [9] for instance for the details).

4.3. Numerical experiments

We present here numerical evidences for illustrating the validity of the numerical strategy that we have proposed in previous section. For that, we consider system (31) where N is taken equal to 3 and the corresponding three internal energies are associated with polytropic ideal gases (thermally and calorically perfect). More precisely, introducing N constant adiabatic exponents $\gamma_i > 1$ for all $i = 1,\ldots,3$, we set

$$
\rho\varepsilon_i^\epsilon = \frac{p_i^\epsilon}{\gamma_i - 1}, \quad i = 1,\ldots,3.
$$

For simplicity we make the choice of constant viscosity laws with a Reynolds number equal to 10^5. As initial data, we propose a step function made of two constant states, called left state and right state in the following, separated by a discontinuity located at $x = 0$ and we approximate the solution on a uniform grid with $\Delta x = 1/300$.

Experiment 1 We set $(\gamma_1,\gamma_2,\gamma_3) = (1.4, 1.6, 1.8)$ and $(\mu_2/\mu_1, \mu_3/\mu_1) = (1., 1.)$, while the left ($l$) and right ($r$) states of initial data read: $(\rho, u, p_1, p_2, p_3)_l = (4., 1., 1.2, 1.4, 1.6)$, $(\rho, u, p_1, p_2, p_3)_r = (2.5568, -1.4305, 0.5162, 0.5103, 0.4999)$.

Experiment 2 We set $(\gamma_1,\gamma_2,\gamma_3) = (1.4, 1.4, 1.4)$ and $(\mu_2/\mu_1, \mu_3/\mu_1) = (1., 1.)$ Left and right states of initial data now read: $(\rho, u, p_1, p_2, p_3)_l = (3., 1., 1., 1.2, 1.4)$, $(\rho, u, p_1, p_2, p_3)_r = (2.6529, -1.1153, 0.8160, 0.9844, 1.1528)$.

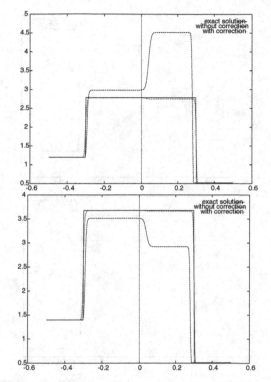

FIGURE 1. Experiment 1 – Pressures 1 and 2

In Figures 1 and 2, we compare some of the corresponding pressure profiles for the exact solutions together with the numerical solutions generated by an usual Godunov approach and our prediction-correction like scheme. As expected, we observe that the classical approach (without correction) fails in capturing the correct solution while the correction step provides us with a numerical solution in good agreement with the exact one. We refer for instance the reader to [6], [9], [10] for additional numerical experiments.

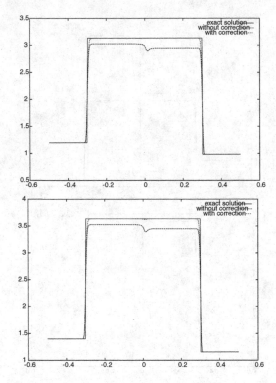

FIGURE 2. Experiment 2 – Pressures 2 and 3

References

[1] Berthon C., *Contribution à l'analyse numérique des équations de Navier-Stokes compressibles à deux entropies spécifiques. Application à la turbulence compressible*, PhD Thesis, Université Paris VI (1999).

[2] Berthon C. and Coquel F., *Travelling wave solutions of a convective diffusive system with first and second order terms in nonconservation form*, Hyperbolic Problems. Theory, Numerics, Applications, Internat. Ser. Numer. Math., 129, Birkhäuser, Basel, pp. 47–54 (1999).

[3] Berthon C. and Coquel F., *Nonlinear projection methods for multi-entropies Navier-Stokes systems* Innovative methods for numerical solutions of partial differential equations, World Sci. Publishing, River Edge, NJ, pp. 278–304 (1998).

[4] Berthon C. and Coquel F., *Travelling wave solutions for the Navier-Stokes equations with several specific entropies*, preprint.

[5] Berthon C., Coquel F. and LeFloch P.G., *Entropy dissipation measure and kinetic relation associated with nonconservative hyperbolic systems*, work in preparation.

[6] Chalons C., *Bilans d'entropie discrets dans l'approximation numérique des chocs non classiques. Application aux équations de Navier-Stokes multi-pression 2D et à quelques systèmes visco-capillaires*, PhD Thesis, Ecole Polytechnique (2002).

[7] Chalons C. and Coquel F., *Numerical approximation of the Navier-Stokes equations with several independent specific entropies*, Proceedings of the Ninth International Conference on Hyperbolic Problems. Theory, Numerics, Applications, T.Y. Hou and E. Tadmor (Eds), pp. 407–418 (2003).

[8] Chalons C. and Coquel F., *The Riemann problem for the multi-pressure Euler system*, to appear in Journal of Hyperbolic Differential Equations (JHDE) (2005).

[9] Chalons C. and Coquel F., *Navier-Stokes equations with several independent pressure laws and explicit predictor-corrector schemes*, to appear in Numerische Math. (2005).

[10] Chalons C. and Coquel F., *Euler equations with several independent pressure laws and entropy satisfying explicit projection scheme*, submitted.

[11] Colombeau J.F., **Elementary Introduction to New Generalized Functions**, North-Holland, Amsterdam (1985).

[12] Colombeau J.F. and Leroux A.Y., *Multiplications of distributions in elasticity and hydrodynamics*, J. Math. Phys., vol. 29, pp. 315–319 (1988).

[13] Coquel F. and LeFloch P.G., *Convergence of finite volumes schemes for scalar conservation laws in several space dimensions: a general theory*, SIAM J. Numer. Anal. 30, pp. 676–700 (1993).

[14] Coquel F. and Marmignon C., *Numerical Methods for weakly ionized gases*, Astrophys. Space Sci., 260 (1998).

[15] Després B. and Lagoutière F., *Numerical resolutions of a two-component compressible fluid model with interfaces*, preprint (2004).

[16] Federer H., **Geometric measure theory**, Grund. der Math. Wiss., vol. 153, Springer-Verlag (1969).

[17] Cordier S., Degond P., Markowich P. et Schmeiser C., *Travelling waves analysis and jump relations for Euler-Poisson model in quasi-neutral limit*, Asymptotic Analysis, vol. 11, pp. 209–240 (1995).

[18] Dal Maso G., LeFloch P.G. and Murat F., *Definition and weak stability of a non conservative product*, J. Math. Pures Appl., vol. 74, pp. 483–548 (1995).

[19] Godlewsky E. and Raviart P.A., **Numerical approximation of hyperbolic systems of conservation laws**, Springer (1995).

[20] Hou T.Y. and LeFloch P.G., *Why nonconservative schemes converge to wrong solutions: error analysis*, Math. of Comp., vol. 62(206), pp. 497–530 (1994).

[21] LeFloch P.G., *Shock waves for nonlinear hyperbolic systems in nonconservative form*, Institute for Math. and its Appl., Minneapolis, Preprint # 593 (1989).

[22] LeFloch P.G., *Entropy weak solutions to nonlinear hyperbolic systems in nonconservative form*, Comm. Partial Differential Equations, vol. 13, pp. 669–727 (1988).

[23] LeFloch P.G. and Liu T.P., *Existence theory for nonlinear hyperbolic systems in nonconservative form*, Forum Math., vol. 5, pp. 261–280 (1993).

[24] LeFloch P.G. and Tzavaras A., *Representation of weak limits and definition of nonconservative products*, SIAM J. Math. Anal., vol. 30, pp. 1309–1342 (1999).

[25] Majda A. and Pego R.L., *Stable Viscosity Matrices for Systems of Conservation Laws*, Journal of Differential Equations, vol. 56, pp. 229–262 (1985).

[26] Raviart P.A. and Sainsaulieu L., *A non conservative hyperbolic system modelling spray dynamics. I. Solution of the Riemann problem*, Math. Models Methods Appl. Sci., vol. 5, pp. 297–333 (1995).

[27] Sainsaulieu L., *An Euler system modelling vaporizing sprays,*in Dynamic of Heterogeneous Combustion and Reacting Systems, Progress Astr. Aero., 152, pp. 280–305 (1993).

[28] Tadmor E., *The numerical viscosity of entropy stable schemes for systems of conservation laws*, Math. Comp., 49, pp. 91–103 (1987).

[29] Volpert A.I., *The space BV and quasilinear equations*, Math. Sbornik, vol. 73 (115) no. 2, pp. 225–267 (1967).

Christophe Chalons
Université Paris 7 & Laboratoire JLL
U.M.R. 7598, Boîte courrier 187
F-75252 Paris Cedex 05, France
e-mail: `chalons@math.jussieu.fr`

Frédéric Coquel
Centre National de la Recherche Scientifique & Laboratoire JLL
U.M.R. 7598, Boîte courrier 187
F-75252 Paris Cedex 05, France
e-mail: `coquel@ann.jussieu.fr`

Analysis and Simulation of Fluid Dynamics
Advances in Mathematical Fluid Mechanics, 69–88
© 2006 Birkhäuser Verlag Basel/Switzerland

Domain Decomposition Algorithms for the Compressible Euler Equations

V. Dolean and F. Nataf

Abstract. In this work we present an overview of some classical and new domain decomposition methods for the resolution of the Euler equations. The classical Schwarz methods are formulated and analyzed in the framework of first order hyperbolic systems and the differences with respect to the scalar problems are presented. This kind of algorithms behave quite well for bigger Mach numbers but we can further improve their performances in the case of lower Mach numbers. There are two possible ways to achieve this goal. The first one implies the use of the optimized interface conditions depending on a few parameters that generalize the classical ones. The second is inspired from the Robin-Robin preconditioner for the convection-diffusion equation by using the equivalence via the Smith factorization with a third order scalar equation.

1. Introduction

When solving the compressible Euler equations by an implicit scheme the nonlinear system is usually solved by Newton's method. At each step of this method we have to solve a linear system which is non-symmetric and very ill conditioned. The necessity of a domain decomposition became more and more obvious. In a previous paper [DLN04] we formulated a Schwarz algorithm (interface iteration which relies on the successive solving of the local decomposed problems and the transmission of the result at the interface) involving transmission conditions that are derived naturally from a weak formulation of the underlying boundary value problem (first formulated in [QS96]). As far as these algorithms are concerned, when dealing with supersonic flows, whatever the space dimension is, imposing the appropriate characteristic variables as interface conditions leads to a convergence of the algorithm which is optimal with regards to the number of subdomains. The only case of interest remains the subsonic one where this property is lost except in the one-dimensional case. We recall briefly these results in order to introduce more general and performant methods such as the optimized interface conditions and the preconditioning methods. The former were widely studied and analyzed for scalar problems such as elliptic equations in [Lio90, EZ98], for the Helmholtz equation in [BD97, CN98] convection-diffusion problems in [JNR01]. For time dependent problems and local times steps, see for instance [GHN01]. The preconditioning methods have also known a wide development in the last decade. The Neumann-Neumann algorithms for symmetric second order problems [RT91] has been the subject of numerous works, see [TW04] and references therein. An extension of these algorithms to non-symmetric scalar problems (the so-called Robin-Robin algorithms) has been done in [ATNV00, GGTN04] for advection-diffusion problems.

In Section 2 we first formulate the Schwarz algorithm for a general linear hyperbolic system of PDEs with general interface conditions inspired by Clerc [DN04b] built in order to have a well-posed problem. The convergence rate is computed in the Fourier space as a function of some parameters. We will further estimate the convergence rate at the discrete level. We will find the optimal parameters of the interface conditions at the discrete level. We will then use the new optimal interface conditions in Euler computations which illustrate the improvement over the classical interface conditions (first described in [QS96]).

As far as preconditioning methods are concerned, to our knowledge, no extension to the Euler equations was done. In Section 2 we will first show the equivalence between the 2D Euler equations and a third order scalar problem, which is quite natural by considering a Smith factorization of this system, see [Gan66] then we define an optimal algorithm for the third order scalar equation inspired from the idea of the Robin-Robin algorithm [ATNV00] applied to a convection-diffusion problem. Afterwards we back-transform it and define the corresponding algorithm applied to the Euler system. All the previous results have been obtained at the continuous level and for a decomposition into 2 unbounded subdomains. After a discretization in a bounded domain we cannot expect that these properties to be preserved exactly. Still we can show by a discrete convergence analysis that the expected results should be very good.

2. A Schwarz algorithm with general interface conditions

In this section we introduce a Schwarz algorithm which is based on general transmission conditions at subdomain interfaces that take into account the hyperbolic nature of the problem. In addition, we recall some existing results concerning the convergence of the algorithm. We consider here a general non-linear system of conservation laws. Under the hypothesis that the solution is regular, we can also write it under a non-conservative (or quasi-linear) equivalent form:

$$\frac{\partial W}{\partial t} + \sum_{i=1}^{d} A_i(W)\frac{\partial W}{\partial x_i} = 0 \tag{1}$$

where the A_i are the Jacobian matrices of the flux vectors. Assume that we first proceed to an integration in time of (1) using a backward Euler implicit scheme involving a linearization of the flux functions and eventually we symmetrize it (we know that when the system admits an entropy it can be symmetrized by multiplying it by the Hessian matrix of this entropy). This operation results in the linearized system:

$$\mathcal{L}(\delta W) \equiv \frac{\mathrm{Id}}{\Delta t} \delta W + \sum_{i=1}^{d} A_i \frac{\partial \delta W}{\partial x_i} = f \tag{2}$$

In the following we will define the boundary conditions that have to be imposed when solving the problem on a domain $\Omega \subset \mathbb{R}^d$. We denote by $A_n = \sum_{i=1}^{d} A_i n_i$, the linear combination of jacobian matrices by the components of the outward normal vector at the boundary of the domain $\partial\Omega$. This matrix is real, symmetric and can be diagonalized

$$A_n = T\Lambda_n T^{-1}, \Lambda_n = \mathrm{diag}(\lambda_i)$$

It can also be split in negative and positive part using this diagonalization

$$A_n = A_n^+ + A_n^-, A_n^\pm = T\Lambda_n^\pm T^{-1}, \Lambda_n^+ = \mathrm{diag}(\max(\lambda_i, 0)), \Lambda_n^- = \mathrm{diag}(\min(\lambda_i, 0))$$

This corresponds to a decomposition with local characteristic variables. A more general splitting in negative(positive) definite parts, A_n^{neg} and A_n^{pos} of A_n can be done such that these matrices satisfy the following properties:

$$A_n = A_n^{\mathrm{neg}} + A_n^{\mathrm{pos}}, \mathrm{rank}(A_n^{\mathrm{neg,pos}}) = \mathrm{rank}(A_n^\pm), A_{-n}^{\mathrm{pos}} = -A_n^{\mathrm{neg}} \tag{3}$$

In the scalar case the only possible choice is $A_n^{\mathrm{neg}} = A_n^-$. Using the previous formalism we can define the following boundary condition:

$$A_n^{\mathrm{neg}} W = A_n^{\mathrm{neg}} g, \text{ on } \partial\Omega$$

Remark 1. *In the case of a classical decomposition in negative and positive part this boundary condition has the physical meaning of the incoming flux in domain Ω. By extension of the properties found in this case we call the last equality of (3) conservation property because it insures that the "out-flow" quantity (given by the positive part of the jacobian flux matrix with opposite direction of the normal) is retrieved out of the "in-flow" quantity imposed by the boundary condition (given the negative part of the Jacobian flux matrix).*

2.1. Schwarz algorithm with general interface conditions

We consider a decomposition of the domain Ω into N overlapping or non-overlapping subdomains $\bar{\Omega} = \bigcup_{i=1}^{N} \bar{\Omega}_i$. We denote by n_{ij} the outward normal to the interface Γ_{ij} between Ω_i and a neighboring subdomain Ω_j. Let $W_i^{(0)}$ denote the initial approximation of the solution in subdomain Ω_i. A general formulation of a Schwarz algorithm for computing $(W_i^{p+1})_{1 \le i \le N}$ from $(W_i^p)_{1 \le i \le N}$ (where p defines the iteration of the Schwarz algorithm) reads:

$$\begin{cases} \mathcal{L}W_i^{p+1} & = f & \text{in } \Omega_i \\ A_{\boldsymbol{n}_{ij}}^{\text{neg}}W_i^{p+1} & = A_{\boldsymbol{n}_{ij}}^{\text{neg}}W_j^p & \text{on } \Gamma_{ij} = \partial\Omega_i \cap \Omega_j \\ A_{\boldsymbol{n}_{ij}}^{\text{neg}}W_i^{p+1} & = A_{\boldsymbol{n}_{ij}}^{\text{neg}}g & \text{on } \partial\Omega \cap \partial\Omega_i \end{cases} \qquad (4)$$

where $A_{\boldsymbol{n}_{ij}}^{\text{neg}}$ and $A_{\boldsymbol{n}_{ij}}^{\text{pos}}$ satisfy (3). We have the following result concerning the convergence of the Schwarz algorithm in the non-overlapping case, due to([Cle98]):

Theorem 1. *If we denote by $E_i^p = W_i^p - W_i$ the error vector associated to the restriction to the ith subdomain of the global solution of the problem. Then, the Schwarz algorithm converges in the following sense:*

$$\begin{cases} \lim_{p\to\infty} \|E_i^p\|_{L^2(\Omega_i)^q} = 0 \\ \lim_{p\to\infty} \| \sum_{j=1}^d A_j \partial_j E_i^p \|_{L^2(\Omega_i)^q} = 0 \end{cases}$$

The convergence rate of the algorithm defined by (4) depends of the choice of the decomposition of $A_{\boldsymbol{n}_{ij}}$ into a negative and a positive part satisfying (3). In order to choose the decomposition (3) we need to relate this choice to the convergence rate of (4).

2.2. Convergence rate of the algorithm with general interface conditions

We consider a two-subdomain non-overlapping or overlapping decomposition of the domain $\Omega = \mathbb{R}^d$, $\Omega_1 =]-\infty, \gamma[\times\mathbb{R}^{d-1}$ and $\Omega_2 =]\beta, \infty[\times\mathbb{R}^{d-1}$ with $\beta \le \gamma$ and study the convergence of the Schwarz algorithm in the subsonic case. A Fourier analysis applied to the linearized equations allows us to derive the convergence rate of the "ξ"th Fourier component of the error. We will first briefly recall the technique of Fourier transform which was already described in detail in [DLN04]. The vector of Fourier variables is denoted by $\boldsymbol{\xi} = (\xi_j, j = 2, \ldots, d)$. Let $(E_i^p)(x) = (W_i^p - W_i)(x)$ be the error vector in the ith subdomain at the pth iteration of the Schwarz algorithm and:

$$\hat{E}(x_1, \xi_2, \ldots, \xi_d) = \mathcal{F}E(x_1, \xi_2, \ldots, \xi_d)$$

$$= \int_{\mathbb{R}^{d-1}} e^{-i\xi_2 x_2 - \ldots - i\xi_d x_d} E(x_1, \ldots, x_d) dx_2 \ldots dx_d$$

the Fourier symbol of the error vector. This transformation is useful only if the A_i matrices are constant which is the case here because we have considered the linearized form of the Euler equations around a constant state $\bar{W}$. The Schwarz algorithm in the Fourier space ($\xi \in \mathbb{R}^{d-1}$) can be written as follows:

$$
\begin{cases}
\dfrac{d}{dx_1} \hat{E}_1^{p+1} = -M(\boldsymbol{\xi})\hat{E}_1^{p+1}, \; x < \gamma \\
\mathcal{A}^{\text{neg}}(\hat{E}_1^{p+1}) = \mathcal{A}^{\text{neg}}(\hat{E}_2^{p}), \; \text{on } x = \gamma
\end{cases}
$$
$$
\begin{cases}
\dfrac{d}{dx_1} \hat{E}_2^{p+1} = -M(\boldsymbol{\xi})\hat{E}_2^{p+1}, \; x > \beta \\
\mathcal{A}^{\text{pos}}(\hat{E}_2^{p+1}) = \mathcal{A}^{\text{pos}}(\hat{E}_1^{p}), \; \text{on } x = \beta
\end{cases}
\tag{5}
$$

where we denoted by $\mathcal{A}^{\text{neg}} = A_n^{\text{neg}}$, $\mathcal{A}^{\text{pos}} = A_n^{\text{pos}}$ with $\boldsymbol{n} = (1,0)$ the outward normal to the domain Ω_1 and:

$$
M(\boldsymbol{\xi}) = A_1^{-1}\left(\frac{1}{\Delta t}\text{Id} + \sum_{i=2}^{d} A_i \xi_{i-1}\right)
\tag{6}
$$

We thus obtain local problems that for a given ξ are very simple ODEs whose solutions can be expressed as linear combinations of the eigenvectors of $M(\boldsymbol{\xi})$ (we denote by $\lambda_j(\boldsymbol{\xi})$ the eigenvalues of $M(\boldsymbol{\xi})$). Here we require that these solutions are bounded at infinity ($-\infty$ and $+\infty$ respectively). We deduce that in the decomposition of $\hat{E}_1(x_1,\boldsymbol{\xi})$ (respectively $\hat{E}_2(x_1,\boldsymbol{\xi})$) we must keep only the eigenvectors corresponding to the negative (respectively the positive) real parts of the eigenvalues. Taking into account these considerations we replace the expressions of the local solutions into the interface conditions (5) to obtain the interface iterations on the α coefficients:

$$
\begin{cases}
(\alpha_j^{1,p+1})_{j,\Re(\lambda_j)<0}(\boldsymbol{\xi}) = T_1\left[(\alpha_j^{2,p})_{j,\Re(\lambda_j)>0}(\boldsymbol{\xi})\right] \\
(\alpha_j^{2,p+1})_{j,\Re(\lambda_j)>0}(\boldsymbol{\xi}) = T_2\left[(\alpha_j^{1,p})_{j,\Re(\lambda_j)<0}(\boldsymbol{\xi})\right]
\end{cases}
$$

Then, the convergence rate of the $\boldsymbol{\xi}$th component of the error vector of the Schwarz algorithm can be computed as the spectral radius of one of the iteration matrices $T_1 T_2(\boldsymbol{\xi})$ or $T_2 T_1(\boldsymbol{\xi})$:

$$
\rho_2^2 \equiv \rho_{\text{Schwarz2}}^2 = \rho(T_1 T_2) = \rho(T_2 T_1)
$$

2.3. The 2D Euler equations

After having defined in a general frame the well-posedness of the boundary value problem associated to a general equation and the convergence of the Schwarz algorithm applied to this class of problems, we will concentrate ourselves on the conservative Euler equations in two-dimensions:

$$
\frac{\partial W}{\partial t} + \nabla.F(W) = 0 \; , \quad W = (\rho, \, \rho\boldsymbol{V}, \, E)^T \; , \quad \nabla = \left(\frac{\partial}{\partial x}, \, \frac{\partial}{\partial y}\right)^T .
\tag{7}
$$

In the above expressions, ρ is the density, $\boldsymbol{V} = (u, \, v)^T$ is the velocity vector, E is the total energy per unit of volume and p is the pressure. In equation (7), $W = W(\mathbf{x}, \mathbf{t})$ is the vector of conservative variables, $\mathbf{x}$ and t respectively denote the space and time variables and $F(W) = (F_1(W) , F_2(W))^T$ is the conservative flux vector whose components are given by

$$F_1(W) = \begin{pmatrix} \rho u \\ \rho u^2 + p \\ \rho uv \\ u(E+p) \end{pmatrix}, \quad F_2(W) = \begin{pmatrix} \rho v \\ \rho uv \\ \rho v^2 + p \\ v(E+p) \end{pmatrix}.$$

The pressure is deduced from the other variables using the state equation for a perfect gas $p = (\gamma_s - 1)(E - \frac{1}{2}\rho \parallel V \parallel^2)$ where γ_s is the ratio of the specific heats ($\gamma_s = 1.4$ for the air). ·

2.4. A new type of interface conditions

We will apply now the method described previously to the computation of the convergence rate of the Schwarz algorithm applied to the two-dimensional subsonic Euler equations. In the supersonic case there is only one decomposition satisfying (3), that is: $\mathcal{A}^{\text{pos}} = A_n$ and $\mathcal{A}^{\text{neg}} = 0$ and the convergence follows in 2 steps. Therefore the only case of interest is the subsonic one. The starting point of our analysis is given by the linearized form of the Euler equations (7) which are of the form (2) where we replace δW by W and to whom we applied a change of variable $\tilde{W} = T^{-1}W$ based on the eigenvector factorization of $A_1 = T\tilde{A}_1 T^{-1}$. In the following we will abandon the $\tilde{}$ symbol):

$$\frac{W}{c\Delta t} + A_1 \partial_x W + A_2 \partial_y W = 0$$

characterized by the Jacobian matrices A_1 and A_2 depending on $M_n = \frac{u}{c}$, $M_t = \frac{v}{c}$ which denote respectively the normal and the tangential Mach number. Before estimating the convergence rate we will derive the general transmission conditions at the interface by splitting the matrix A_1 into a positive and negative part. We have the following general result concerning this decomposition:

Lemma 1. Let $\lambda_1 = M_n - 1$, $\lambda_2 = M_n + 1$, $\lambda_3 = \lambda_4 = M_n$. Suppose we deal with a subsonic flow: $0 < u < c$ so that $\lambda_1 < 0$, $\lambda_{2,3,4} > 0$. Any decomposition of $A_1 = A_n$, $n = (1,0)$ which satisfies (3) has to be of the form:

$$\mathcal{A}^{\text{neg}} = \frac{1}{a_1}\mathbf{u} \cdot \mathbf{u}^t, \ \mathbf{u} = (a_1, a_2, a_3, a_4)^t$$
$$\mathcal{A}^{\text{pos}} = A_n - \mathcal{A}^{\text{neg}}.$$

where $(a_1, a_2, a_3, a_4) \in \mathbb{R}^4$ satisfies $a_1 \leq \lambda_1 < 0$ and $\frac{a_1}{\lambda_1} + \frac{a_2^2}{a_1\lambda_2} + \frac{a_3^2}{a_1\lambda_3} + \frac{a_4^2}{a_1\lambda_4} = 1$.

For the details of the proof see [DN04b].

We will proceed now to the estimation of the convergence rate using some results from [DLN04]. Following the technique described in Section 2.2 we estimate the convergence rate in the non-overlapping case and we use the non-dimensioned wave-number $\bar{\xi} = c\Delta t\xi$ and if we drop the bar symbol, we get for the general interface conditions the following:

$$
\begin{cases}
\rho^2_{2,\text{novr}}(\xi) = \left| 1 - \dfrac{4M_n(1-M_n)(1+M_n)R(\xi)a_1^2(a+M_nR(\xi))}{D_1 D_2} \right| \\[2ex]
D_1 = R(\xi)[a_1(1+M_n) - a_2(1-M_n)] \\
\quad\quad + a[a_1(1+M_n) + a_2(1-M_n)] - i\sqrt{2}a_3\xi(1-M_n^2) \\[2ex]
D_2 = M_n a_1 [R(\xi)[a_1(1+M_n) - a_2(1-M_n)] \\
\quad\quad + a[a_1(1+M_n) + a_2(1-M_n)]] \\
\quad\quad + a_3(1-M_n^2)[a_3(R+a) - iM_n a_1\xi\sqrt{2}]
\end{cases}
\tag{8}
$$

Remark 2. *The expression (8) gives the convergence rate in the classical case for $a_1 = -(1-M_n) = \lambda_1(0)$ and $a_2 = a_3 = a_4 = 0$, which corresponds to the classical transmission conditions. Moreover, Theorem 1 proves that this quantity is always strictly inferior to 1 as the algorithm is convergent.*

In order to simplify our optimization problem we will take $a_3 = 0$, we can thus reduce the number of parameters to 2, a_1 and a_2, as we can see that a_4 can be expressed as a function of a_1, a_2 and a_3. We can also see that the convergence rate is a real quantity when the flow is normal to the interface $M_t = 0$. In the same time for the optimization purpose only we introduce the parameters: $b_1 = -a_1/(1-M_n)$ and $b_2 = a_2/(1+M_n)$ which provide a simpler form of the convergence rate:

$$
\rho^2_{2,\text{novr}}(\xi) = \left| 1 - \frac{4b_1(a + M_nR(\xi))R(\xi)}{(R(\xi)(b_1 + b_2) + a(b_1 - b_2))^2(Mn + 1)} \right|
\tag{9}
$$

Before proceeding to the analysis of the general case we recall some results found in the classical case obtained in [DLN04]. The asymptotic convergence rate in the non-overlapping case:

$$
\lim_{k\to+\infty} \rho_{2,\text{novr}}(k) = \sqrt{\left(\frac{1-3M_n}{1+M_n}\right)^2 + \frac{8M_nM_t^2}{(1+M_n)^3}} < 1
\tag{10}
$$

is always strictly inferior to 1. Moreover, in the particular case $M_n^* = 1/3$ and $M_t = 0$, this limit becomes null. The inequality (10) has a numerical meaning. For a given discretization, let $\xi_{\max}$ denote the largest frequency supported by the numerical grid. This largest frequency is of the order π/h with h a typical mesh size. The convergence rate in a numerical computation made on this grid can be estimated by $\rho_2^h = \max_{|\xi|<\xi_{\max}} \rho_2(\xi)$. From (10), we have that $\rho_2^h \le \max_{|\xi|<\xi_{\max}} \rho_2(\xi) < 1$. This means that for finer and finer grids, the number of iterations may increase slightly but should not go to infinity. Thus the optimization problem with respect to the parameters b_1 and b_2, makes sense:

$$
\min_{(b_1,b_2)\in\mathcal{I}_1\times\mathcal{I}_2(b_1)} \max_{\xi\ge 0} \rho(\xi)
\tag{11}
$$

The solving of this problem is quite a tedious task even in the non-overlapping case, where we can obtain analytical expression of the parameters only for some values of the Mach number. In the same time, we have to analyze the convergence of the overlapping algorithm. Indeed, standard discretizations of the interface conditions correspond to overlapping decompositions with an overlap of size $\delta = h$, h being the mesh size, as seen in [CFS98] and [DLN04]. Analytic optimization with respect to b_1 and b_2 seems out of reach. We will have to use numerical procedures of optimization.

In order to get closer to the numerical simulations we will estimate the convergence rate for the discretized equations with general transmission conditions, both in the non-overlapping and the overlapping case and then optimize numerically this quantity in order to get the best parameters for the convergence. Following a similar procedure as that described in [DLN04] and [DN04b] we will first discretize the problem by a finite volume scheme then we formulate a Schwarz algorithm whose convergence rate is estimated at the discrete level. Therefore, we will get the theoretical optimized parameters at the discrete level by means of a numerical algorithm, by calculating the following

$$\rho(b_1, b_2) = \max_{k \in \mathcal{D}_h} \rho_2^2(k, \Delta x, M_n, M_t, b_1, b_2)$$

$$\min_{(b_1, b_2) \in \mathcal{I}_h} \rho(b_1, b_2) \tag{12}$$

where $\mathcal{D}_h$ is a uniform partition of the interval $[0, \pi/\Delta x]$ and $\mathcal{I}_h \subset \mathcal{I}$ a discretization by means of a uniform grid of a subset of the domain of the admissible values of the parameters. This kind of calculations are done once for all for a given pair (M_n, M_t) before the beginning of the Schwarz iterations. An example of such a result is given in the Figure 1 Mach number $M_n = 0.2$. The computed parameters from the relation (12) will be further referred to with a superscript th. The theoretical estimates are compared afterwards with the numerical ones obtained by running the Schwarz algorithm with different pairs of parameters which lie in a an interval such that the algorithm is convergent. We are thus able to estimate the optimal values for b_1 and b_2 from these numerical computations. These values will be referred to by a superscript num.

2.5. Implementation and numerical results

We present here a set of results of numerical experiments that are concerned with the evaluation of the influence of the interface conditions on the convergence of the non-overlapping Schwarz algorithm of the form. The computational domain is given by the rectangle $[0, 1] \times [0, 1]$. The numerical investigation is limited to the resolution of the linear system resulting from the first implicit time step using a Courant number CFL=100. In all these calculations we considered a model problem: a flow normal to the interface (that is when $M_t = 0$). In Figures 1 and 2 we can see an example of a theoretical and numerical estimation of the reduction factor of the error. We illustrate here the level curves which represent the log of the precision after 20 iterations for different values of the parameters (b_1, b_2), the minimum being attained in this case for $b_1^{th} = 1.3$ and $b_2^{th} = -0.5$, $b_1^{num} = 1.4$ and $b_2^{num} = -0.6$. We can see that we have good theoretical estimates of these parameters we can therefore use them in the interface conditions of the Schwarz algorithm.

M_n	b_1^{th}	b_2^{th}	b_1^{num}	b_2^{num}
0.1	1.6	−0.8	1.6	−0.9
0.2	1.3	−0.5	1.4	−0.6
0.3	1.25	−0.3	1.25	−0.45
0.4	1.08	−0.15	1.08	−0.28
0.5	1.03	−0.08	1.02	−0.23
0.6	1.0	0.0	1.0	0.0
0.7	1.02	0.06	1.01	0.04
0.8	1.03	0.08	1.02	0.06
0.9	1.06	0.08	1.04	0.06

TABLE 1. Overlapping Schwarz algorithm:
Numerical vs. theoretical parameters

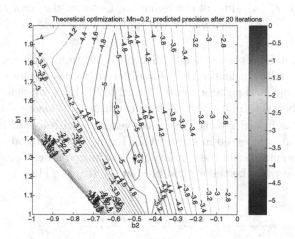

FIGURE 1. Isovalues of the predicted reduction factor of the
error after 20 iterations via formula (12)

M_n	IT_0^{num}	$IT_{\text{op}}^{\text{num}}$	M_n	IT_0^{num}	$IT_{\text{op}}^{\text{num}}$
0.1	48	19	0.5	22	18
0.2	41	20	0.7	20	16
0.3	32	20	0.8	22	15
0.4	26	19	0.9	18	12

TABLE 2. Overlapping Schwarz algorithm: Classical vs.
optimized counts for different values of M_n

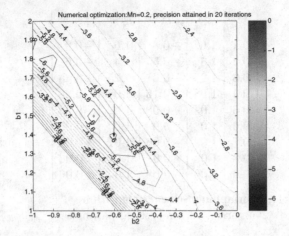

FIGURE 2. Isovalues of the reduction factor of the error after
20 iterations for the finite volume code

Table 2 summarizes the number of Schwarz iterations required to reduce the initial linear residual by a factor 10^{-6} for different values of the reference Mach number with the optimal parameters b_1^{num} and b_1^{num}. Here we denoted by IT_0^{num} and IT_{op}^{num} the observed (numerical) iteration number for classical and optimized interface conditions in order to achieve a convergence with a threshold $\varepsilon = 10^{-6}$. The same results are presented in Figure 4. In Figure 3 we compare the theoretical estimated iteration number in the classical and optimized case. Comparing Figures 3 and 4 we can see that the theoretical prediction are very close to the numerical tests.

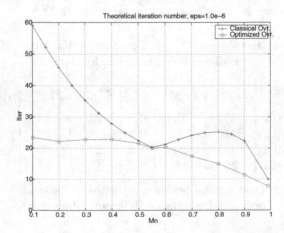

FIGURE 3. Theoretical iteration number: Classical vs.
optimized conditions

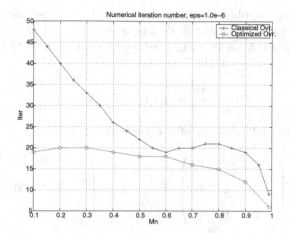

FIGURE 4. Numerical iteration number: Classical vs. optimized conditions

The conclusion of these numerical tests is, on one hand, that the theoretical prediction is very close to the numerical results: we can get by a numerical optimization (12) a very good estimate of optimal parameters (b_1, b_2)). On the other hand, the gain, in number of iterations, provided by the optimized interface conditions, is very promising for low Mach numbers, where the classical algorithm doesn't give optimal results. We can note that the optimized convergence rate is monotone with respect to the normal Mach number while the classical one isn't. For bigger Mach numbers, for instance, those who are close to 1, the classical algorithm already has a very good behavior so the optimization is less useful. In the same time we studied here the zero order and therefore very simple transmission conditions. The use of higher order conditions (see [GMN02]) is a possible way that can be further studied to obtain even better convergence results.

3. A new preconditioning method

In this section we will show the equivalence between the linearized Euler system and a third order scalar equation. The motivation for this transformation is that a new algorithm is easier to design for a scalar equation than for a system of partial differential equations.

3.1. Equivalence of the Euler system to a scalar equation

The starting point of our analysis is given by the linearized form of the Euler equations (7) written in primitive variables (p, u, v, S). In the following we suppose that the flow is isentropic, which allows us to drop the equation of the entropy (which is totally decoupled with respect to the others). We denote by $W = (P, U, V)^T$ the vector of unknowns and by A and B the jacobian matrices of the fluxes $F_i(w)$

to whom we already applied the variable change from conservative to primitive variables. In the following, we shall denote by $\bar{c}$ the speed of the sound and we consider the linearized form (we will mark by the bar symbol, the constant state around which we linearize) of the Euler equations:

$$\mathcal{P}W \equiv \frac{W}{\Delta t} + A\partial_x W + B\partial_y W = f \tag{13}$$

characterized by the following Jacobian matrices:

$$A = \begin{pmatrix} \bar{u} & \bar{\rho}\bar{c}^2 & 0 \\ 1/\bar{\rho} & \bar{u} & 0 \\ 0 & 0 & \bar{u} \end{pmatrix} \qquad B = \begin{pmatrix} \bar{v} & 0 & \bar{\rho}\bar{c}^2 \\ 0 & \bar{v} & 0 \\ 1/\bar{\rho} & 0 & \bar{v} \end{pmatrix} \tag{14}$$

We can re-write the system (13) by denoting $\beta = \frac{1}{\Delta t} > 0$ under the form

$$\mathcal{P}W \equiv (\beta I + A\partial_x + B\partial_y)\, W = f \tag{15}$$

We will study this system with the help of the Smith factorization.

3.1.1. Smith factorization.

We first recall the definition of the Smith factorization of a matrix with polynomial entries and apply it to systems of PDEs, see [Gan66] and references therein.

Theorem 2. *Let n be an integer and C an invertible $n \times n$ matrix with polynomial entries in the variable $\lambda : C = (c_{ij}(\lambda))_{1 \le i,j \le n}$.*
Then, there exist three matrices with polynomial entries E, D and F with the following properties:

- $\det(E) = \det(F) = 1$.
- *D is a diagonal matrix.*
- $C = EDF$.

Moreover, D is uniquely defined up to a reordering and multiplication of each entry by a constant by a formula defined as follows. Let $1 \le k \le n$,

- *S_k is the set of all the submatrices of order $k \times k$ extracted from C.*
- *$\text{Det}_k = \{\text{Det}(B_k) \backslash B_k \in S_k\}$*
- *LD_k is the largest common divisor of the set of polynomials Det_k.*

Then,

$$D_{kk}(\lambda) = \frac{LD_k(\lambda)}{LD_{k-1}(\lambda)}, \quad 1 \le k \le n \tag{16}$$

(by convention, $LD_0 = 1$).

Application to the Euler system. We first take formally the Fourier transform of the system (15) with respect to y (the dual variable is ξ). We keep the partial derivatives in x since in the sequel we shall consider a domain decomposition with an interface whose normal is in the x direction. We note

$$
\hat{\mathcal{P}} = \begin{pmatrix} \beta + \bar{u}\partial_x + i\xi\bar{v} & \bar{\rho}\bar{c}^2\partial_x & i\bar{\rho}\bar{c}^2\xi \\ \frac{1}{\bar{\rho}}\partial_x & \beta + \bar{u}\partial_x + i\xi\bar{v} & 0 \\ \frac{i\xi}{\bar{\rho}} & 0 & \beta + \bar{u}\partial_x + i\bar{v}\xi \end{pmatrix} \tag{17}
$$

We can perform a Smith factorization of $\hat{\mathcal{P}}$ by considering it as a matrix with polynomials in ∂_x entries. We have

$$
\hat{\mathcal{P}} = EDF \tag{18}
$$

where

$$
D = \begin{pmatrix} 1 & 0 & 0 \\ 0 & 1 & 0 \\ 0 & 0 & \hat{\mathcal{L}}\hat{\mathcal{G}} \end{pmatrix} \tag{19}
$$

and

$$
E = \frac{1}{(\bar{u}(\bar{c}^2 - \bar{u}^2))^{1/3}} \begin{pmatrix} i\bar{\rho}\bar{c}^2\xi & 0 & 0 \\ 0 & \bar{u} & 0 \\ \beta + \bar{u}\partial_x + i\bar{v}\xi & E_2 & \frac{\bar{c}^2 - \bar{u}^2}{i\xi\bar{\rho}\bar{c}^2} \end{pmatrix}
$$

and

$$
F = - \begin{pmatrix} \frac{\beta + \bar{u}\partial_x + i\xi\bar{v}}{i\xi\bar{\rho}\bar{c}^2} & \frac{\partial_x}{i\xi} & 1 \\ \frac{\partial_x}{\bar{\rho}\bar{u}} & \frac{\beta + \bar{u}\partial_x + i\xi\bar{v}}{\bar{u}} & 0 \\ \frac{\bar{u}}{\beta + i\xi\bar{v}} & \frac{\bar{\rho}\bar{u}^2}{\beta + i\xi\bar{v}} & 0 \end{pmatrix} \tag{20}
$$

where

$$
E_2 = \bar{u}\frac{(-\bar{u}\bar{c}^2 + \bar{u}^3)\partial_{xx} + (2\bar{u}^2 - \bar{c}^2)(\beta + i\xi\bar{v})\partial_x + \bar{u}((\beta + i\xi\bar{v})^2 + \xi^2\bar{c}^2)}{\bar{c}^2(i\beta + i\xi\bar{v})},
$$

$$
\hat{\mathcal{G}} = \beta + \bar{u}\partial_x + i\xi\bar{v} \tag{21}
$$

and

$$
\hat{\mathcal{L}} = \beta^2 + 2i\xi\bar{u}\bar{v}\partial_x + 2\beta(\bar{u}\partial_x + i\xi\bar{v}) + (\bar{c}^2 - \bar{v}^2)\xi^2 - (\bar{c}^2 - \bar{u}^2)\partial_{xx} \tag{22}
$$

Equation (19) suggests that the derivation of a domain decomposition method (DDM) for the third order operator $\mathcal{L}G$ is a key ingredient for a DDM for the compressible Euler equations.

3.2. A new algorithm applied to a scalar third order problem

In this section we will describe a new algorithm applied to the third order operator found in Section 3.1.1. We want to solve

$$\mathcal{L}\mathcal{G}(Q) = g \tag{23}$$

where Q is scalar unknown function and g is a given right-hand side. The algorithm will be based on the Robin-Robin algorithm [ATNV00] for the convection-diffusion problem. Then we will prove its convergence in 2 iterations. We first note that the elliptic operator $\mathcal{L}$ can also be written as:

$$\mathcal{L} = -\operatorname{div}(A\nabla) + \mathbf{a}\nabla + \beta^2, \; A = \begin{pmatrix} \bar{c}^2 - \bar{u}^2 & -\bar{u}\bar{v} \\ -\bar{u}\bar{v} & \bar{c}^2 - \bar{v}^2 \end{pmatrix} \text{ where } \mathbf{a} = 2\beta(\bar{u}, \bar{v}) \tag{24}$$

Without loss of generality we assume in the sequel that the flow is subsonic and that $\bar{u} > 0$ and thus we have $0 < \bar{u} < \bar{c}$.

3.2.1. The algorithm for a two-domain decomposition. We consider now a decomposition of the plane $\mathbb{R}^2$ into two non-overlapping sub-domains $\Omega_1 = (-\infty, 0) \times \mathbb{R}$ and $\Omega_2 = (0, \infty, 0) \times \mathbb{R}$. The interface is $\Gamma = \{x = 0\}$. The outward normal to domain Ω_i is denoted $\mathbf{n_i}$, $i = 1, 2$. Let $Q^{i,k}$, $i = 1, 2$ represent the approximation to the solution in subdomain i at the iteration k of the algorithm. We define the following algorithm:

ALGORITHM 1. *We choose the initial values $Q^{1,0}$ and $Q^{2,0}$ such that $\mathcal{G}Q^{1,0} = \mathcal{G}Q^{2,0}$ and we compute $(Q^{i,k+1})_{i=1,2}$ from $(Q^{i,k})_{i=1,2}$ by the following iterative procedure:*
Correction step *We compute the corrections $\tilde{Q}^{1,k}$ and $\tilde{Q}^{2,k}$ as solution of the homogeneous local problems:*

$$\begin{cases} \mathcal{L}\mathcal{G}\tilde{Q}^{1,k} = 0 \text{ in } \Omega_1, \\ (A\nabla - \tfrac{1}{2}\mathbf{a})\mathcal{G}\tilde{Q}^{1,k} \cdot \mathbf{n_1} = \gamma^k, \text{ on } \Gamma. \end{cases} \qquad \begin{cases} \mathcal{L}\mathcal{G}\tilde{Q}^{2,k} = 0 \text{ in } \Omega_2, \\ (A\nabla - \tfrac{1}{2}\mathbf{a})\mathcal{G}\tilde{Q}^{2,k} \cdot \mathbf{n_2} = \gamma^k, \text{ on } \Gamma, \\ \tilde{Q}^{2,k} = 0, \text{ on } \Gamma. \end{cases}$$

$$\tag{25}$$

where $\gamma^k = -\tfrac{1}{2}\left[A\nabla\mathcal{G}Q^{1,k} \cdot \mathbf{n_1} + A\nabla\mathcal{G}Q^{2,k} \cdot \mathbf{n_2}\right]$.
Update step. *We update $Q^{1,k+1}$ and $Q^{2,k+1}$ by solving the local problems:*

$$\begin{cases} \mathcal{L}\mathcal{G}Q^{1,k+1} = g, \text{ in } \Omega_1, \\ \mathcal{G}Q^{1,k+1} = \mathcal{G}Q^{1,k} + \delta^k, \text{ on } \Gamma. \end{cases} \qquad \begin{cases} \mathcal{L}\mathcal{G}\tilde{Q}^{2,k+1} = g, \text{ in } \Omega_2, \\ \mathcal{G}Q^{2,k+1} = \mathcal{G}Q^{2,k} + \delta^k, \text{ on } \Gamma, \\ Q^{2,k+1} = Q^{1,k} + \tilde{Q}^{1,k}, \text{ on } \Gamma. \end{cases} \tag{26}$$

where $\delta^k = \tfrac{1}{2}\left[\mathcal{G}\tilde{Q}^{1,k} + \mathcal{G}\tilde{Q}^{2,k}\right]$.

Proposition 1. *Algorithm 1 converges in 2 iterations.*

For the details of the proof see [DN04a].

3.3. A new algorithm applied to the Euler system

After having found an optimal algorithm which converges in two steps for the third order model problem, we focus on the Euler system by translating this algorithm into an algorithm for the Euler system. It suffices to replace the operator $\mathcal{LG}$ by the Euler system and Q by the last component $F(W)_3$ of $F(W)$ in the boundary conditions. This algorithm is quite complex since it involves second order derivatives of the unknowns in the boundary conditions on $\mathcal{G}F(W)_3$. It is possible to simplify it. By using the Euler equations in the subdomain, we have lowered the degree of the derivatives in the boundary conditions. After lengthy computations that we omit here, we find a simpler algorithm. We write it for a decomposition in two subdomains with an outflow velocity at the interface of domain Ω_1 but with an interface not necessarily rectilinear. In this way, it is possible to figure out how to use for a general domain decomposition. In the sequel, $\mathbf{n} = (n_x, n_y)$ denotes the outward normal to domain Ω_1, $\partial_n = \nabla \cdot \mathbf{n}$ the normal derivative at the interface, $\partial_\tau = \nabla \cdot \tau$ the tangential derivative, $U_n = U n_x + V n_y$ and $U_\tau = -U n_y + V n_x$ are respectively the normal and tangential velocity at the interface between the subdomains. Similarly, we denote $\bar{u}_n$ (resp. $\bar{u}_\tau$) the normal (resp. tangential) component of the velocity around which we have linearized the equations.

ALGORITHM 2. *We choose the initial values* $W^{i,0} = (P^{i,0}, U^{i,0}, V^{i,0})$, $i = 1, 2$ *such that* $P^{1,0} = P^{2,0}$ *and we compute* $W^{i,k+1}$ *from* $W^{i,k}$ *by the iterative procedure with two steps:*
Correction step *We compute the corrections* $\tilde{W}^{1,k}$ *and* $\tilde{W}^{2,k}$ *as solution of the homogeneous local problems:*

$$\begin{cases} \mathcal{P}\tilde{W}^{1,k} = 0, & in \ \Omega_1, \\ -(\beta + \bar{u}_\tau \partial_\tau)\tilde{U}_n^{1,n} + \bar{u}_n \partial_\tau \tilde{U}_\tau^{1,k} = \gamma^k, & on \ \Gamma. \end{cases}$$

$$\begin{cases} \mathcal{P}\tilde{W}^{2,k} = 0, & in \ \Omega_2, \\ (\beta + \bar{u}_\tau \partial_\tau)\tilde{U}_n^{2,k} - \bar{u}_n \partial_\tau \tilde{U}_\tau^{2,k} = \gamma^k, & on \ \Gamma \\ \tilde{P}^{2,k} + \bar{\rho}\bar{u}_n \tilde{U}_n^{2,k} = 0, & on \ \Gamma. \end{cases} \tag{27}$$

where $\gamma^k = -\frac{1}{2}\left[(\beta + \bar{u}_\tau \partial_\tau)(U_n^{2,k} - U_n^{1,k}) + \bar{u}_n \partial_\tau(\tilde{U}_\tau^{1,k} - \tilde{U}_\tau^{2,k})\right]$.

Update step. *We compute the update of the solution* $W^{1,k+1}$ *and* $W^{2,k+1}$ *as solution of the local problems:*

$$\begin{cases} \mathcal{P}W^{1,k+1} = f_1, & in \ \Omega_1, \\ P^{1,k+1} = P^{1,k} + \delta^k, & on \ \Gamma. \end{cases}$$

$$\begin{cases} \mathcal{P}W^{2,k+1} = f_2, & in \ \Omega_2, \\ P^{2,k+1} = P^{2,k} + \delta^k, & on \ \Gamma, \\ (P + \bar{\rho}\bar{u}_n U_n)^{2,k+1} = (P + \bar{\rho}\bar{u}_n U_n)^{1,k} + (\tilde{P} + \bar{\rho}\bar{u}_n \tilde{U}_n)^{1,k}, & on \ \Gamma. \end{cases} \tag{28}$$

where $\delta^k = \frac{1}{2}\left[\tilde{P}^{1,k} + \tilde{P}^{2,k}\right]$.

M_n	Classical iter.	Classical prec. GMRES	New DDM iter.	New prec. GMRES
0.001	32	18	16	16
0.01	30	18	16	16
0.1	28	17	14	14
0.2	24	17	14	14
0.3	20	17	14	14
0.4	18	16	14	14
0.5	16	16	12	12
0.6	15	16	12	12
0.7	14	16	12	12
0.8	14	17	14	14

TABLE 3. Subdomain solves counts for different values of M_n, $M_t(y)$

Proposition 2. *For a domain $\Omega = \mathbb{R}^2$ divided into two non-overlapping half-planes, algorithm 2 converges in two iterations.*

For the details of the proof see [DN04a].

3.4. Numerical results

We compare the proposed method and the classical method defined in [DLN04] involving interface conditions that are derived naturally from a weak formulation of the underlying boundary value problem. We present here a set of results of numerical experiments on a model problem. We consider a decomposition into different number of subdomains and for a linearization around a constant or non-constant flow. The computational domain is given by the rectangle $[0 , 4] \times [0 , 1]$ with a uniform discretization using 80×20 points. The numerical investigation is limited to the solving of the linear system resulting from the first implicit time step using a Courant number CFL=100. For all tests, the stopping criterion was a reduction of the maximum norm of the error by a factor 10^{-6}. In the following, for the new algorithm, each iteration counts for 2 as we need to solve twice as much local problems than with the classical algorithm. For an easier comparison of the algorithms, the figures shown in the tables are the number of subdomains solves.

In Table 3, we consider a linearization around a variable state where the tangential velocity is given by $M_t(y) = 0.1(1 + \cos(\pi y))$ and we vary the normal Mach number.

In Figure 5, we linearize the equations around a variable state for a normal flow to the interface ($M_t = 0.0$), where the initial normal velocity is gives by $M_n(y) = 0.5(0.2 + 0.04 \tanh(y/0.2))$. The sensitivity to the mesh size is shown in the Table 4. We can see that for the new algorithm the growth in the number

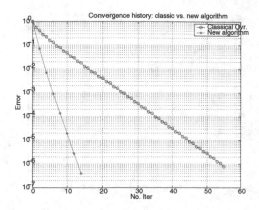

FIGURE 5. Convergence curves for the classical
and the new algorithms

h ($M_n = 0.001$)	Classical	New DDM	h ($M_n = 0.1$)	Classical	New DDM
1/10	65	18	1/10	56	12
1/20	67	18	1/20	57	14
1/40	70	18	1/40	59	16

TABLE 4. Subdomain solves counts for different mesh size

of iterations is very weak as the mesh is refined, the same property being already known for the classical one.

In Figure 6, we consider for a three subdomains decomposition a linearization around a variable state for a normal flow to the interface. The normal velocity is given by $M_n(y) = 0.5(0.2 + 0.04 \tanh(y/0.2)))$ (the same as for the 2 subdomain case).

These tests show that the new algorithm is very stable with respect to various parameters such as the mesh size and the Mach number. We see that the convergence in two iterations of the continuous algorithm is lost at the discrete level although the subdomain solves are very reasonable. Moreover, a stabilization was necessary for the discretization of the interface condition (27) in order to keep the algorithm converging. The optimal discretization of this interface condition is not yet quite well known. The comparison with the classical algorithm is favorable for Mach numbers smaller than 0.5 and especially very low Mach numbers by a factor of almost 4.

The next set of tests concerns a decomposition into 4 subdomains using a 2×2 decomposition of a $40 \times 40 = 1600$ point mesh. Table 5 summarizes the number of GMRES iterations required to reduce the initial linear residual by a factor 10^{-6} for different values of the reference Mach number for the classical algorithm and

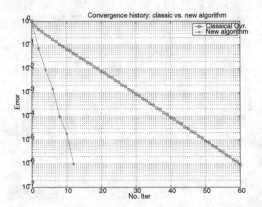

FIGURE 6. Convergence curves for the classical
and the new algorithm

M_n	Classical (iter.)	Classical prec GMRES	New DDM prec. GMRES
0.001	75	36	24
0.01	70	36	24
0.1	56	36	30
0.2	44	36	32
0.3	34	34	34

TABLE 5. Iteration count for different values of M_n

the number of GMRES iteration necessary to achieve convergence when solving the interface system.

We now consider a linearization around a variable state for a normal flow to the interface, where the normal velocity is gives by the expression $M_n(y) = 0.5(0.2 + 0.04 \tanh(y/0.2)))$ (the same as for the 2 subdomain case) and we are solving the homogeneous equations verified by the error vector at the first time step. The convergence history is given in Figure 7.

3.5. Conclusion

We designed a new domain decomposition for the Euler equations inspired by the idea of the Robin-Robin preconditioner applied to the advection-diffusion equation. We used the same principle after reducing the system to scalar equations via a Smith factorization. The resulting algorithm behaves very well for low Mach numbers, where usually the classical algorithm doesn't give very good results. We reduce the number of subdomain solves by almost a factor 4 for linearization around a constant and variable state as well. A general theoretical study and more comprehensive numerical tests have to be done in order to firmly assess the applicability of the proposed algorithm to large scale computations.

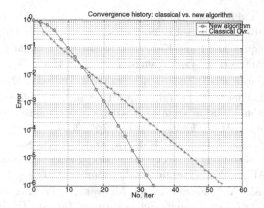

FIGURE 7. Comparison between the classical and the new algorithm

This work can also be seen as a first step for deriving new domain decomposition methods for the 2D or 3D compressible Navier-Stokes equations. Indeed, the derivation of the algorithm which is based on the Smith factorization is in fact general and can be applied to arbitrary systems of partial differential equations.

References

[ATNV00] Yves Achdou, Patric Le Tallec, Frédéric Nataf, and Marina Vidrascu. A domain decomposition preconditioner for an advection-diffusion problem. *Comput. Methods Appl. Mech. Engrg.*, 184:145–170, 2000.

[BD97]   J.D. Benamou and B. Deprés. A domain decomposition method for the Helmholtz equation and related optimal control. *J. Comp. Phys.*, 136:68–82, 1997.

[CFS98]   X.-C. Cai, C. Farhat, and M. Sarkis. A minimum overlap restricted additive Schwarz preconditioner and applications to 3D flow simulations. *Contemporary Mathematics*, 218:479–485, 1998.

[Cle98]   S. Clerc. Non-overlapping Schwarz method for systems of first order equations. *Cont. Math*, 218:408–416, 1998.

[CN98]   Philippe Chevalier and Frédéric Nataf. Symmetrized method with optimized second-order conditions for the Helmholtz equation. In *Domain decomposition methods, 10 (Boulder, CO, 1997)*, pages 400–407. Amer. Math. Soc., Providence, RI, 1998.

[DLN04]   V. Dolean, S. Lanteri, and F. Nataf. Convergence analysis of a schwarz type domain decomposition method for the solution of the euler equations. *Appl. Num. Math.*, 49:153–186, 2004.

[DN04a]   V. Dolean and F. Nataf. A new domain decomposition method for the compressible euler equations. Technical Report 567, CMAP - Ecole Polytechnique, 2004.

[DN04b]   V. Dolean and F. Nataf. An optimized schwarz algorithm for the compressible
            euler equations. Technical Report 556, CMAP - Ecole Polytechnique, 2004.

[EZ98]   Bjorn Engquist and Hong-Kai Zhao. Absorbing boundary conditions for do-
            main decomposition. *Appl. Numer. Math.*, 27(4):341–365, 1998.

[Gan66]   Felix R. Gantmacher. *Theorie des matrices*. Dunod, 1966.

[GGTN04]   L. Gerardo-Giorda, P. Le Tallec, and F. Nataf. A robin-robin preconditioner
            for advection-diffusion equations with discontinuous coefficients. *Comput.
            Methods Appl. Mech. Engrg.*, 193:745–764, 2004.

[GHN01]   Martin J. Gander, Laurence Halpern, and Frédéric Nataf. Optimal Schwarz
            waveform relaxation for the one-dimensional wave equation. Technical Report
            469, CMAP, Ecole Polytechnique, September 2001.

[GKM+91]   R. Glowinski, Y.A. Kuznetsov, G. Meurant, J. Periaux, and O.B. Widlund,
            editors. *Fourth International Symposium on Domain Decomposition Methods
            for Partial Differential Equations*, Philadelphia, 1991. SIAM.

[GMN02]   M.-J. Gander, F. Magoulès, and F. Nataf. Optimized Schwarz methods with-
            out overlap for the Helmholtz equation. *SIAM J. Sci. Comput.*, 24-1:38–60,
            2002.

[JNR01]   Caroline Japhet, Frédéric Nataf, and Francois Rogier. The optimized order
            2 method. application to convection-diffusion problems. *Future Generation
            Computer Systems FUTURE*, 18, 2001.

[Lio90]   Pierre-Louis Lions. On the Schwarz alternating method. III: a variant for
            nonoverlapping subdomains. In Tony F. Chan, Roland Glowinski, Jacques
            Périaux, and Olof Widlund, editors, *Third International Symposium on Do-
            main Decomposition Methods for Partial Differential Equations, held in Hous-
            ton, Texas, March* 20–22, 1989, Philadelphia, PA, 1990. SIAM.

[QS96]   A. Quarteroni and L. Stolcis. Homogeneous and heterogeneous domain de-
            composition methods for compressible flow at high Reynolds numbers. Tech-
            nical Report 33, CRS4, 1996.

[RT91]   Y.H. De Roeck and P. Le Tallec. Analysis and Test of a Local Domain De-
            composition Preconditioner. In *R. Glowinski et al.* [GKM+91], 1991.

[TW04]   A. Toselli and O. Widlund. *Domain Decomposition Methods – Algorithms
            and Theory*. Springer Series in Computational Mathematics. Springer Verlag,
            2004.

V. Dolean
Univ. de Nice and INRIA
Sophia Antipolis
F-06108 Nice Cedex 02, France
e-mail: dolean@math.unice.fr

F. Nataf
CMAP, CNRS UMR 7641
Ecole Polytechnique
F-91128 Palaiseau Cedex, France
e-mail: nataf@cmap.polytechnique.fr

Analysis and Simulation of Fluid Dynamics
Advances in Mathematical Fluid Mechanics, 89–108

The Two-Jacobian Scheme for Systems of Conservation Laws

Rosa Donat and Pep Mulet

1. Introduction

Shock capturing techniques for the computation of discontinuous solutions to hyperbolic conservation laws are based on an old (by now) theorem of Lax and Wendroff establishing that the limit solutions of a consistent scheme in conservation form are in fact weak solutions to the PDE and, thus, their discontinuities will propagate at the right speeds.

Over the years, it has become clear that one of the most successful strategies for designing a shock capturing scheme is to follow Godunov's lead and use the solution to the Riemann problem (the only initial-value problem easy enough to be solved explicitly) as an essential building block of the scheme.

Godunov assumed that a flow solution could be represented by a series of piecewise constant states with discontinuities at the cell interfaces. A piecewise constant function is a reasonable numerical representation of the solution in regions of smooth flow and it is specially well suited near discontinuities. The discretized flow solution is evolved by considering the nonlinear interaction between its component states. Viewed in isolation, each pair of neighboring states constitutes a Riemann problem, which can be solved exactly. If there is no interaction between neighboring Riemann problems, the global solution is easily found by piecing together these Riemann solutions. The approximate solution at the next time level is then obtained averaging over each cell this global solution.

Godunov's method becomes a finite volume scheme in conservation form because it uses solutions to Riemann problems, which are themselves exact solutions of the conservation laws. Solving Riemann problems at interfaces serves the purpose of correctly discriminating between information which should propagate with different speeds. The scheme mimics much of the relevant physics, hence it leads to an accurate and well-behaved treatment of shock waves.

Research supported by DGICYT grant MTM2005-07214.

For gas dynamics simulations, Godunov's method computes the exact solution to a Riemann problem at each cell interface. However, most of the structure of the exact solution is lost in the averaging process used to update each cell value. This observation suggests that it may not be worthwhile calculating the Riemann solution exactly. In fact, one may be able to obtain equally good numerical results with an approximate Riemann solution obtained by some less expensive means.

Economy is the chief motivation in Roe's approximate Riemann solver. It exploits the fact that the Riemann problem for any linear hyperbolic system of equations can easily be solved. Roe replaces the original nonlinear system of conservation laws

$$U_t + [\mathcal{F}(U)]_x = 0 \tag{1}$$

by a local linearization

$$U_t + A(U_L, U_R)U_x = 0 \tag{2}$$

where $A(U_L, U_R)$ is a constant matrix, constructed to have the crucial property

$$A(U_L, U_R)(U_R - U_L) = \mathcal{F}(U_R) - \mathcal{F}(U_L). \tag{3}$$

Thanks to this property, the solution to the linearized problem (2) coincides with the solution to the exact problem (1) whenever this involves merely a single shock or contact discontinuity. In addition, (3) guarantees that the resulting scheme can be written in conservation form (see, e.g., [13]). On the other hand since rarefaction waves do not appear in linear systems, the scheme can (and does) produce non-physical expansion shocks in the computed flows, unless appropriate *entropy corrections* are enforced.

Other approximate Riemann solvers, based on Roe's simplification, have emerged over the years. Their basic design principle is (as in Roe's scheme) that it might be sufficient to find only an approximate solution to a Riemann problem, provided that this approximate solution still describes important nonlinear behavior ([9, 6]).

Godunov-type schemes are indeed very robust in most situations. However, they can, on occasions, fail quite spectacularly. Reports on approximate Riemann solver failures and their respective corrections are abundant in the literature. Well-known examples in one dimension include shock reflection problems, where most shock capturing schemes produce an unphysical 'overheating' near the reflecting wall [20], oscillations in slowly moving shock waves [27, 1, 12, 30], non-linearizable Riemann problems [6], etc. In two dimensions, it is well known that Roe's method can sometimes admit solutions with an inexplicably kinked Mach stem, or an unphysical *carbuncle* at the head of a bow shock (see, e.g., [25, 22, 5, 24] and references therein).

It should be noted that the failures of a specific Riemann solver may usually be repaired by the judicious use of a small amount of artificial dissipation. However, this technique often implies the tuning and re-tuning of various parameters, which degrades the automatic character of Godunov type schemes. Moreover, the type and amount of viscosity to be added in each particular deficiency is, usually, not the same, further aggravating the user.

An interesting strategy, proposed by Quirk [22], is to combine two or more approximate Riemann solvers. With this approach, it is possible to control certain instabilities by changing the flavor of the dissipation mechanism rather than increasing the absolute level of dissipation. While this approach is very attractive, and seems to work properly in the examples shown in [22], it still has a problem dependent parameter left: when and where to use one Riemann solver in preference to another.

In [5], a new numerical flux formula was proposed. The resulting numerical scheme shares some of the flavor of Quirk's approach in that it can be interpreted as a combination of two *solvers*: Roe's scheme and a Lax-Friedrichs-type scheme. However, the numerical flux formulas of these two schemes are intertwined in a non-linear way. The resulting flux formula leads to an entropy satisfying, shock capturing scheme in which there are no adjustable parameters left.

In this paper, we review various scenarios where the flux formula of [5] has been used with success and show some of the advantages that it offers with respect to more classical alternatives, such as Roe's scheme. The layout of the paper is as follows: in Section 2 we describe the basic features of characteristic based schemes for systems of conservation laws, which is the natural framework for the Two-Jacobian (2J) numerical flux proposed in [5]. The 2J scheme is described in Section 3, as well as the necessary steps to obtain a High Resolution Shock Capturing scheme from the basic first order numerical flux formula. Section 4 is devoted to applications and numerical examples.

2. Characteristic-based schemes for systems of hyperbolic conservation laws

A fully discrete numerical scheme in conservation form for (1) has the form

$$U_i^{n+1} - U_i^n + \frac{\Delta t}{\Delta x}[F_{i+\frac{1}{2}} - F_{i-\frac{1}{2}}] = 0 \qquad (4)$$

where $U_i^n \approx U(x_i, t_n)$, $x_i = i\Delta x$, $t_n = n\Delta t$, $F_{i+\frac{1}{2}} = F(U_{i-r}, \ldots, U_{i+r+1})$ $(r \geq 0)$ and F is the *numerical flux function* of the scheme. If $F(U, \ldots, U) = \mathcal{F}(U)$, i.e., the numerical flux function is consistent, Lax-Wendroff's theorem ensures that discontinuities are captured by scheme (4), i.e., they move at the right speed even if they are unresolved (see, e.g., [13]).

In fact, (4) guarantees that the rate of change of the vector of conserved quantities, U, is equal to a difference of fluxes, in analogy with the integral form of the system of conservation laws, hence the *conservation form* is very natural if we view U_i^n as an approximation to the average of the solution $U(x, t)$ in the cell $[x_{i-\frac{1}{2}}, x_{i+\frac{1}{2}}]$ (note that $U(x_i, t_n) = \frac{1}{\Delta x} \int_{x_{i-1/2}}^{x_{i+1/2}} U(x, t_n)dx + O(\Delta x^2)$). Within this finite-volume approach, $F_{i+\frac{1}{2}}$ plays the role of an average flux through the cell boundary over the time interval $[t_n, t_{n+1}]$:

$$F_{i+\frac{1}{2}} \approx \frac{1}{\Delta t} \int_{t_n}^{t_{n+1}} \mathcal{F}(U(x_{i+\frac{1}{2}}, t))dt.$$

The design of high order, High Resolution Shock-Capturing (HRSC henceforth) schemes in conservation form for systems of equations is somewhat involved if one insists on using the fully discrete form (4), specially if the order of accuracy is larger than 2 [10]. The semi-discrete conservative approximation to (1)

$$U_t + \frac{1}{\Delta x}[F_{i+\frac{1}{2}} - F_{i-\frac{1}{2}}] = 0 \tag{5}$$

allows to naturally decouple the time and space approximations and it provides a better setting to design higher order schemes. Notice that the numerical flux $F_{i+\frac{1}{2}}$ in (5) is now naturally interpreted as an approximation to the flux through the cell boundary $x_{i+\frac{1}{2}}$ at a given time,

$$F_{i+\frac{1}{2}} \approx \mathcal{F}(U(x_{i+\frac{1}{2}}, t))$$

In Godunov's scheme, the numerical flux function is computed by solving a Riemann problem at each cell boundary. One has to keep in mind, however, that the main purpose served by introducing a Riemann solver (either exact or approximate) into a numerical scheme for conservation laws is that of providing physical realism by correctly discriminating between information which should propagate with different speeds. This is a recurrent theme when solving hyperbolic equations [9, 10, 13, 21, 28, 29, 32], since the direction in which information propagates is determinant to construct stable upwind finite difference schemes capable of approximating their exact solutions.

The local upwind directions at each cell boundary can also be obtained by analyzing the local characteristic structure obtained by diagonalizing the Jacobian matrix $J(U) = \partial \mathcal{F}/\partial U$, rather than by solving directly a Riemann problem. This approach has been used in flux-vector splitting schemes (e.g., [29]) and it is the general technique used in characteristic based schemes.

The characteristic based approach has been extensively used in the design of finite-difference Essentially Non Oscillatory (ENO) schemes [28], which could be considered as the natural framework for the 2J scheme. In what follows, we explain in some detail the basic mechanisms underlying this technique (see also [7])

Let us consider a system of m convective conservation laws in one spatial dimension,

$$U_t + [\mathcal{F}(U)]_x = 0. \tag{6}$$

In a smooth region of the flow, we can get a better understanding of the structure of the system by expanding out the derivative term as

$$U_t + JU_x = 0$$

where $J = \frac{\partial F}{\partial U}$ is the Jacobian matrix of the system. In a hyperbolic system this matrix is diagonalizable. If L is the matrix whose rows are the left eigenvectors of J and R is the matrix whose columns are the right eigenvectors of J we have

$$LJR = \text{Diag}(\lambda^p)$$

and the eigenvectors λ^p are all real.

Let us now fix a state U_0 and consider the linear system

$$U_t + R_0 J_0 L_0 U_x = 0 \tag{7}$$

where $L_0 = L(U_0)$, $R_0 = R(U_0)$, $J_0 = J(U_0)$. System (7) can be equivalently written as follows

$$W_t^0 + J_0 W_x^0 = 0 \tag{8}$$

where $W^0 = L_0 U$. This is a diagonal system, each equation being of the form

$$w_t + \lambda w_x = 0$$

and we can discretize each scalar equation independently in a λ-upwind biased fashion.

Clearly, when $U \approx U_0$, the systems (6) and (8) are very *close*, hence the local propagation of information mechanisms in (6) can be conveniently approximated by those of (8). The local upwind directions at the cell boundary $x_{i+1/2}$ could, thus, be obtained by analyzing the Jacobian matrix at an appropriately selected *interface state*.

Let us postpone the discussion of the selection of the interface state for the moment, and assume that $U_{i+1/2}$ is the interface state at the cell boundary $x_{i+1/2}$. The rationale behind the flux computation for $F_{i+1/2}$ put forward in finite-difference Essentially Non Oscillatory (ENO) schemes [28] goes as follows: multiply the entire system by the *constant* left eigenvector matrix $L_{i+1/2} = L(U_{i+1/2})$ to obtain

$$[L_{i+1/2}U]_t + [L_{i+1/2}\mathcal{F}(U)]_x = 0. \tag{9}$$

According to the local linearization (7), (8), it is approximately true that the pth component of this system, i.e., *pth local characteristic field*, rigidly translates in space at the corresponding *characteristic velocity* $\lambda_{i+1/2}^p$. Hence, we proceed to discretize the $p = 1, \ldots, m$ scalar components of this system independently, using upwind biased differencing with the upwind direction for the pth equation determined by the sign of $\lambda_{i+1/2}^p$. The corresponding discretization expressed in the original variables is obtained by pre-multiplying the resulting spatially discretized system of equations by $R_{i+1/2} = R(U_{i+1/2})$:

$$U_t + R_{i+1/2}\Delta(L_{i+1/2}\mathcal{F}(U)) = 0$$

where Δ stands for the upwind biased discretization operator.

Thus, if $w_s^p = L_{i+1/2}^p U_s$, and $\mathcal{F}_s^p = L_{i+1/2}^p \mathcal{F}(U_s)$, $p = 1, \ldots, m$, $s = i - r, \ldots, i + r + 1$, are the *characteristic variables* and *characteristic fluxes* at the

$x_{i+1/2}$ cell boundary, the numerical flux function at this location is obtained as

$$F_{i+1/2} = \sum_{i=1}^{p} F_{i+1/2}^{p} R_{i+1/2}^{p} \tag{10}$$

where the *characteristic numerical fluxes* $F_{i+1/2}^{p}$ are obtained from appropriate upwind discretizations of the components of (9).

It should be obvious that the choice of interface state is quite important to any characteristic based scheme. It defines the local linearization of the non-linear problem, determining the transformation to the local characteristic fields and, thus, what the upwind directions are, as well as what quantities are to be upwind differenced.

In standard ENO schemes it was thought that the precise form of this interface state was not so important. The standard ENO schemes [28] use the linear average of the states at nodes adjacent to the interface,

$$J_{i+\frac{1}{2}} = J(U_{i+1/2}) = J\left(\frac{U_i + U_{i+1}}{2}\right),$$

essentially because this centered linear approximation is second order accurate in smooth regions. However, recent developments show that this choice can be relevant in causing certain numerical pathologies [7, 5].

Clearly, in smooth regions it makes little difference whether the derivatives are computed in an upwind biased fashion or in some combination of upwind and downwind. The precise determination of the Jacobian (and the transformation to characteristic fields) is, thus, not critically important. Between nodes in an unresolved steep gradient the situation can be completely different. There, a centrally averaged Jacobian, or *any* artificially constructed averaged Jacobian, can differ significantly from the left and right Jacobian matrices interpolated from left and right nodal state values, and there is no clear reason why any averaged Jacobian would be the right choice for a proper transformation to characteristic variables at a cell boundary.

There is an even more basic reason to avoid the use of an averaged Jacobian at the interface when using ENO techniques. ENO schemes were developed to address the special difficulties that arise in the numerical solution of systems of nonlinear conservation laws with very high resolution. The philosophy underlying an ENO method is simple: when reconstructing a profile for use in a convective flux term, one should not use high order polynomial interpolation across a steep gradient in the data. Such an interpolant would be highly oscillatory and ultimately corrupt the computed solution. ENO schemes solve this problem by using local adaptive stencils in order to obtain information automatically from smooth regions of the solution. For physical consistency and stability, the polynomial reconstructions should also be biased to extrapolate data from the direction in which information propagates, the *upwind* direction.

However, if a basic ENO goal is to avoid interpolation across discontinuities (steep profiles) in the data, it seems highly incoherent to consistently mix information at each cell boundary in order to compute the numerical flux functions.

Near an unresolved steep gradient in the flow, in which the states may vary by a large amount from one node to the next, it makes sense to try to use *directly* the two interface states, obtained from ENO extrapolation of nodal values, and the characteristic information contained in the two Jacobian matrices evaluated at these states, rather than attempt to define a single representative midpoint Jacobian.

3. The 2J scheme

As described in [5], the starting point of the 2J-numerical flux is a combination of Roe's flux and a Local-Lax-Friedrichs (LLF from now on) flux for scalar conservation laws $u_t + f(u)_x = 0$. It was first proposed in [28], where it was labelled F^{RF}:

$$F^{RF}(u_l, u_r) = \begin{cases} f(u_l) & \text{if} \quad f' > 0 \text{ in } [u_l, u_r] \\ f(u_r) & \text{if} \quad f' < 0 \text{ in } [u_l, u_r] \\ \frac{1}{2}(f(u_l) + f(u_r) - \alpha(u_r - u_l)) & \text{else} \end{cases} \tag{11}$$

$$\alpha = \max_{u \in [\{u_l, u_r\}]} |f'(u)| \tag{12}$$

where $[\{u_l, u_r\}]$ should be understood as the range of u-values that lie between u_l and u_r, the states to the left and to the right of the interface.

If f' does not change sign in $[\{u_l, u_r\}]$, then (11) is equivalent to Godunov's flux formula (see, e.g., [13])

$$F^G(u_l, u_r) = \begin{cases} \min_{u_l \leq u \leq u_r} f(u) & \text{if} \quad u_l \leq u_r \\ \max_{u_r \leq u \leq u_l} f(u) & \text{if} \quad u_l > u_r. \end{cases} \tag{13}$$

If, on the contrary, f' changes sign on $[\{u_l, u_r\}]$, then F^{RF} is obtained by switching to the more viscous, entropy satisfying Lax-Friedrichs scheme,

$$F^{LLF}(u_l, u_r) = \frac{1}{2}\left(f(u_l) + f(u_r) - \alpha(u_r - u_l)\right).$$

It is well known that for convex f, the schemes of Roe and Godunov differ only at transonic rarefactions. Since the LLF flux is monotone (see [28]), F^{RF} is an 'entropy fix' for Roe's flux. Notice that for convex f one could switch to LLF only when $f'(u_l) < 0 < f'(u_r)$. However, we keep (11) as established because it is more general and it works properly also for non-convex conservation laws.

Notice that the scheme that results from using the F^{RF} flux formula can be understood as the combination of two *solvers*: Godunov's (or Roe's) method and the (local) Lax

Friedrichs scheme. The experiments reported in [28] and our own experimentation confirm that conservative schemes whose numerical flux function is F^{RF} always approximate the physically relevant solution of scalar conservation laws, even for non convex f. Moreover, local pathologies, like the dog-leg effect, either do not show up in numerical approximations, or are reduced to $O(\Delta x)$ glitches in the first order version of the scheme. Higher order versions completely eliminate the pathology.

The numerical flux function of the 2J scheme for systems of conservation laws follows the basic characteristic-based strategy of ENO schemes, as described in the previous section, in order to extend to systems a numerical flux formula designed specifically for scalar conservation laws. As mentioned in the previous section, the essential difference lies in the direct use of the two interface values U_l and U_r and the two associated Jacobian matrices $J(U_l)$ and $J(U_r)$. The spectral information of these two matrices serves to compute two sets of characteristic variables and fluxes at each cell interface.

Let $L^p(U_l)$, $L^p(U_r)$, $(R^p(U_l)$, $R^p(U_r))$, be the (normalized so that $L^p \cdot R^q = \delta_{p,q}$) left (resp. right) eigenvectors of the Jacobian matrices $J(U_l),J(U_r)$. Let $\lambda_p(U_l),\lambda_p(U_r)$, $p = 1,\ldots,m$ be their corresponding eigenvalues.

The first step of the algorithm proposed in [5] is as follows: compute

$$\text{For } p = 1,2\ldots,m \begin{cases} \omega_l^p = L^p(U_l) \cdot U_l & \mathcal{F}_l^p = L^p(U_l) \cdot \mathcal{F}(U_l) \\ \omega_r^p = L^p(U_r) \cdot U_r & \mathcal{F}_r^p = L^p(U_r) \cdot \mathcal{F}(U_r). \end{cases} \tag{14}$$

The second step proceeds as described in Figure 1.

For any hyperbolic system in which the fields are either genuinely nonlinear or linearly degenerate, like the Euler equations of gas dynamics, we test the possible

> **for** $p = 1,\ldots,m$
>> **if** $\lambda_p(U_l) \cdot \lambda_p(U_r) \geq 0$
>>> **if** $\lambda_p(U_l) > 0$ **then**
>>>> $F_+^p = \mathcal{F}_l^p, \qquad F_-^p = 0$
>>> **else**
>>>> $F_+^p = 0, \qquad F_-^p = \mathcal{F}_r^p$
>>> **endif**
>> **else**
>>> $\alpha_p = \max\{|\lambda_p(U_l)|, |\lambda_p(U_r)|\}$
>>> $F_+^p = .5(\mathcal{F}_l^p + \alpha_p\omega_l^p), \qquad F_-^p = .5(\mathcal{F}_r^p - \alpha_p\omega_r^p)$
>> **endif**
> **endfor**

$$F^M(U_l,U_r) = \sum_{p=1}^m \left(F_+^p R^p(U_l) + F_-^p R^p(U_r)\right)$$

FIGURE 1. First order 2J flux splitting algorithm to compute $F^M(U_l,U_r)$.

sign changes of $\lambda_p(U)$ as in [28], by checking the sign of $\lambda_p(U_l) \cdot \lambda_p(U_r)$. In case of existence of a field neither genuinely nonlinear nor linearly degenerate, the definition of α_p should, most likely, be changed. Any convenient definition would necessarily involve an appropriately defined curve in phase space, say $\Gamma(U_l, U_r)$, connecting U_l and U_r. We will not pursue this issue any further in this paper.

The structure of the 2J numerical flux formula,

$$F^M(U_l, U_r) = \sum_{p=1}^{m} \left(F_+^p R^p(U_l) + F_-^p R^p(U_r) \right) \tag{15}$$

is similar to that of (10). Here, the *sided* characteristic information is re-assembled by using the normalized right eigenvectors from the appropriate side.

From the algorithm described in Figure 1, we see that if $U_l = U_r = U$, then $F_+^p + F_-^p = \mathcal{F}^p = L^p(U)\mathcal{F}(U)$. Then

$$F^M(U, U) = \sum_{p=1}^{m} \mathcal{F}^p(U) R^p(U) = \mathcal{F}(U),$$

i.e., the 2J numerical flux function is consistent. Hence, a first order scheme based on (15) is obtained by taking $U_l = U_i$ and $U_r = U_{i+1}$ to compute the flux at the $x_{i+1/2}$ cell-interface, i.e.,

$$U_i^{n+1} = U_i^n - \frac{\Delta t}{\Delta x}(F^M(U_i^n, U_{i+1}^n) - F^M(U_{i-1}^n, U_i^n))$$

We remark here that the design of the 2J numerical flux involves a nonlinear combination of the schemes of Roe and LLF, performed in each 'local' characteristic field. There are no adjustable parameters left in the scheme, although the 'artificial dissipation' can still be modified via the parameter α.

3.1. The flux-vector splitting structure of F^M

Let us define

$$F^+(U, V) = \sum F_+^p(U, V) R^p(U), \qquad F^-(U, V) = \sum F_-^p(U, V) R^p(V), \tag{16}$$

where F_+^p and F_-^p are as described in Figure 1. Observe that both $F^\pm = F^\pm(U, V)$ depend (in a nonlinear way) on the left and right states. From the algorithm in Figure 1, we see that F^+ collects the contribution of the *right wind* driven information (i.e., positive eigenvalues for each local characteristic field), while F^- collects the contribution of the *left wind* driven information. Notice that if $\lambda^p(U) > 0$, $\lambda^p(V) > 0$ for all p, then $F_-^p = 0$, $F_+^p = L^p(U)\mathcal{F}(U)$, thus

$$F^M(U, V) = F^+(U, V) = \mathcal{F}(U).$$

Analogously, if $\lambda^p(U) < 0$, $\lambda^p(V) < 0$ for all p, then $F_+^p = 0$, $F_-^p = L^p(V)\mathcal{F}(V)$, thus

$$F^M(U, V) = F^-(U, V) = \mathcal{F}(V).$$

The 2J numerical flux formula

$$F^M(U, V) = F^+ + F^-$$

displays, thus, a clear *flux-splitting* structure, since F^+ is associated to the *right-moving* waves at the interface, i.e., to the positive eigenvalues for each local characteristic field, while F^- is associated to the negative eigenvalues and, hence, to the left-moving information.

It is illuminating to examine the particular case in which the flux function is homogeneous of degree one, i.e., $\mathcal{F}(U) = J(U) \cdot U$, as in the Euler equations of gas dynamics. Then we get

$$\mathcal{F}_l^p = L^p(U_l) \cdot \mathcal{F}(U_l) = \lambda_p(U_l) \, \omega_l^p$$

$$\mathcal{F}_r^p = L^p(U_r) \cdot \mathcal{F}(U_r) = \lambda_p(U_r) \, \omega_r^p.$$

Let us assume that for the pth field $\lambda_p(U_l) \cdot \lambda_p(U_r) \geq 0$. The computations in the previous section would immediately lead to

$$F_+^p = \max(\lambda_p(U_l), 0) \cdot \omega_l^p = \lambda_p^+(U_l) \cdot \omega_l^p$$

$$F_-^p = \min(\lambda_p(U_r), 0) \cdot \omega_r^p = \lambda_p^-(U_r) \cdot \omega_r^p.$$

Now, if this is the case *for all* p, we readily get

$$F^+ = \sum F_+^p R^p(U_l) = \sum \lambda_p^+(U_l) \, \omega_l^p \cdot R^p(U_l) = J^+(U_l) \cdot U_l$$

$$F^- = \sum F_-^p R^p(U_r) = \sum \lambda_p^-(U_r) \, \omega_r^p \cdot R^p(U_r) = J^-(U_r) \cdot U_r$$

with

$$J^+(U) = R(U)\Lambda^+ L(U), \quad J^-(U) = R(U)\Lambda^- L(U); \quad \Lambda^\pm = \text{Diag}(\lambda_p^\pm(U)). \quad (17)$$

The computations above reveal that F^M is equivalent to the flux-vector splitting formula described by Steger and Warming in [29], when there is no change in sign in any of the eigenvalues.

When the eigenvalue corresponding to a characteristic field changes sign across a given interface, F^+ and F^- in (16) will depend on both the left and right states and the flux formula of the 2J scheme will differ from that of the Steger and Warming. Notice that, as opposed to the Steger and Warming construction, (15) does not assume homogeneity. It could, thus, be considered a generalization of the flux-vector splitting construction in [29] which can equally be applied to non homogeneous fluxes, such as the flux in Burgers equation or the equations of gas dynamics for real gases [31], as long as the eigen-structure of the Jacobian matrix is known.

3.2. The 1J scheme: Shu Osher finite difference schemes

Let us assume that we insist on defining an interface state U^*, obtained from the values U_l and U_r, which will be used to perform the computation of the local characteristic fields, as described in Section 2.

The first step of this 1J (for 1-Jacobian) flux computation substitutes (14) by

$$\text{For } p = 1, 2 \ldots, m \begin{cases} \omega_l^p = L^p(U^*) \cdot U_l & \mathcal{F}_l^p = L^p(U^*) \cdot \mathcal{F}(U_l) \\ \omega_r^p = L^p(U^*) \cdot U_r & \mathcal{F}_r^p = L^p(U^*) \cdot \mathcal{F}(U_r). \end{cases} \quad (18)$$

If we perform the second step, as specified in Figure 1, we end up with the following numerical flux formula

$$F^{OJ}(U_l, U_r) = \sum_{p=1}^{m}(F_+^p + F_-^p)R^p(U^*) = \sum_{p=1}^{m} F^p R^p(U^*).$$

It is easy to check that the characteristic numerical flux

$$F^p = (F_+^p + F_-^p) = \begin{cases} \mathcal{F}_l^p & \text{if } \lambda_p(U_l) > 0, \lambda_p(U_r) > 0 \\ \mathcal{F}_r^p & \text{if } \lambda_p(U_l) < 0, \lambda_p(U_r) < 0 \\ \frac{1}{2}(\mathcal{F}_l^p + \mathcal{F}_r^p - \alpha_p(\omega_r^p - \omega_l^p)) & \text{if } \lambda_p(U_l) \cdot \lambda_p(U_r) \leq 0 \\ \alpha_p = \max(|\lambda_p(U_l)|, |\lambda_p(U_r)|). \end{cases}$$

Hence, the 1J numerical flux is absolutely equivalent to the first order numerical flux formula obtained from the ENO-RF algorithm in [28].

The 2J numerical flux can also be interpreted as an extension of the Shu-Osher numerical flux for finite difference schemes developed in [28]. As mentioned in Section 2, the authors in [28] use $U^* = (U_l + U_r)/2$ as the interface state, but any other average state, like the Roe mean when known, can also be used. The 2J strategy seems to be better suited to the ENO upgrading of the basic first order scheme, since it insists on non-mixing information at interfaces. As observed in [7], this fact can explain the absence of certain numerical inaccuracies observed when using Shu-Osher ENO schemes in some scenarios.

3.3. Improving the order of accuracy

The semi-discrete conservative formulation (5) is particularly useful in developing HRSC schemes with order of accuracy greater than two, since it allows for an easy decoupling of the issues of spatial and temporal accuracy. In particular, high order accuracy in time is obtained in ENO schemes by applying an adequate ODE solver, such as the special family of Runge-Kutta time integration schemes developed by Shu and Osher [28]. They are easy to implement, have good stability properties and also have a 'Total variation Diminishing' (TVD) property, which prevents the time stepping scheme from introducing spurious spatial oscillations into upwind-biased spatial discretizations.

In ENO schemes, the spatial accuracy is improved by using an appropriate ENO reconstruction of the data at each time step. In the case of a system, it is agreed [10, 28] that this interpolation should be done in the local characteristic fields, since it is these quantities, and not the primitive conserved variables such as mass, momentum and energy, that are properly thought as propagating in various directions.

In [28], Shu and Osher use the moving-stencil idea directly on numerical fluxes to get high order finite-difference ENO schemes. For the scalar problem, the numerical flux formula of the high order ENO scheme recommended in [28] can be expressed as follows:

$$F_{i+1/2} = F^{RF}(U_i, U_{i+1}) + HOT_{i+1/2} \tag{19}$$

where $F^{RF}(U_i, U_{i+1})$ is given by (11) and HOT_{i+1} correspond to the higher order terms necessary to improve the order of accuracy. These HOT terms involve
divided differences of the values $f_i = f(U_i)$ and U_i, hence their scheme uses only
nodal values of the conserved variables and it is somewhat faster and easier to
implement than the original cell-averaged ENO schemes of [10].

• For systems, the same scalar ENO reconstruction leading to (19) is carried
out in each local characteristic field. The process needs, thus, to enlarge the number
of local characteristic quantities to be computed in the first step of the algorithm.
In [28], where one interface state $U_{i+1/2}$ is used to transform to local characteristic
variables at the $x_{i+1/2}$ interface, the first step for the high order ENO Shu-Osher
(or 1J) $F_{i+1/2}$ flux computation substitutes (18) by

For $p = 1, \ldots, m$.   For $s = -r, \ldots, r+1$

$$\begin{cases} \omega_{i+s}^p = L^p(U_{i+1/2}) \cdot U_{i+s} \\ \mathcal{F}_{i+s}^p = L^p(U_{i+1/2}) \cdot \mathcal{F}(U_{i+s}). \end{cases} \tag{20}$$

These $2r + 2$ values are used to perform a high order accurate upwind ENO reconstruction, which is then evaluated at the cell-interface to obtain the numerical
characteristic flux $F_{i+1/2}^p$. We refer the reader to [28] for details.

To upgrade the 2J scheme, one needs to compute first a high order approximation to $U_l = U_{i+1/2}^l$ and $U_r = U_{i+1/2}^r$, the interface values. This is accomplished
by constructing an upwind-biased ENO interpolant of the U values on each side
of the interface. Once the two interface states have been computed, (14) is substi-
tuted by

For $p = 1, \ldots, m$.   For $s = -r, \ldots, r+1$

$$\begin{cases} \omega_{i+s}^{p,l} = L^p(U_{i+1/2}^l) \cdot U_{i+s} & \omega_{i+s}^{p,r} = L^p(U_{i+1/2}^r) \cdot U_{i+s} \\ \mathcal{F}_{i+s}^{p,l} = L^p(U_{i+1/2}^l) \cdot \mathcal{F}(U_{i+s}) & \mathcal{F}_{i+s}^{p,r} = L^p(U_{i+1/2}^r) \cdot \mathcal{F}(U_{i+s}). \end{cases} \tag{21}$$

Each *sided-set* of characteristic information is ENO-reconstructed and evaluated as specified in [28] in order to obtain the high order version of the sided
characteristic fluxes $(F_+^p)_{i+1/2}$ and $(F_-^p)_{i+1/2}$. The characteristic information is
re-assembled together as specified in formula (15). The third order ENO extension
of the 2J scheme has been carefully described in [7].

The greatest advantage of the finite-difference ENO schemes developed by
Shu and Osher in [28] is that it extends to higher dimensions in a 'dimension by
dimension' fashion, so that the 1D method applies unchanged to higher dimensional problems (see [19]).

In the finite volume ENO framework developed in [10], in which the numerical values U_i^n are approximations to the cell-averages of the solution, the spatial

upgrading involves a piecewise polynomial reconstruction, $\mathbf{R}(\cdot, U^n)$, of the solution from its cell-averages. If $U(x, t_n) = \mathbf{R}(\cdot, U^n) + O(\Delta x^r)$, an rth order finite-volume version of the scheme (4) can be obtained following the semi-discrete formulation,

$$\frac{d}{dt}U_i(t) = -\frac{1}{\Delta x}[\tilde{F}_{i+1/2} - \tilde{F}_{i-1/2}] \qquad (22)$$

where

$$\tilde{F}_{i+1/2} = F^M\left(\mathbf{R}(x_{i+1/2} - 0; U(t)), \mathbf{R}(x_{i+1/2} + 0; U(t))\right).$$

This was the approach described in [5]. We recall that the reconstruction procedure computes first ENO piecewise polynomial reconstructions for the *local characteristic fields*, and this applies also to the computation of the high order ENO extrapolated values U_l and U_r at each interface. The HRSC schemes that result from this procedure are a bit more expensive than those obtained following the Shu-Osher flux-difference approach. Throughout our numerical experimentation, we have not observed any significant difference between the two options, and the finite difference version of the scheme is the one we have currently in use.

4. Numerical examples

In [5, 4, 7, 15, 2, 26, 14, 31], the reader can find many examples illustrating the use of the 2J scheme in different scenarios.

Here, we display two cases that involve non-trivial scenarios. Our numerical examples concern the equations that describe the dynamics of relativistic flows and an extension of the Euler equations for multifluid gas dynamics. In both cases, the shock capturing technology had been previously used with various degrees of success.

It is worth recalling that the 2J scheme, only requires the spectral decomposition of the Jacobian matrices of the fluxes. If this ingredient is known this flux formula can easily substitute the computation of the numerical flux in an existing hydrodynamical code.

4.1. A 2D simulation of a supersonic relativistic jet

HRSC schemes are nowadays routinely used in Relativistic Fluid Dynamics (RFD henceforth). The term applies to flows in which the velocities (of the individual particles or of the fluid as a whole) approach c, the velocity of light. It also applies to those flows where the effects of the background gravitational field (or that generated by the matter itself) are so important that a description in terms of Einstein theory of gravity becomes necessary.

Explicit shock capturing codes for relativistic simulations appear first in the early nineties and are mainly Godunov-type schemes in which the use of an approximate Riemann solver is an essential ingredient

In [4], an explicit, ready to-use formulation of the full eigenstructure of the Jacobian matrices associated with the fluxes of the 3D hyperbolic system that models relativistic fluid flow is given for the first time. This knowledge allows for the use of the ENO-type schemes mentioned in the previous sections, as well as the 2J scheme reviewed in this paper.

In [17], high resolution shock capturing schemes were systematically used to study the morphology and dynamics of *relativistic jets* encountered in some astrophysical scenarios. We have chosen one particular model of the large sample pool in [17] to illustrate the performance gains of the 2J scheme with respect to another Roe-like linearized solver previously employed in relativistic simulations (see [18]).

The numerical simulation shows the evolution of a relativistic fluid injected supersonically into the computational domain through a small nozzle. The computational domain, which is discretized in cylindrical coordinates (r, z), is 25 units long in the z-direction and 7 units wide in the r-direction. It is covered by a uniform numerical grid consisting of 500×140 zones. The beam fluid is injected into the grid parallel to the symmetry axis (the z axis) through a nozzle at the bottom $(r = 0)$ of the left boundary of the grid $(z = 0)$, which is 20 zones wide (i.e., of length unity). Outflow boundary conditions are used at all boundaries except at the symmetry axis $(r = 0$ boundary) where reflection conditions are imposed, and at the nozzle, where fixed inflow beam conditions are used.

The initial model that we consider for the injected beam fluid corresponds to a Mach 6 flow with $\Gamma = 5/3$. The density outside the beam is $\rho = 1$ and the velocity of the fluid at the nozzle is $v^r = 0$ and $v^z = 0.99c$, i.e., well in the *ultrarelativistic* regime.

Figure 2 shows a snap-shot of the simulation obtained with two different codes applied to this initial set-up. Both codes use the same piecewise-parabolic reconstruction [16], and time-stepping procedures, but one incorporates the 2J recipe in the numerical flux computation while the other one uses the Roe-type linearized solver used in [18] (essentially the 1J scheme with the mean $U_{i+1/2} = (U_i + U_{i+1})/2$ at the interface).

In both figures, the leading bow shock, the internal contact discontinuity which is Kelvin-Helmholtz unstable and the beam, the innermost internal channel are clearly identified. At this time of the evolution the beam presents an internal conical shock and a Mach disk at its head, which slows down the material inside the beam.

The most significant difference between both plots is the small protuberance ahead of the bow shock which appears when using the 1J scheme. This is purely a numerical artifact, since the Mach disk prevents the material inside the beam from pushing other material ahead of the bow shock. This local pathological behavior is well known in blunt body simulations in gas dynamics (see [22, 24] and references therein) and receives the name of *carbuncle*. As pointed out by Quirk [22], Roe's scheme admits sometimes this spurious solution, being the effect more likely to appear for high Mach number flows and the more closely the grid is aligned to the bow shock. This is precisely what we have here. As can be seen in Figure 2 the pathology seems to disappear when the 2J scheme is used.

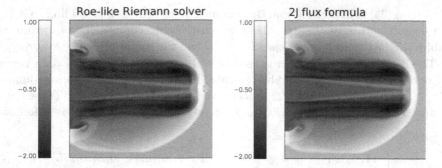

FIGURE 2. Simulation of a supersonic relativistic jet: the logarithm of the rest-mass density – gray-scale – obtained with a PPM reconstruction procedure and
Left: a linearized Roe-like solver. Notice the "carbuncle phenomenon" ahead of the bow shock
Right: The 2J flux formula. The behavior of the numerical solution agrees with the physics of the problem

4.2. A mass-fraction multifluid model

In this section we analyze the computational performance of the numerical scheme for a test problem consisting of a 1.22 Mach shock wave impinging a R22 refrigerant gas bubble. This problem was proposed and studied by Haas and Sturtevant [8] and numerically simulated by Quirk and Karni [23]. A similar simulation for Helium instead of R22 gas has been carried out in [8, 23, 15].

The gas dynamics for a fluid composed by the mixture of two perfect gasses in thermal equilibrium can be modelled by the Euler equations with an additional equation for the conservation of the first gas, hence of both gasses due to mass conservation. In one space dimension, this model becomes:

$$\begin{pmatrix} \rho \\ \rho u \\ E \\ \rho\phi \end{pmatrix}_t + \begin{pmatrix} \rho u \\ \rho u^2 + p \\ u(E + p) \\ \rho\phi u \end{pmatrix}_x = 0, \tag{23}$$

where the last equation expresses the conservation of the volume fraction $\rho\phi$ of the first gas. In these equations ρ represents the density of the mixture, ϕ the mass fraction of the first gas, u is the velocity, E is the total energy per unit volume and p is the pressure, related to the density ρ, the specific internal energy ϵ and the mass fraction by the equation of state:

$$p = (\gamma - 1)\rho\epsilon, \quad \gamma = \frac{C_{p_1}\phi + C_{p_2}(1 - \phi)}{C_{v_1}\phi + C_{v_2}(1 - \phi)}, \tag{24}$$

with C_{p_i} (resp. C_{v_i}) being the specific heats at constant pressure (resp. volume) of fluid i.

System (23) is hyperbolic and the unique requirements for the 2J flux splitting is the eigenstructure of the Jacobians, which is easily computed from primitive variables [15] (actually, it only depends on u, ϕ and the sound velocity $c = \sqrt{\gamma p/\rho}$).

The computational domain is the interval $[0, 0.445]$. The initial data represents a one-dimensional R22 "bubble", located in the interval $[0.2, 0.25]$ surrounded by air. A left-travelling Mach 1.22 shock wave is located at $x = 0.275$. These data, normalized so that both the density and the pressure of quiescent air is 1, follows:

$$(\rho, u, p, \phi)(x, t = 0)$$
$$= \begin{cases} (1, 0, 1, 1) & \text{if } x \in [0, 0.2] \cup [0.25, 0.275] \\ (3.1538, 0, 1, 0) & \text{if } x \in (0.2, 0.25) \\ (1.3764, -0.3947, 1.5698, 0) & \text{if } x \in [0.25, 0.445]. \end{cases} \quad (25)$$

In Figure 3 a high resolution (8000 cells) approximate solution at $t = 0.18$ time units is computed by the 2J scheme with a CFL number of 0.9. This solution is taken as reference for lower resolution simulations (800 cells) computed by the 2J scheme and the 1J scheme with fifth order weighted ENO (see [11]) flux reconstructions.

This time, the plots in Figure 3 do not show any significant differences with respect to the 1J scheme. Multifluid simulations are notoriously hard, due to various numerical pathologies that appear when using shock capturing schemes. In previous works (e.g., [7, 15]) we have observed that the importance of the observed pathological behavior of numerical nature seems to be diminished when using the 2J scheme.

To conclude, we show, in Figure 4, a two-dimensional R22 bubble simulation, computed at $t = 0.31$ time units on a computational domain of the same length covered with a 8000×800 uniform computational mesh. The 2J scheme is applied to the system

$$\begin{pmatrix} \rho \\ \rho u \\ \rho v \\ E \\ \rho\phi \end{pmatrix}_t + \begin{pmatrix} \rho u \\ \rho u^2 + p \\ \rho u v \\ u(E + p) \\ \rho\phi u \end{pmatrix}_x + \begin{pmatrix} \rho v \\ \rho u v \\ \rho v^2 + p \\ v(E + p) \\ \rho\phi v \end{pmatrix}_y = 0, \quad (26)$$

with initial conditions given by

$$(\rho, u, p, \phi)(x, y, t = 0)$$
$$= \begin{cases} (1, 0, 1, 1) & \text{if } x \in [0, 0.2] \cup [0.25, 0.275] \\ (3.1538, 0, 1, 0) & \text{if } (x, y) \in B \\ (1.3764, -0.3947, 1.5698, 0) & \text{if } x \in [0.25, 0.445] \end{cases} \quad (27)$$
$$B = \{(x, y)/(x - 0.225)^2 + y^2 < 0.0225^2\}$$

and vertical velocity $v = 0$ everywhere. Reflecting boundary conditions are applied at the top and bottom edges, outflow applied elsewhere.

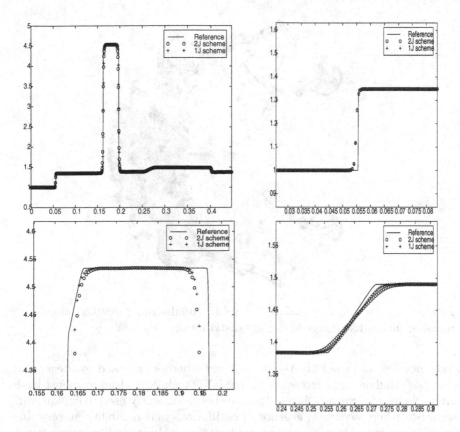

FIGURE 3. One-dimensional R22 bubble simulation, $t = 0.18$ time units: density plots of reference solution (solid line, 8000 cells), 2J solution (*circ*, 800 cells) and 1J solution (+, 800 cells)

As noted by Quirk [22], in performing very fine computations, when the numerical viscosity inherent to the scheme becomes very low, some numerical schemes for systems of conservation laws tend to show an anomalous behavior. Our own experimentation confirms this fact [3, 15]. As we see in Figure 4, the very low numerical viscosity allows for the full development of the Richtmyer-Meshkov instabilities developed at the bubble interface. Similar behavior has been observed at certain two-dimensional contact discontinuities [3].

Conclusions

In this paper, we review the numerical scheme initially proposed in [5] for hyperbolic conservation laws, paying special attention to the close relationship between this scheme and other more classical methods like the Shu-Osher finite-difference

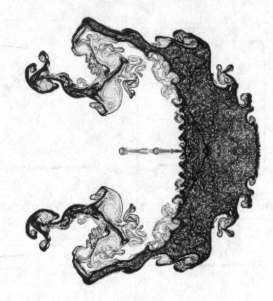

FIGURE 4. Two-dimensional R22 bubble simulation, $t = 0.31$ time
units: schlieren-like image of density distribution

ENO schemes [28] and the Flux-Vector splitting schemes proposed by Steger and
Warming [29]. Higher order extensions, using ENO techniques lead to robust High
Resolution Shock Capturing schemes that are being currently used to explore various scenarios where accuracy, absence of oscillations and reliability at very fine
resolution is needed. Areas of current applications include shallow water flows,
interfaces in acoustics problems and penalization techniques for the simulation of
fluid flow with obstacles.

Acknowledgements

The authors would like to thank J.M. Martí for providing the simulations of
the ultrarelativistic jets. Support from the Spanish MEC, through the project
MTM2005-07214 is gratefully acknowledged.

References

[1] M. Arora and P.L. Roe *On post-shock oscillations due to shock capturing schemes in unsteady flows* J. Comput. Phys., v. **130**, (1997) pp. 25.

[2] G. Chiavassa and R. Donat, *Point Value Multiscale Algorithms for 2D Compressible Flows,* SIAM J. Sci. Comp. **23**, 805–823 (2001)

[3] G. Chiavassa and R. Donat and A. Marquina, *Fine-Mesh Numerical Simulations for 2D Riemann Problems with a Multilevel scheme,* Intl. Series of Numerical Mathematics **140** 247–256 (2001).

[4] R. Donat, J.A. Font, J.M. Ibañez and A. Marquina *A Flux-Split Algorithm applied to Relativistic Flows*. To appear in J. Comput. Phys.

[5] R. Donat and A. Marquina, *Capturing Shock Reflections: An improved Flux Formula* J. Comput. Phys., v. **125**, (1996) pp. 42–58.

[6] B. Einfeldt, *On Godunov type methods for Gas Dynamics*, SINUM v. **25**, (1988) pp. 294–318.

[7] R. Fedkiw, B. Merriman, R. Donat and S. Osher, *The Penultimate scheme for Systems of Conservation Laws* Innovative methods for numerical solutions of partial differential equations (Arcachon, 1998), 49–85, World Sci. Publishing, River Edge, NJ, (2002).

[8] J.-F. Haas and B. Sturtevant, *Interaction of weak shock waves with cylindrical and spherical gas inhomogeneities*, J. Fluid. Mech., 181 (1987), pp. 41–76.

[9] A. Harten, P.D. Lax and B. van Leer, *On Upstream differencing and Godunov type Schemes for Hyperbolic Conservation Laws*, SIAM Review, **25**, pp. 35–61 (1983).

[10] A. Harten, B. Engquist, S. Osher and S. Chakravarthy, *Uniformly High Order Accurate Essentially Non-oscillatory Schemes III*, J. Comput. Phys., v. **71** No. 2, (1987), pp. 231–303.

[11] G. Jiang and C. Shu, *Efficient implementation of weighted ENO schemes*, J. Comp. Phys., 126 (1996), pp. 202–228.

[12] S. Karni and S. Canic *Computations of Slowly Moving Shocks* J. Comput. Phys., v. **136**, (1997) pp. 132–139.

[13] R.J. Leveque *Numerical methods for Conservation Laws*, Birkhäuser Verlag, Zürich, (1990).

[14] B. Lombard, R. Donat, *The explicit simplified interface method for compressible multicomponent flows* SIAM J. of Sci. Comp. **27**, 208–230 (2005)

[15] A. Marquina and P. Mulet *A flux-split algorithm applied to conservative models for multicomponent compressible flows*, J. Comput. Phys., v. **185** , (2003) pp. 120–138.

[16] J.M. Martí and E. Müller, *J. Comp. Phys.* **123**, 1 (1996).

[17] J.M. Martí, E. Müller, J.A. Font, J.M. Ibáñez and A. Marquina, *Astrophys. J.*, **479** 151–163 (1997)

[18] J.M. Martí, E. Müller and J.M. Ibáñez, *Astron. Astrophys.*, **281**, L9 (1994).

[19] B. Merriman *Understanding the Shu-Osher conservative finite difference form*, J. Sci. Comput., v. **19**, (2003) pp. 309–322.

[20] W.F. Noh, *Errors for the Calculations of Strong Shocks using an artificial viscosity and an artificial heat flux*, J. Comput. Phys., v. **72**, (1987) pp. 78–120.

[21] S. Osher, F. Solomon, *Upwind difference schemes for hyperbolic systems of Conservation Laws*, Math. Comput., v. **38**, (1982) pp. 339–374.

[22] J. Quirk, *A contribution to the great Riemann Solver debate*, Intl. J. Num. Meth. Fluids, v. **18** (1994) pp. 555–574. Also ICASE Report 92–64 (1992).

[23] J. Quirk and S. Karni, *On the dynamics of a shock-bubble interaction*, J. Fluid. Mech., 318 (1996), pp. 129–163.

[24] M. Pandolfi and D. D'Ambrosio *Numerical instabilities in upwind methods: analysis and cures for the "carbuncle" phenomenon*, J. Comput. Phys., v. **166**, (2001) pp. 271–301.

[25] K.M. Peery and S.T. Imlay, *Blunt Body Flow Simulations*, AAIA paper 88-2904.

[26] A. Rault, G. Chiavassa and R. Donat, *Shock-Vortex Interactions at High Mach Numbers* Journal of Scientific Computing, **19** 347–371 (2003).

[27] T.W. Roberts, *The behavior of flux-difference splitting schemes near slowly moving shock waves,* J. Comput. Phys. v. **90** (1990) pp. 141–160.

[28] C.W. Shu and S.J. Osher, *Efficient Implementation of Essentially Non-Oscillatory Shock Capturing Schemes II,* J. Comput. Phys., v. **83**, (1989) pp. 32–78.

[29] J. Steger and R.F. Warming *Flux Vector Splitting of the Inviscid Gasdynamics Equations with Application to Finite Difference Methods,* J. Comput. Phys., v. **40**, (1981) pp. 263–293.

[30] Y. Stiriba, R. Donat, *A Numerical Study of Post Shock Oscillations in Slowly Moving Shock Waves,* Computer and Mathematics with Applications, **46**, pp. 719–739 (2003).

[31] Y. Stiriba, A. Marquina and R. Donat *'Equilibrium real gas computations using Marquina's scheme* International J. for Numerical Methods in Fluids, **41** 275–301 (2003).

[32] B. Van Leer, *Flux-vector splitting for the Euler equations,* presented at the 8th international Conference on Numerical Methods for Engineering, Aachen, June 1982.

Rosa Donat and Pep Mulet
Departament de Matemática Aplicada
Universitat de València
E-46100-Burjassot (València) Spain

Analysis and Simulation of Fluid Dynamics
Advances in Mathematical Fluid Mechanics, 109–127
© 2006 Birkhäuser Verlag Basel/Switzerland

Do Navier-Stokes Equations Enable to Predict Contact Between Immersed Solid Particles?

M. Hillairet

Abstract. We present here a short overview of recent results on a paradox appearing in the area of fluid-solid interactions. This paradox states that, in two space dimensions, strong solutions to viscous models describing fluid-solid interactions do not permit rigid solids inside the fluid to collide.

1. Introduction

1.1. Paradox after paradox

Since the very beginning of the study of fluid-solid interactions, several paradoxical situations revealed the difficulties to make theoretical models meet experimental reality. The first example worth mentioning must be the d'Alembert paradox. This result states that, under suitable assumptions, a perfect fluid does not exert any force on an immersed solid body. As an example, a bird should not be able to fly in the air. The difference between this result and experimental intuition motivates looking for the way in which the hypothesis that the fluid is perfect do not comply with the reality. Actually, it is known that, in order to describe precisely interactions between a solid and a fluid, there is a need to take into account viscous effects of the fluid at the interface. However, as emphasized by the Stokes paradox, this modification is not sufficient. Indeed, this paradox states that an infinite cylinder moving with no dynamics in a viscous fluid is necessarily at rest. Once again, the theoretical result is not in agreement with the physical intuition of the problem. Actually, the lack of precision of the viscous model is of a quite different nature. Indeed, the existence of the Stokes paradox relies on a mathematical difficulty in dealing with two-dimensional unbounded stationary flows.

Our interest lives here in a third case where the study of the evolution of solid bodies in a fluid is expected to lead to a controversial theoretical result. Indeed, several authors inferred that strong solutions to systems coupling partial differential equations and ordinary differential equations, and derived in order to model fluid-solid interactions, do not permit collisions between several solid bodies in the fluid [2, 13]. However, the question is still open at the moment in two and in three space dimensions. On the contrary, V.N. Starovoitov constructed examples of weak solutions to these same systems for which two solid bodies collide [13]. But, this also leads to further open problems. We review here several results motivated by these problems.

We do not claim to discuss the modelization aftermaths of such a paradox. Indeed, some might argue that Navier-Stokes equations is not the right model to consider when two solid bodies come too close one with the other. Others might ask if we cannot consider that, when the distance between two particles is negligible with respect to the typical size of bodies, then, there is contact, even if this distance does not vanish mathematically. However, this, as for the older paradoxes cited above, represents a reflection to lead once the paradox is proved to exist. In particular, we hope that rigorous no-collision proofs lead to a better understanding of the mathematical limits of the model (and, consequently, on the limits of its applications).

1.2. Models

Typical fluid-structure model under consideration reads as follows. For simplicity, we present the full system only for the constant-density, incompressible, bidimensional case. A cavity $\Omega \subset \mathbb{R}^m$ ($m = 2$) is filled with a newtonian incompressible viscous fluid and n rigid bodies (see Figure 1). We refer indifferently to the rigid bodies as "the particles" or "the solid bodies".

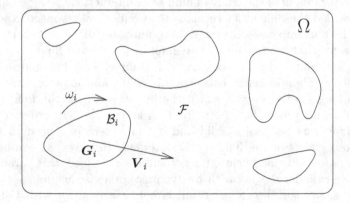

FIGURE 1. n rigid bodies in a viscous incompressible fluid

The fluid (without the particles) fills a domain denoted $\mathcal{F}(t)$. Hence, the boundary $\partial\mathcal{F}(t)$ of $\mathcal{F}(t)$ is built up with the boundary of the cavity and the boundaries of the different particles. The behavior of the fluid is described by $(\rho, \boldsymbol{u}, p)$ a density/velocity/pressure field satisfying the classical incompressible

Navier-Stokes equations. Hence, we obtain a system of partial differential equations in $(\rho, \boldsymbol{u}, p)$ over the domain $\mathcal{F}(t)$. This system is complemented with no-slip boundary conditions over $\partial \mathcal{F}(t)$. These conditions are a first way of interaction between the fluid and the solid bodies.

Concerning the particles, the domain occupied by the ith particle is denoted $\mathcal{B}_i(t)$. To the ith particle, we associate its center of mass $\boldsymbol{G}_i$ and a pair translational/rotational speeds $(\boldsymbol{V}_i, \omega_i)$ computed with respect to $\boldsymbol{G}_i$. By definition, $\dot{\boldsymbol{G}}_i = \boldsymbol{V}_i$, the angular speed ω_i is a scalar quantity and the velocity-field of the particle reads $\boldsymbol{V}_i + \omega_i(\boldsymbol{x} - \boldsymbol{G}_i)^\perp$, where $\boldsymbol{x}$ is a fixed system of coordinates associated to Ω and $\perp$ denotes the direct rotation with angle $\pi/2$. These quantities, $\boldsymbol{V}_i$ and ω_i, satisfy the classical solid mechanics relations (balance of forces and of momentum), applied to the ith solid body. In these relations are involved the stress and torque applied by the fluid onto the ith solid. This is the second way of interaction between the rigid bodies and the fluid.

In the incompressible constant-density case, the full system reads, with adimensionalized coefficients:

$$
\begin{cases}
\partial_t \boldsymbol{u} + \operatorname{Re} \boldsymbol{u} \cdot \nabla \boldsymbol{u} = \operatorname{div}(\mathbb{T}(\boldsymbol{u}, p)), \\
\operatorname{div}(\boldsymbol{u}) = 0,
\end{cases} \quad \text{in } \mathcal{F}(t), \tag{1}
$$

with boundary conditions:

$$
\begin{cases}
\boldsymbol{u}_{|\partial\Omega} = 0, \\
\boldsymbol{u}_{|\partial\mathcal{B}_i(t)} = \boldsymbol{V}_i + \omega_i(\boldsymbol{x} - \boldsymbol{G}_i)^\perp, \quad \forall i = 1, \ldots, n
\end{cases} \tag{2}
$$

and

$$
\begin{cases}
-\displaystyle\int_{\partial\mathcal{B}_i(t)} \mathbb{T}(\boldsymbol{u}, p)\boldsymbol{n}_i \, d\sigma_i = \dot{\boldsymbol{V}}_i, \\
-\displaystyle\int_{\partial\mathcal{B}_i(t)} (\boldsymbol{x} - \boldsymbol{G}_i)^\perp \cdot (\mathbb{T}(\boldsymbol{u}, p)\boldsymbol{n}_i) \, d\sigma_i = \dot{\omega}_i,
\end{cases} \quad \forall i = 1, \ldots, n. \tag{3}
$$

Here, the vector $\boldsymbol{n}_i$ stands for the normal to $\partial\mathcal{B}_i(t)$ directed inside $\mathcal{B}_i(t)$ and $d\sigma_i$ is the element of length of $\partial\mathcal{B}_i$. As the ith body is rigid, they do not depend on time. The notation $\mathbb{T}(\boldsymbol{u}, p)$ stands for the stress tensor of the fluid. This is an $m \times m$ matrix. In the Newtonian case, we have, precisely:

$$
\mathbb{T}(\boldsymbol{u}, p) := \nabla \boldsymbol{u} + \nabla^\perp \boldsymbol{u} - p\,\mathrm{I}_m = 2\mathrm{D}(\boldsymbol{u}) - p\,\mathrm{I}_m,
$$

with $\mathrm{D}(\boldsymbol{u})$ the symmetric part of $\nabla \boldsymbol{u}$, and I_m the (m, m) identity matrix. Recall that m is the dimension of the flow, and that we focus on the case $m = 2$. When writing $\operatorname{div}(\mathbb{T}(\boldsymbol{u}, p))$, we apply the divergence operator to each line of the matrix $\mathbb{T}(\boldsymbol{u}, p)$. In this way we obtain a column vector. Finally, the symbol Re stands for the Reynolds number. It measures the ratio between the inertia and viscous forces of the fluid. The full system $(1, 2, 3)$ is referred to as (FNS) for full Navier-Stokes. This system is complemented with initial data on the velocity field of the fluid $\boldsymbol{u}$, the positions of the solid bodies $\mathcal{B}_i$ and their speeds $(\boldsymbol{V}_i, \omega_i)$.

Some barotropic compressible versions of this problem have also been studied. In this case, broadly speaking, we replace the divergence-free condition by a pressure law $p = \mathrm{p}(\rho)$ and we add the continuity equation for the density. Typically, these studies considered isentropic pressure law: $p = a\rho^\gamma$ with $\gamma > m/2$.

1.3. Organization of paper

In the next section, we present the concept of strong and weak solutions of (FNS) and the state of the art concerning the Cauchy theory for these solutions. In particular, we emphasize the importance of the question of collision occurrence in this Cauchy theory. First, we explain the importance that the particles do not meet in order that the weak solutions are reliable. Then, we describe the difficulties induced by collisions preventing from further existence of strong solution.

Before attacking the question of collision occurrence directly in several dimensions, the first attempts on that question were naturally tried on baby models. Namely, in **Section 3**, we present one constant-density model studied in one dimension and describe the result obtained on such models and the techniques used therein.

Finally, in the last section, we envisage extensions of this technique in several dimensions for a particular geometry. We exhibit some particular cases where there cannot be any collision between a rigid body and the boundary of the cavity and we propose some intermediate steps in order to obtain rigorously the no-collision result with two-dimensional flows.

2. Cauchy theory and collisions

2.1. Weak solutions

2.1.1. Energy estimate. For simplicity, we assume $\mathrm{Re} = 1$. Before introducing the concept of weak solution, let us justify the minimum regularity we might ask for a solution. To this end, let multiply (1_a) by u. We obtain:

$$\int_{\mathcal{F}(t)} \partial_t u \cdot u + \int_{\mathcal{F}(t)} (u \cdot \nabla) u \cdot u - \int_{\mathcal{F}(t} (\Delta u - \nabla p) \cdot u = 0. \tag{4}$$

In the first term, we recognize the differentiation of the $\mathrm{L}^2(\mathcal{F})$-norm of u with respect to time. Nevertheless, as the domain occupied by the fluid is moving, we have to compensate with the right integral over the boundary:

$$\int_{\mathcal{F}(t)} \partial_t u \cdot u = \frac{\mathrm{d}}{\mathrm{dt}} \left[\int_{\mathcal{F}(t)} \frac{|u|^2}{2} \right] - \sum_{i=1}^{n} \int_{\partial \mathcal{B}_i(t)} \frac{|u|^2}{2} u \cdot n.$$

But, integrating by parts the second integral in (4), we obtain:

$$\int_{\mathcal{F}(t)} (u \cdot \nabla) u \cdot u = \sum_{i=1}^{n} \int_{\partial \mathcal{B}_i(t)} \frac{|u|^2}{2} u \cdot n.$$

Consequently, aggregating the first and second integrals in (4), the derivative of the $L^2(\mathcal{F})$-norm of u with respect to time yields. Then, integrating by parts the last integral in (4), and applying the divergence-free condition and the solid mechanics relations, we obtain:

$$\int_{\mathcal{F}(t} (\Delta u - \nabla p) \cdot u = -V_i \cdot \dot{V}_i - \omega_i \dot{\omega}_i - 2 \int_{\mathcal{F}(t)} |\mathrm{D}(u)|^2,$$

with $\mathrm{D}(u)$ the symmetric part of ∇u. Finally, the first estimate is the classical decrease of kinetic energy, augmented in our case with the kinetic energy of the particles:

$$\frac{\mathrm{d}}{\mathrm{dt}} \left[\int_{\mathcal{F}(t)} |u|^2 + |V_i|^2 + |\omega_i|^2 \right] + 4 \int_{\mathcal{F}(t)} |\mathrm{D}(u)|^2 = 0. \tag{5}$$

2.1.2. Concept of weak solutions. There are several ways to define weak solutions. However, all these formulations lead to the same misunderstanding when collision occurs. In order to make things the most clear as possible, we present here the ideas of one concept precisely. Namely, we make precise the ideas in [11].

First of all, as the fluid sticks to its interfaces with the rigid bodies, we prolong its velocity-field u by the rigid velocity of the bodies inside the domain they occupy. We obtain in this way a function defined over Ω which belongs to $\mathrm{H}_0^1(\Omega)$ and we still denote u. Given the above estimate, the concept of weak solution requires that this prolonged function belongs to the space $\mathrm{L}^\infty(0, T; \mathrm{L}^2(\Omega)) \cap \mathrm{L}^2(0, T; \mathrm{H}_0^1(\Omega))$.

Once the velocity field has been prolonged up to Ω, a fundamental problem arises. Indeed, we have now to determine where the solid bodies are in the fluid and to prescribe that the velocity-field u has rigid-velocity on their domains. To this end, two facts come into play. The first observation is that the rigid bodies move along trajectories associated to the flow of u. Consequently, denoting η the characteristic function of the domain occupied by the solid bodies, it satisfies the system:

$$\begin{cases} \eta^0(x) = \begin{cases} 1 & \text{if } x \in \bigcup_{i=1}^{n} \mathcal{B}_i^0, \\ 0 & \text{else}, \end{cases} \\ \partial_t \eta + u \cdot \nabla \eta = 0. \end{cases} \tag{6}$$

The second observation is a classical lemma which states that, if the symmetric gradient of some velocity field u vanishes in some open connected subset $\Omega' \subset \Omega$, then it has rigid velocity in Ω'. Consequently, one way to invert the prolongation to Ω of the velocity field is to add:

- a new unknown, namely η,
- a new equation, namely (6).
- a compatibility condition, namely that the symmetric gradient $\mathrm{D}(u)$ of u vanishes on the support of η.

Concerning test-functions in the weak formulation of (1,3), the basic observation is that, taking a divergence-free $w \in \mathrm{H}_0^1(\Omega)$ with rigid velocity $W_i + O_i(x - G_i)^\perp$ in $\mathcal{B}_i(t)$, we obtain :

$$\int_{\mathcal{F}(t)} \mathrm{div}(\mathbb{T}(u,p)) \cdot w = -2 \int_{\mathcal{F}(t)} \mathrm{D}(u) : \mathrm{D}(w)$$

$$+ \sum_{i=1}^n \left(\int_{\partial\mathcal{B}_i} (\mathbb{T}(u,p)n_i) \cdot W_i + \int_{\partial\mathcal{B}_i} ((\mathbb{T}(u,p)n_i) \cdot O_i(x - G_i)^\perp. \right)$$

Consequently, the set of test-functions reads:

$w \in \mathrm{H}^1((0,T) \times \Omega)$ with $w(T) = 0$, w is divergence-free and

$\qquad w$ has rigid velocity at almost any time t in the support of $\eta(t, \cdot)$.   (7)

As the reader may note, the definition of the set of test-functions depends itself on the solution. To conclude, a weak solution is a pair (u, η) such that:

- η is a characteristic function satisfying (6) in a weak sense,
- $u \in \mathrm{L}^\infty(0,T; \mathrm{L}^2(\Omega)) \cap \mathrm{L}^2(0,T; \mathrm{H}_0^1(\Omega))$, and $\mathrm{D}(u)$ vanishes on the support of η,
- we have, for all test-functions w in the space made precise in (7):

$$\int_{\Omega \times (0,T)} u \cdot (\partial_t w + u \cdot \nabla w) - 2\mathrm{D}(u) : \mathrm{D}(w) = - \int_\Omega u^0 \cdot w(0, \cdot). \qquad (8)$$

2.1.3. Weak solutions with collisions. With such a concept of weak solutions, J.A. San Martin, M. Tucsnak and V.N. Starovoitov prove in [11] global existence of solutions regardless collisions. Corresponding results were obtained for weak solutions to the three-dimensional compressible and incompressible case by E. Feireisl in [3, 4]. In their study, J.A. San Martin, M. Tucsnak and V.N. Starovoitov describe collisions between solid bodies inside the fluid for weak solutions. They prove:

Lemma 1. *Given a weak solution to* (1, 3), *denoting* $h_{i,j} = dist(\mathcal{B}_i(t), \mathcal{B}_j(t))$, *then, if* $h_{i,j}(t_0) = 0$ *for* $t_0 < \infty$, *we have:*

(i) $\mathcal{B}_i$ *and* $\mathcal{B}_j$ *have the same rigid velocity in* t_0, *i.e., the two bodies move as one bigger rigid body,*

(ii) *we have the following local behavior of* $h_{i,j}$:

$$\lim_{t \to t_0} \frac{|h_{i,j}(t)|}{|t - t_0|^2} = 0.$$

Further considerations by E. Feireisl [3] state also that we have an impermeability principle. Namely, two particles in the fluid cannot penetrate each other. Such a result is satisfactory, because we assumed the particle to be rigid ! Nonetheless V.N. Starovoitov proved in [13] that, choosing suitable source-terms, the particles in the fluid collide, in the bidimensional case. Hence, collision is apparently possible for weak solutions.

For weak solutions, the difficulties are rejected on the meaning of weak solutions in presence of contact between rigid bodies. Indeed, one interpretation of the above lemma is that, if two particles collide in a weak solution, then they become glued at their contact point and evolve later on as one bigger particle. However, the second part **(ii)** of this lemma suggests that there is a narrow space for collision. The drawback of this result comes from its first part. Indeed, in order to obtain **(i)**, the authors observe that, if two bodies, say $\mathcal{B}_i$ and $\mathcal{B}_j$, have a contact point on their boundaries, then necessarily both bodies have the same rigid velocity. Hence, not only the bodies move with the same velocity, but also both bodies have thee same rigid-velocity for any test-function in the weak formulation. Then, in the weak formulation we cannot test the solid mechanics relations on both particles independently. For example, assume that $\mathcal{B}_i$ touches $\partial\Omega$ of the cavity in t_0. Then, as the boundary of the cavity is at rest, any test-function $\boldsymbol{w}$ will satisfy $\boldsymbol{w}_{|\mathcal{B}_i} = 0$. Consequently, even if we integrate formally by parts our weak formulation, we do not have enough test-function in order to obtain the solid mechanics relations for $\mathcal{B}_i$.

To conclude, notice that the concept of weak solution is underdetermined in case of collision. In particular, V.N. Starovoitov gave a non-uniqueness result in [12]. E. Feireisl mentions that the weak solutions constructed up to now (in two and in three dimensions) are obtained adding the "sticky" condition. This means that the colliding particles remain glued at their contact point since the collision event. In order to complete the study, it remains in particular to identify if this is the condition expected in order to describe physically relevant configurations. As for the shock for hyperbolic problems some time ago, we still lack an entropy condition identifying the right collision law.

2.2. Strong solutions

In order to study the regularity of weak solutions with smooth initial data, B. Desjardins and M.J. Esteban use $\partial_t \boldsymbol{u}$ as test-function in the weak formulation [1]. They obtain:

$$\int_\Omega |\nabla \boldsymbol{u}|^2(t) + \int_0^t \int_\Omega |\partial_t \boldsymbol{u}|^2 \leqslant \int_\Omega |\nabla \boldsymbol{u}^0|^2 + \int_0^t \int_\Omega |\boldsymbol{u} \cdot \nabla \boldsymbol{u}|^2, \qquad (9)$$

where one can dominate:

$$\int_0^t \int_\Omega |\boldsymbol{u} \cdot \nabla \boldsymbol{u}|^2 \leqslant |\boldsymbol{u}|_4^2 |\nabla \boldsymbol{u}|_4^2.$$

Using the ellipticity of the Stokes operator with Dirichlet boundary conditions, as long as the distances between particles remain greater than a strictly positive constant, they then obtain, for arbitrary $\varepsilon > 0$:

$$\int_0^t \int_\Omega |\boldsymbol{u} \cdot \nabla \boldsymbol{u}|^2 \leqslant |\boldsymbol{u}|_4^2 |\nabla \boldsymbol{u}|_4^2 \leqslant \varepsilon \int_0^t \int_\Omega |\partial_t \boldsymbol{u}|^2 + C_\varepsilon \int_0^T |\nabla \boldsymbol{u}|_2^4.$$

Finally, it yields a control on

$$\sup_{0,t} |\nabla \boldsymbol{u}|_2^2, \quad \text{and} \quad \int_0^t \int_{\mathcal{F}(t)} |\nabla^2 \boldsymbol{u}|^2.$$

That is the notion of strong solution. Consequently, B. Desjardins and M.J. Esteban prove existence of solutions satisfying the above regularity for smooth initial data up to collision between particles in the fluid. As a bypass, this furnishes a weak-strong uniqueness principle, *i.e.*, weak solutions with smooth initial data are actually strong solutions. Similar results were obtained independently, and for other variants of the model, by T. Takahashi [14] and M.D. Gunzburger, H.C. Lee and G.A. Seregin [6]. In the three-dimensional case, B. Desjardins and M.J. Esteban showed in [1] that the collision between particles is a new mode for explosion of strong solutions which can be added to the already-suspected ones for Navier Stokes equations. In the two-dimensional case, T. Takahashi and M. Tucsnak gave an example emphasizing that if no collision occurs then the strong solution is global. Indeed, they prove that, if a single particle evolves in an unbounded fluid, then the strong solution is global. The question of long-time existence of strong solution to the fluid-solid interaction problem thus reduces, in two dimensions, to proving that solid bodies cannot collide in strong solutions.

3. One-dimensional baby models

Facing the difficulty to deal with the full complexity of the problem of collision in several dimensions, J.-L. Vàzquez and E. Zuazua proposed to give a try on some easier cases. The first natural way to simplify the context is to work in lower dimension. Thus, in [16], J.L. Vàzquez and E. Zuazua introduce the following baby-model. They consider n punctual particles in a Burgers–Hopf fluid. The positions of the particles are denoted by $(h_i)_{i=1,\ldots,n} \in \mathbb{R}^n$ and the velocity-field of the fluid is $u(t, x) \in \mathbb{R}$. These quantities are assumed to satisfy:

$$\begin{cases} u_t + \kappa(u^2)_x - \mu u_{xx} = 0, & \text{in } \mathbb{R} - \{h_i\}_{i=1,\ldots,n} \\ u(t, h_i) = \dot{h}_i, & i = 1,\ldots,n, \\ [u_x](t, h_i) = m_i \ddot{h}_i, & i = 1,\ldots,n. \end{cases} \tag{10}$$

Here $[u_x](t, h_i)$ denotes the jump of u_x in the position (t, h_i). The equation (10_a) is a 1D-approximation to the Navier-Stokes system while (10_c) is a 1D-approximation to the solid mechanics relations. As we are in one dimension, there is no way to take into account the compressibility properties of the fluid. Moreover, particles are points. Thus, their movements reduce to translations and the definition of the force exerted by the fluid on the particle is somewhat problematic. However, it is possible to extend the definition of the stress exerted by the fluid on the particle, requiring it is the trace of the elliptic operator in (10_a). As the elliptic operator into account is simply u_{xx}, its trace is the jump of u_x in the site of the particle.

We would like to emphasize that the parameter κ in (10_a) has a dimension. In particular, we set all other physical parameters to 1 in the sequel, but we keep κ because it has an influence on the dynamics (see **Section** 3.3 and [8]).

3.1. Energy estimate

As in the previous section, we justify formally the regularity we might expect for strong solutions deriving a priori estimates. So, assume the positions of the particles satisfy: $h_1 < h_2 < \cdots < h_n$. For technical purposes we add two particles h_0 and h_{n+1}. The first one is sited in $-\infty$ and the second one in $+\infty$. In the whole estimate derivation, we assume that quantities computed in h_0 and h_{n+1} vanish.

Let multiply (10_a) by $\beta'(u)$, where: $\beta : \mathbb{R} \to [0, \infty)$ is a convex function with $\beta(0) = 0$. It yields after integration by parts:

$$\frac{d}{dt} \left[\int_{\mathbb{R}} \beta(u) + \sum_{i=1}^{n} \beta(\dot{h}_i) \right] \leqslant 0. \tag{11}$$

Choosing suitable β, we obtain:

- conservation of any $L^p(\mathbb{R})$ norms of the solution,
- conservation of the momentum of the solution: $M := \int_{\mathbb{R}} u + \sum_{i=1}^{n} \dot{h}_i$,
- positivity of the solution.

In the particular case $p = 2$, we can measure the decay of the $L^2(\mathbb{R})$-norm, we obtain:

$$\int_{\mathbb{R}} u^2(t, \cdot) + \sum_{i=1}^{n} |\dot{h}_i|^2 + 2 \int_0^t \int_{\mathbb{R}} u_x^2 = \int_{\mathbb{R}} u^2(0, \cdot) + \sum_{i=1}^{n} |\dot{h}_i(0)|^2. \tag{12}$$

Before looking for further regularity of the solution, we recall that, in (10), we prescribe jump conditions on u_x. Consequently, it is hopeless to look for solutions u such that u_x is continuous. In particular, we might not expect that $u \in H^2(\mathbb{R})$. Such a regularity is only expectable in any of the (h_i, h_{i+1}). Hence, we multiply (10_a) by u_{xx} on any interval (h_i, h_{i+1}) for $i = 0, \ldots, n$. We obtain, after integration by parts:

$$\int_{h_i}^{h_{i+1}} u_t u_{xx} = u_x(t, h_{i+1}) \left(u_t(t, h_{i+1}) + \frac{\dot{h}_2}{2} u_x(t, h_{i+1}) \right)$$

$$- u_x(t, h_i) \left(u_t(t, h_i) + \frac{\dot{h}_i}{2} u_x(t, h_1) \right) - \frac{d}{dt} \left[\int_{h_i}^{h_{i+1}} \frac{(u_x)^2}{2} \right], \tag{13}$$

where, actually, $u_x(t, h_i)$ stands for the limit of $u_x(t, y)$ when y tends to h_i inside (h_i, h_{i+1}). Moreover, rewriting $(u^2)_x u_{xx} = u((u_x)^2)_x$:

$$\int_{h_i}^{h_{i+1}} (u^2)_x u_{xx} = u(t, h_{i+1})(u_x(t, h_{i+1}))^2 - u(t, h_i)(u_x(t, h_i))^2 - \int_{h_i}^{h_{i+1}} (u_x)^3. \tag{14}$$

Recall that, differentiating (10_b) with respect to time, we obtain:

$$\ddot{h}_i(t) = u_t(t, h_i) + \dot{h}_i(t)u_x(t, h_i), \quad \text{for } i = 1, \ldots, n. \tag{15}$$

We replace in (13). It yields:

$$\frac{d}{dt}\left[\int_{h_i}^{h_{i+1}} \frac{(u_x)^2}{2}\right] + \int_{h_i}^{h_{i+1}} (u_{xx})^2 + \int_{h_i}^{h_{i+1}} \frac{(u_x)^3}{2} + \ddot{h}_i u_x(t, h_{i+1}) - \ddot{h}_{i+1} u_x(t, h_{i+1})$$
$$= \frac{(2\kappa - 1)}{2} \int_{h_i}^{h_{i+1}} (u^2)_x u_{xx}, \tag{16}$$

Summing these equalities when the index i ranges 0 to n, we obtain:

$$\frac{d}{dt}\left[\int_{\mathbb{R}} \frac{(u_x)^2}{2}\right] + \int_{\mathbb{R}} (u_{xx})^2 + \int_{\mathbb{R}} \frac{(u_x)^3}{2} + \sum_{i=1}^{n} |\ddot{h}_i|^2 = \frac{(2\kappa - 1)}{2} \int_{\mathbb{R}} (u^2)_x u_{xx}. \tag{17}$$

In this equality, the right-hand side is simply dominated, making use of the injection: $H^1(\mathbb{R}) \subset \mathcal{C}(\mathbb{R}) \cap L^\infty(\mathbb{R})$. Thus, for $H^1(\mathbb{R})$ initial data, we make use of the decay of the $L^\infty(\mathbb{R})$ norm of the solution and:

$$|u(t, \cdot)|_\infty \leqslant \sup\left\{|u^0|_\infty, \sup_{i=1,\ldots,n} |\dot{h}_i^0|\right\}.$$

Consequently, the only annoying term in (17) is the integral of u_x^3. To get rid of it, J.L. Vàzquez and E. Zuazua prove a generalization of the above injection in the case when the functions have a finite set of jumps. They prove:

Lemma 2. *Given $h_0 = -\infty < h_1 < h_2 < \cdots < h_n < h_{n+1} = \infty$ and $\varepsilon > 0$. For any $v \in L^2(\mathbb{R})$ such that $v \in H^1(h_i, h_{i+1})$ for all $i = 0, \ldots, n$, we have $v \in L^\infty(\mathbb{R})$, and there exists a constant C depending exclusively on ε for which:*

$$|v|_\infty^2 \leqslant C |v|_2^2 + \varepsilon \left(\sum_{i=0}^{n} |v_x|_{2,(h_i,h_{i+1})}^2 + \sum_{i=1}^{n} |[v](h_i)|^2\right).$$

Applying suitably this lemma to u_x, it yields finally:

$$\frac{d}{dt}\left[\int_{\mathbb{R}} \frac{(u_x)^2}{2}\right] + \left\{\int_{\mathbb{R}} \frac{(u_{xx})^2}{2} + \sum_{i=1}^{n} |\ddot{h}_i|^2\right\}$$
$$\leqslant C\left(1 + (|u^0|_\infty + |\dot{h}_0^1| + |\dot{h}_2^0|)^2 + \int_{\mathbb{R}} (u_x)^2\right) \int_{\mathbb{R}} (u_x)^2,$$

with a constant C depending exclusively on κ. In this inequality it is still annoying to have $|u_x|_2^4$ in the right-hand side. It still seems that the $H^1(\mathbb{R})$ norm of the solution can blow up in finite time. However, denoting

$$\phi := C\left(1 + (|u^0|_\infty + |\dot{h}_0^1| + |\dot{h}_2^0|)^2 + |u_x|_2^2\right)$$

we have, because of first order estimate: $\phi \in L^1(0,T)$. Hence, applying the Gronwall lemma, we may integrate the above inequality into:

$$\int_{\mathbb{R}} \frac{(u_x)^2(t,\cdot)}{2} + \frac{1}{2} \int_0^t \exp\left\{\int_s^t \phi(\alpha)\mathrm{d}\alpha\right\} \left(\int_{\mathbb{R}} |u_{xx}|^2 + \sum_{i=1}^n |\ddot{h}_i|^2(s)\mathrm{d}s\right)$$
$$\leqslant \int_{\mathbb{R}} \frac{(u_x^0)^2}{2} \exp\left\{\int_0^t \phi(\alpha)\mathrm{d}\alpha\right\}. \quad (18)$$

Finally, the $H^1(\mathbb{R})$ norm of the solution is locally finite. Consequently, we say that $(u, (h_i)_{i=1,\ldots,n})$ is a finite-energy solution to (10), if and only if:

(i) $(h_i)_{i=1,\ldots,n} \in \left[H^2(0,T)\right]^n$,
(ii) $u \in \mathcal{C}([0,T], H^1(\mathbb{R}))$ with $u \in H^2((h_i, h_{i+1}))$ for any $i = 0,\ldots,n$ and a.a. $t \in (0,T)$ and:

$$\int_0^T \sum_{i=0}^n \int_{h_i}^{h_{i+1}} |u_{xx}|^2 < \infty.$$

(iii) each equation of (10) is satisfied almost everywhere,
(in $(0,T) \times \mathbb{R} - \{(t, h_i(t))\}_{t\in(0,T), i=1,\ldots,n}$ and $(0,T)$ respectively).

3.2. Cauchy theory for solutions

In [16], J.L. Vàzquez and E. Zuazua prove

Theorem 1. *Suppose* $(u^0, (h_i^0)_{i=1,\ldots,n}, (\dot{h}_i^0)_{i=1,\ldots,n}) \in H^1(\mathbb{R}) \times \mathbb{R}^{(2n)}$ *satisfies the compatibility condition:*

$$\begin{cases} h_1^0 < h_2^0 < \ldots < h_n^0, \\ u^0(h_i^0) = \dot{h}_i^0, \quad \forall i = 1,\ldots,n. \end{cases} \quad (19)$$

Then, there exists a unique global finite-energy solution to (10).

The same result has been obtained by the author independently [7]. The first step of the proof of this theorem is to reduce the problem to a system in a fixed geometry using a change of spatial variable fixing the position of the particles. As the speeds of the particles are controlled via a priori estimates, this change of variable is non-singular at least locally in time. However, this introduces non-linear terms. Thus, we conclude to local existence and uniqueness of solutions applying classical tools in non-linear pdes.

In order to obtain a long-time result, notice that the above a priori estimates imply that the only difficulty leaves in proving that particles do not collide in finite time. Indeed, under this assumption, we may apply the above estimates and $\|u\|_1$ is locally bounded in time, and we can thus prolong any solution to an arbitrary lengthy of time. Notice that estimating $\|u\|_1$, we obtain a control on

$$\int_0^t \sum_{i=1}^n \left[\|u\|_{2,(h_i, h_{i+1})}^2 + |[u](h_i)|^2 \right].$$

But, applying **Lemma** 2, this induces that u is Lipschitz at almost any time, with a locally-integrable (in time) Lipschitz constant. Consequently, the Cauchy-Lipschitz theory implies that the dynamical system: $\dot{X} = u(t, X)$, admits a unique flow. Regarding boundary condition (10$_b$), the position of the particles remains on trajectories of this flow. Thus, applying the Cauchy-Lipschitz theorem, we obtain that there can be no collision in finite time.

J.L. Vàzquez and E. Zuazua prove in [16] that the regularizing properties of the underlying heat equation enable to construct "strong" solutions for $L^2(\mathbb{R})$ initial data with no compatibility conditions. Indeed, the regularizing properties imply that the solution is instantaneously in $H^1(\mathbb{R})$, in order that the meaning of diverse boundary conditions become clear. Hence, for this baby model there is no deep difference between strong and weak solutions contrary to the multi-dimensional case.

3.3. Asymptotic of solutions

To end up with (10), we mention some results on the asymptotic behavior of the distances between particles. Namely, our interest leaves in determining whether the particles tend to collide asymptotically or not. In [16], J.L. Vàzquez and E. Zuazua give a precise description of the asymptotic behavior of the velocity-field of the fluid:

Theorem 2. *Given initial data in* $H^1(\mathbb{R}) \cap L^1(\mathbb{R})$. *The unique solution*

$$(u, (h_i)_{i=1,\ldots,n})$$

to (10) *satisfies:*

$$t^{\frac{1}{2}(1-\frac{1}{p})} |u(t) - U(t)|_p \to 0, \;\; as \; t \to \infty, \tag{20}$$

for all $1 \leqslant p \leqslant \infty$, *where*

$$U(x,t) = \frac{1}{\sqrt{t}} f_M\left(\frac{x}{\sqrt{t}}\right),$$

is the self-similar solution to the Burgers equation with initial data $M d_0$, *and* d_0 *is the Dirac measure located in 0.*

Moreover, if $M > 0$, *we have:*

$$t^{-\frac{1}{2}} \left| h_i(t) - c\sqrt{t} \right| \to 0, \;\; as \; t \to \infty, \; i = 1, \ldots, n$$

where $c > 0$ *is the unique solution to* $f_M(c) = \frac{c}{2}$.

This result is obtained applying the same technique used by J.L. Vàzquez and E. Zuazua for the one-particle case in [15], namely, a precise use of the scale invariance of the system. Indeed, one may notice that if $(u, (h_i)_{i=1,\ldots,n})$ is a solution to (10) with masses m_i then:

$$u_\lambda(t, x) = \lambda u(\lambda^2 t, \lambda x), \quad h_{i,\lambda}(t) = \frac{h_i(\lambda^2 t)}{\lambda}, \quad \forall (t, x) \in (0, \infty) \times \mathbb{R},$$

is also a solution to (10) with masses m_i/λ. Consequently, the energy of the particles is completely absorbed by the fluid and the particles tend to follow trajectories

of the flow of the velocity-field solution to the viscous Burgers equation. In particular, the particles follow asymptotically the same trajectory. However, this does
not imply that they collide asymptotically. Indeed, they reach the same trajectory,
but within the scale of the system, namely $\sqrt{t}$. This lets space for any behavior.

In particular, J.L. Vàzquez and E. Zuazua proved that,

- if $\kappa < 1$, the particles collide asymptotically,
- if $\kappa > 1$, the particles diverge asymptotically.

They also compute the rate at which the particles collide or diverge.

We obtained independently a last asymptotic behavior in [8]. Namely, as
classical with Burgers–like equations, the asymptotic behavior is deeply depending
on the first moment M of the initial data. In particular, previous results were
obtained when this first moment is not null. In the case the first moment is null
or $\kappa = 1$, the third following behavior yields:

- the particles tend to fixed positions remaining at finite distance one with the
  other.

To obtain it, we used the so-called Hopf–Cole transform:

$$w(t, x) = \exp\left(-\kappa \int_{-\infty}^{x} u\right) \in \mathcal{C}([0, \infty) \times \mathbb{R}, (0, \infty)).$$

Due to the conservation of the $L^1(\mathbb{R})$ norm of u, the distance between h_i and h_{i+1}
is equivalent to any $L^p(h_i, h_{i+1})$ norm of w, where $1 \leqslant p < \infty$. This new unknown
is a solution to:

$$w_t - w_{xx} = \kappa \sum_{1 \leqslant i \leqslant i(x)} m_i \ddot{h}_i w. \qquad (21)$$

On this system the source-like influence of jump conditions in the h_i's appears
clearly. Then, classical energy-estimate techniques enable to give new proofs of
the results obtained by J.L. Vàzquez and E. Zuazua and extend them to the case
$\kappa = 1$ and vanishing initial first moment M. This leads in particular to obtain the
third asymptotic behavior mentioned above.

4. Two-dimensional aspects

With this one-dimensional experience at hand, the new challenge is to attack the
2D case. Unfortunately, no result has been obtained yet with the full Navier-Stokes
system. We present here some approaches on a simplified version and explain why
they give hope that the result with the full non-linear system is at hand.

We consider the most simple geometry in order to catch the collision phe-
nomenon. An homogeneous disk $\mathcal{B}$ is moving above a ramp $\mathcal{P}$ in a viscous incompressible fluid. Notations and physical descriptions are summarized in Figure 2.
We keep G for the center of mass, h for the distance between $\mathcal{B}$ and $\mathcal{P}$ and (V, ω)
for the dynamics of the particle. Such geometries have been studied in the 3D case
in [9]. In that study, the authors prove that, given a translational speed V parallel

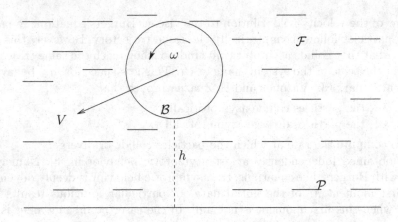

FIGURE 2. The particle $\mathcal{B}$ falls along the ramp $\mathcal{P}$.

to $\mathcal{P}$, for some masses of the solid and some orientations of the ramp (which is no longer assumed horizontal), there exist stationary solutions to (FNS).

4.1. Some no collision criteria. Part I

The first task in order to adapt the one-dimensional arguments is to obtain an estimate linking $\dot{h}$ and h providing a criterion for the lack of collision. In one dimension, we did not mention the estimate itself, however, recall that the idea of the Cauchy-Lipschitz argument in order to prove that trajectories for a flow attached to a Lipschitzian vector field do not cross, is that, if u is k-Lipschitz between h_i and h_{i+1} then

$$\left|\frac{\mathrm{d}}{\mathrm{dt}}[h_{i+1} - h_i]\right| = |u(h_{i+1}) - u(h_i)| \leqslant k\,|h_{i+1} - h_i|\,.$$

Consequently, $|h_{i+1} - h_i|$ remains strictly positive, due to a Gronwall-like inequality. Hence, our first objective is to prove an inequality like:

$$\dot{h} \leqslant Ch^{\beta}\|u\|,$$

with $\beta \geqslant 1$ and $\|u\|$ a suitable norm.

Applying the result in [11] underlying **Lemma** 1, we merely get:

$$|\dot{h}| \leqslant Kh^{\frac{3}{4}}\left(\int_{\mathcal{F}}|\nabla u|^2\right)^{\frac{1}{2}}.$$

But $3/4 < 1$ and this is not enough to conclude. V.N. Starovoitov proposed several improvement of β in such an inequality, using $L^p(\mathcal{F})$-norm of ∇u with $p \neq 2$ [13]. Nonetheless, in these estimates, either β is lower than 1, either we do not have enough information on the involved norm of ∇u in order to conclude. So, we propose another method using directly second order derivatives of u.

Before going further with the case when $\mathcal{B}$ is a disk, we justify that this is the annoying case. To this end, we take into account more general shapes than disks for the particle. The main point is that all these shapes have an axis symmetry with respect to the direction orthogonal to $\mathcal{P}$. Introducing (e_1, e_2) the orthonormal basis for $\mathbb{R}^2$ such that e_1 is parallel to $\mathcal{P}$, this direction is the direction associated to e_2.

We assume that near the contact point, i.e., the point on $\mathcal{B}$ realizing the distance between $\mathcal{P}$ and $\mathcal{B}$, the distance between the point of $\mathcal{B}$ and its orthogonal projection on $\mathcal{P}$ is $h(r) = h + Kr^\alpha$ with $K > 0$ and r the distance between the projection of the point and the projection of the contact point (see Figure 3). In particular, in the case of a disk, $\alpha = 2$. If $\alpha > 2$, the particle is locally flat around the contact point.

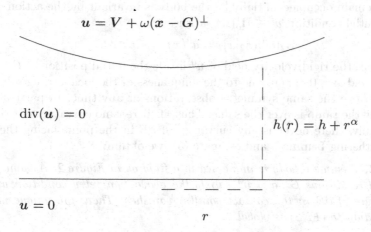

$$u = V + \omega(x - G)^\perp$$

$$\text{div}(u) = 0$$

$$h(r) = h + r^\alpha$$

$$u = 0$$

$$r$$

FIGURE 3. General shape for the particle.

With such a geometry, the method used in [11] implies that, for r sufficiently small:

$$|V \cdot e_2| r^2 \leqslant K r^{\frac{1}{2}} h(r)^{\frac{3}{2}} \left(\int_{\mathcal{F}} |\nabla u|^2 \right)^{\frac{1}{2}}. \tag{22}$$

In order to optimize this inequality, we choose r is $h^{\frac{1}{\alpha}}$. Thus, we obtain:

Lemma 3. *Given a geometry as in Figure 3, assume that $u \in H_0^1(\Omega)$ is divergence-free and has rigid velocity $V + \omega(x - G)^\perp$ in $\mathcal{B}$. Then, for h sufficiently small:*

$$|V \cdot e_2| \leqslant K h^{\frac{3}{2}(1 - \frac{1}{\alpha})} \left(\int_{\mathcal{F}} |\nabla u|^2 \right)^{\frac{1}{2}}, \tag{23}$$

where K is an absolute constant.

Notice that $\dot{h} = V \cdot e_2$. With such an inequality at hand, we observe that the exponent of h is likely to give us a no-collision result as soon as $\alpha \geqslant 3$. Note that the minimum energy estimate on solutions to the fluid-solid interaction problems implies a control on $|\nabla u|_2$ (see (5)) in $L^2(0, T)$ with no dependence upon h. Hence, the above lemma implies that, whenever $\alpha \geqslant 3$, even weak solutions cannot develop collision in this geometry.

However, assuming that the particle is smooth, one should note that it is not possible that the particle is flat at any point of its boundary. Indeed, if that would be the case, its boundary would be a straight line, and our body would not be bounded! Thus, if the particle rotate, at some moment there will be some curvature at the contact point. So, in order to apply only **Lemma** 3 and the first order energy estimate to obtain the lack of collision, we have to restrict a the class of solutions under consideration. Namely, we focus symmetric solutions. We introduce $\mathbb{S}_2$ the symmetry with respect to e_2 and assume that initially we have:

- the domain occupied initially by the body is invariant by the action of $\mathbb{S}_2$,
- the initial condition $u^0 \in \mathrm{H}_0^1(\Omega)$ satisfies:

$$\mathbb{S}_2[u^0](\mathbb{S}_2[x]) = u^0(x), \quad \forall\, x \in \Omega.$$

In particular, the rigid velocity-field of u^0 inside the initial position of $\mathcal{B}$ satisfies: $V \cdot e_1 = 0$ and $\omega = 0$. Then, due to the uniqueness of the local strong solution to (FNS), we have the same symmetry observations at any time. In particular, the point facing the ramp is ever the same. Thus, if there is no curvature of $\partial\mathcal{B}$ in this point initially, there never is any curvature of $\partial\mathcal{B}$ in the point facing the ramp. Finally, gathering **Lemma** 3 and estimate (5), we obtain:

Theorem 3. *Assume a body is immersed in a fluid as in Figure 2. Assume further that initial conditions $\mathcal{B}^0$ and u^0 satisfy the above symmetry condition and that the curvature of $\partial\mathcal{B}$ in the contact point(s) vanishes. Then, the unique maximal strong solution to (FNS) is global.*

4.2. Some no collision criteria. Part II

In order to examine the case where the boundary of $\mathcal{B}$ has curvature in the contact point, we consider the model case $\mathcal{B}$ is a disk with radius 1. We still denote $h(r)$ the distance between one point and its orthogonal projection on the ramp $\mathcal{P}$. Thus, we have: $h(r) = h + 1 - \sqrt{1 - r^2}$.

What does remain of **Lemma** 3 in that case? Of course, we obtain the exponent yielding in **Lemma** 1, i.e., $3/4$, when reaching first order derivatives of u. In order to improve β we reach second order derivatives of u. To this end, let us remind the way (22) is obtained.

The idea is to integrate $\mathrm{div}(u) = 0$ on domains $\Omega_{h,\delta}$ as in Figure 4. We obtain:

$$\int_{\partial\mathcal{B}_h \cap \partial\Omega_h} u \cdot n\,\mathrm{d}\sigma = \int_0^{h(\delta)} (u_1(\delta, x_2) - u_1(-\delta, x_2))\mathrm{d}x_2.$$

But, as $u(\delta, 0) = 0$, for any $x_1 \in (-1, 1)$:

$$|u_1(x_1, x_2) - u_1(-x_1, x_2)| \leqslant \int_0^{x_2} |\partial_2 u_1(x_1, x_2') - \partial_2 u_1(-x_1, x_2)|\mathrm{d}x_2'.$$

Moreover, due to our assumption that u has rigid velocity in $\mathcal{B}$, notice that

$$u_1(x_1, h(x_1)) = u_1(-x_1, h(x_1)).$$

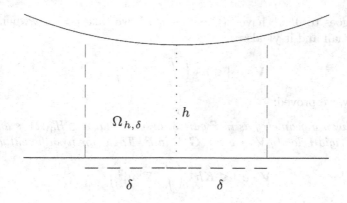

FIGURE 4. Sets $\Omega_{h,\delta}$

Consequently, for any $x_1 \in (-1,1)$, there exists $y_0(x_1)$ such that

$$\partial_2(u_1(x_1, y_0(x_1)) - u_1(-x_1, h(x_1)) = 0.$$

Then:

$$|\partial_2 u_1(x_1, x_2') - \partial_2 u_1(-x_1, x_2')|$$

$$\leqslant \sqrt{h(x_1)} \left\{ \left(\int_0^{h(x_1)} |\partial_{22} u_1(x_1, s)|^2 \mathrm{d}s \right)^{\frac{1}{2}} + \left(\int_0^{h(x_1)} |\partial_{22} u_1(-x_1, s)|^2 \mathrm{d}s \right)^{\frac{1}{2}} \right\}.$$

Integrating this inequality two times, it yields:

$$\left| \int_0^{h(\delta)} (u_1(\delta, x_2) - u_1(-\delta, x_2)) \mathrm{d}x_2 \right|$$

$$\leqslant (h(\delta))^{\frac{5}{2}} \left\{ \left(\int_0^{h(\delta)} |\partial_{22} u_1(\delta, s)|^2 \mathrm{d}s \right)^{\frac{1}{2}} + \left(\int_0^{h(-\delta)} |\partial_{22} u_1(-\delta, s)|^2 \mathrm{d}s \right)^{\frac{1}{2}} \right\}.$$

On the other hand:

$$\int_{\partial \mathcal{B}_h \cap \partial \Omega_{h,\delta}} \boldsymbol{u} \cdot \boldsymbol{n} \mathrm{d}\sigma = \int_{-\delta}^{\delta} u_2(x_1, h(x_1)) \mathrm{d}x_1 = \delta(\boldsymbol{V} \cdot \boldsymbol{e}_2).$$

Integrating this relation for $\delta \in (0, r)$ with r sufficiently small, it yields:

$$|\boldsymbol{V} \cdot \boldsymbol{e}_2| r^2 \leqslant (h(r))^{\frac{5}{2}} r^{\frac{1}{2}} \left(\int_{\Omega_{h,r}} |\partial_{22} u_1|^2 \right)^{\frac{1}{2}}.$$

As, when r goes to 0, we have $h(r) \sim h + r^2/2$, we take $r = \sqrt{h}$ whenever h is sufficiently small and it yields:

$$|V \cdot e_2| \leqslant h^{\frac{7}{4}} \left(\int_{\mathcal{F}} |\nabla^2 u|^2 \right)^{\frac{1}{2}}. \tag{24}$$

Consequently, we proved:

Lemma 4. *Given a geometry as in Figure 3, assume that $u \in H_0^1(\Omega)$ is divergence-free and has rigid velocity $V + \omega(x - G)^\perp$ in $\mathcal{B}$. Then, for h sufficiently small:*

$$|V \cdot e_2| \leqslant K h^{\frac{7}{4}} \left(\int_{\mathcal{F}} |\nabla^2 u|^2 \right)^{\frac{1}{2}}. \tag{25}$$

where K is an absolute constant.

This time, we get an exponent greater than 1. In particular, from this estimate we can state the following result:

Theorem 4. *Given a strong solution to (FNS) with the geometry as depicted in Figure 1, assume that, for any $T > 0$ we have an a priori estimate such that*

$$\int_0^T \int_{\mathcal{F}} |\nabla^2 u|^2 < \infty, \tag{26}$$

then the strong solution is maximal.

It is worth the case to note that, we could actually replace (26) by:

$$\int_0^T h^{\frac{3}{2}} \int_{\mathcal{F}} |\nabla^2 u|^2 < \infty,$$

4.3. Application

Such estimates on the second order derivatives of u are not known for the full Navier-Stokes system. However, in collaboration with J.L. Vàzquez [10, 7] we considered a baby 2D-model with no pressure. It reads as follows:

$$\begin{cases} \partial_t u - \Delta u = 0, & \text{in } \mathcal{F}(t), \\ u_{|\mathcal{P}} = 0, \quad u_{|\infty} = 0, \quad u_{|\partial B(t)} = V + \omega(x - G)^\perp, \end{cases} \tag{27}$$

with

$$\begin{cases} \int_{\partial B} \partial_n u \, d\sigma + \dot{V} = 0, \\ \int_{\partial B} \partial_n u \cdot (x - G)^\perp \, d\sigma + \dot{\omega} = 0, \end{cases} \tag{28}$$

Making use of elliptic estimates for the problem [5]:

$$\begin{cases} \Delta u = g, & \text{in } \mathcal{F} \\ u_{|\partial \mathcal{F}} = 0, \end{cases}$$

we manage to obtain the expected control on second order derivatives of u. However, to our knowledge, such estimates remain to be obtained for the Stokes system.

References

[1] B. Desjardins and M.J. Esteban. Existence of weak solutions for the motion of rigid bodies in a viscous fluid. *Arch. Ration. Mech. Anal.*, 146(1):59–71, 1999.

[2] E. Feireisl. On the motion of rigid bodies in a viscous fluid. *Appl. Math.*, 47(6):463–484, 2002. Mathematical theory in fluid mechanics (Paseky, 2001).

[3] E. Feireisl. On the motion of rigid bodies in a viscous fluid. *Appl. Math.*, 47(6):463–484, 2002. Mathematical theory in fluid mechanics (Paseky, 2001).

[4] E. Feireisl. On the motion of rigid bodies in a viscous compressible fluid. *Arch. Ration. Mech. Anal.*, 167(4):281–308, 2003.

[5] P. Grisvard. *Elliptic problems in nonsmooth domains*, volume 24 of *Monographs and Studies in Mathematics*. Pitman (Advanced Publishing Program), Boston, MA, 1985.

[6] M.D. Gunzburger, H.-C. Lee, and G.A. Seregin. Global existence of weak solutions for viscous incompressible flows around a moving rigid body in three dimensions. *J. Math. Fluid Mech.*, 2(3):219–266, 2000.

[7] M. Hillairet. *Aspects interactifs de la mécanique des fluides*. PhD thesis, Ecole Normale Supérieure de Lyon, 2005.

[8] M. Hillairet. Asymptotic collision between solid particles in a Burgers-Hopf fluid. *Asymptotic analysis*, In press.

[9] M. Hillairet and D. Serre. Chute stationnaire d'un solide dans un fluide visqueux incompressible le long d'un plan incliné. *Ann. Inst. H. Poincaré Anal. Non Linéaire*, 20(5):779–803, 2003.

[10] M. Hillairet and J.L. Vàzquez. A first no-collision result in two dimensions. *In preparation*, 2005.

[11] J.A. San Martín, V. Starovoitov, and M. Tucsnak. Global weak solutions for the two-dimensional motion of several rigid bodies in an incompressible viscous fluid. *Arch. Ration. Mech. Anal.*, 161(2):113–147, 2002.

[12] V.N. Starovoĭtov. On the nonuniqueness of the solution of the problem of the motion of a rigid body in a viscous incompressible fluid. *Zap. Nauchn. Sem. St.-Petersburg. Otdel. Mat. Inst. Steklov. (POMI)*, 306(Kraev. Zadachi Mat. Fiz. i Smezh. Vopr. Teor. Funktsii. 34):199–209, 231–232, 2003.

[13] V.N. Starovoitov. Behavior of a rigid body in an incompressible viscous fluid near a boundary. In *Free boundary problems (Trento, 2002)*, volume 147 of *Internat. Ser. Numer. Math.*, pages 313–327. Birkhäuser, Basel, 2004.

[14] Takéo Takahashi. Analysis of strong solutions for the equations modeling the motion of a rigid-fluid system in a bounded domain. *Adv. Differential Equations*, 8(12):1499–1532, 2003.

[15] J.L. Vàzquez and E. Zuazua. Large time behavior for a simplified 1D model of fluid-solid interaction. *Comm. Partial Differential Equations*, 28(9-10):1705–1738, 2003.

[16] J.L. Vàzquez and E. Zuazua. Lack of collision in a simplified 1-d model for fluid-solid interaction. Preprint, may 2004.

M. Hillairet

Analysis and Simulation of Fluid Dynamics
Advances in Mathematical Fluid Mechanics, 129–154

The Reduced Basis Element Method for Fluid Flows

Alf Emil Løvgren, Yvon Maday and Einar M. Rønquist

Abstract. The reduced basis element approximation is a discretization method for solving partial differential equations that has inherited features from the domain decomposition method and the reduced basis approximation paradigm in a similar way as the spectral element method has inherited features from domain decomposition methods and spectral approximations. We present here a review of the method directed to the application of fluid flow simulations in hierarchical geometries. We present the rational and the basics of the method together with details on the implementation. We illustrate also the rapid convergence with numerical results.

1. Introduction

The numerical simulations of fluid flows is a challenging task on which the computational fluid dynamic and the applied math communities have been working for years. These simulations still fill a large part of the many supercomputers of the planet in a quest for a better reproduction of real life situations. Many applications require a rapid evaluation of the flow picture corresponding to some documented natural phenomenon; among the most prominent applications in this area is the study of internal flows in hierarchical geometries as seen in medical applications. Examples include the analysis of blood flows in arteries (as, e.g., Figure 1, left, which presents a reconstruction of the Willis complex used for numerical simulations [29]), and air flow in the lung (as, e.g., Figure 1, right, which presents a reconstruction of the upper part of the lung used for numerical simulations [5]). A related example from engineering applications is the study of a building's infrastructure for the design of an air conditioning network.

In this range of applications, the challenge of the simulations comes more from the complexity of the geometry and its representation than from the fine structures of the flow itself. Actually it can be noticed that there is some repetitiveness or similarities in the behavior of the flow that allows for the definition of reduced

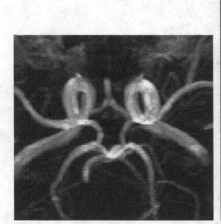
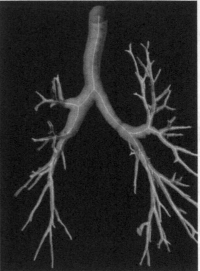

FIGURE 1. The left plot represents the geometry of the Willis complex
that is composed of many blood vessels designing an intricate network
(thanks to [29]). The right plot represents the reconstructed geometry
of the upper part of the lung exhibiting a hierarchical network (thanks
to [5]).

model strategies (see, e.g., [20]). The quite general way of deriving such reduced
models that will be presented in this paper combines three strategies that have
received quite a lot of attention in the computational community:

- the reduced basis method
- the domain decomposition method
- the a posteriori estimations

and has been named the reduced basis element method with rapid certificate of
fidelity.

The reduced basis method is used in the case where we want to solve rapidly
a large number of problems governed by some partial differential equations that
depend on a parameter. The strategy is composed of two stages: the off-line computation during which a few typical problems are solved by classical discretization
methods; the solutions to these problems are the basis in which the on-line computation will be performed through a Galerkin process. This approach can compete
with the best high order discretization methods, and due to the off-line/on-line
strategy, it is very rapid. The extreme high quality and speed of the solution procedure is balanced by the fact that this method is tuned only to particular situations.

The basis is not multipurpose and should only be used in case of a parameter dependent PDE. In some sense this reduced basis method is in the spirit of "learning strategy". We refer to [7, 1, 19, 6, 23, 24, 15].

The domain decomposition method is a "divide and conquer" approach that has benefited from the development of parallel supercomputers. The idea, when a partial differential equation on a given domain has to be solved, is to break the domain into overlapping or non overlapping subdomains and combine the solution strategies of the same PDE over the subdomains, yielding independent smaller tasks, in a proper way to iteratively approximate the solution of the global problem. This method has been very popular and has been developing rapidly over the past twenty years. We refer to, e.g., [26], [30] for a general overview of the problem.

A posteriori analysis is the mathematical equivalent of "precision error bars" that are well known for real experiments and that are attached to any experimental data as a mandatory complement to any measure or output in order to know where the unreachable truth lies. Once a mathematical model is provided, in our case through the definition of the partial differential equation, complemented with the necessary initial and boundary conditions, and once the mathematical analysis allows us to specify in which sense a solution to the problem has to be sought, any numerical method aims at approximating this solution at the price of a certain amount of computation. Most often the larger the computation, the closer the approximated solution is to the "exact solution". The a posteriori analysis complements the computations with computable bounds on the approximation by quantifying the error that has been committed. We refer to [31, 8] for general presentations of these strategies.

In the reduced basis element method we consider the geometry of the computational domain to be the generic parameter. The domain is decomposed into smaller blocks, all of them can be viewed as the deformation of a few reference shapes. Associated with each reference shape are previously computed solutions (typically computed over different deformations of the reference shapes). The precomputed solutions are mapped from the reference shapes to the different blocks of the decomposed domain, and the solution on each block is found as a linear combination of the mapped precomputed solutions. The solutions on the different blocks are glued together using Lagrange multipliers. Our hierarchical flow systems can be decomposed into pipes and bifurcations.

In Section 2, we present the basics of the domain decomposition and the solution attached to it; actually the method can be compared with a numerical plumber toolbox where the elemental domain+attached basis can be hooked together. We first explain the approach on a single domain case and then on a domain decomposed geometry.

In Section 3 we present the reduced basis element approach for simulation of the Laplace, Stokes and Navier-Stokes problem.

In Section 4, we provide the basics of the numerical implementation, including a new method to generate the deformation mappings.

Finally, some numerical results are provided in Section 5 that illustrate the potential of the method together with convergence tables.

We end this introduction by indicating that the work presented here is actually motivated by the simulation of air flows in the lung. The geometry of the respiration tree is indeed of such a complexity that a multiscale/multimodel has to be constructed in order to be able to derive implementable ab initio discretization methods.

What we propose is a decomposition of this tree into 4 stages where different models will be exploited:

- the upper part, including the mouth and nose, that goes down to the first bifurcations
- the medium part, from the second or third bifurcation, down to the 8th or 10th
- the distal part, down to the acini
- the acini

and with all these stages being imbedded in a structural parenchyma.

It is then an easy matter to realize that the exact representation of the flow inside this complete tree governed by the 3D Navier-Stokes equations is currently far from being achievable and will still not be for a long time. This is currently only feasible, though still quite expensive, for the upper part. It is at the level of the second part, that we refer to the reduced basis element method. Concerning the distal and the terminal part, the description of the set of acini evokes easily the reference to homogenization. This will be fractal homogenization for the former and multiphysics (fluid-structure interaction), non stationary homogenization for the latter. We refer to [4] for a first analysis in fractal homogenization and to [2] where a first fluid structure model interaction is considered.

These four different models are hooked together as is explained in [11] on a simpler model, resulting in a viable multiscale/multimodel.

2. Basics of the reduced basis element method

2.1. The monodomain case

In the reduced basis method, there is typically a *parameter dependent problem* to be solved for many instances of the parameter (generally denoted by μ). In the reduced basis element method, the parameter represents the shape of the domain on which some partial differential equation has to be solved. In the single domain case, there exists a "reference domain", denoted as $\hat{\Omega}$, and the problem modelling the phenomenon of interest has to be solved on "deformations" of $\hat{\Omega}$ denoted as $\Omega_\Phi = \Phi(\hat{\Omega})$ where the "parameter" $\mu = \Phi$ is a regular enough, one to one, mapping.

In the reduced basis method, there is typically a *fundamental assumption* that the "dimension" of the set S of all solutions obtained by letting the parameter take all admissible values, is small in the sense that the set of all solutions $u(\mu)$, when

μ varies in the parameter set, can be approximated very well by its projection over a finite- and low-dimensional vectorial space. Then, for well enough chosen μ_i, there exist coefficients $\alpha_i = \alpha_i^N(\mu)$ such that the finite sum $\sum_{i=1}^N \alpha_i u(\mu_i)$ is very close to $u(\mu)$ for any μ. In the reduced basis element method, the parameter being the shape of the domain, analysis of the regular dependency of the solution of a PDE on the domain can be found in, e.g., [21]. In addition to the theoretical analysis that may lead to believe that the assumption of the small dimension of S holds qualitatively true, it is most of the time enlightening to get quantitative information on the confirmation of this *fundamental assumption*. In general, we suggest to perform a preliminary feasibility analysis from which we can get "experimentally" an evaluation of the N-width $d_N(S, X)$ of the set S of all solutions for the different admissible parameters. Following [22], it is defined as

$$d_N(S, X) = \inf_{X_N} \sup_{u \in S} \sup_{u_N \in X_N} \|u - u_N\|_X \ ,$$

(where X is some appropriate normed space, and X_N is a generic finite-dimensional subspace of X with dimension equal to N) and it can be evaluated revealing a potential rapid convergence of this width as N grows (note that we prefer the denomination "grow" to the classic one "goes to infinity" since an accuracy of 10^{-4} is often achieved for N of order tens). This indeed will already provide a good evaluation of the potential of the reduced basis concept to work well. Note that a little bit of intuition may be useful here since it may not be the $u(\mu_i)$ that have a small width but some $\mathcal{F}[u(\mu_i)]$ or even $\mathcal{F}(\mu_i)[u(\mu_i)]$ where $\mathcal{F}$ is some simple transformation; we will see this for the Stokes problem below.

Once these basic considerations are made, the outline of the method may proceed. For some well-chosen instances of the mapping, $\Phi_1, \Phi_2, \ldots, \Phi_N$, the solution to the problem is solved through your preferred numerical scheme (that, in our case is the spectral element method) over the domains $\Phi_i(\hat{\Omega})$, with a good enough accuracy. These solutions, named $u_1, u_2, \ldots, u_N$, are then stored on $\hat{\Omega}$ through an appropriate change of variables involving the mapping Φ_i: $\hat{u}_i = \mathcal{F}_i[u_i]$. This provides (at most) N functions $\hat{u}_1, \hat{u}_2, \ldots, \hat{u}_N$ over $\hat{\Omega}$, linearly independent, selected in the set $\hat{S}$ of all possible solutions mapped back onto $\hat{\Omega}$. An example of such an appropriate change of variables is $\hat{u}_i = u_i \circ \Phi_i$, but we will see that more involved changes of variables may be proposed. For any new problem to be solved over the domain $\Omega_\Phi = \Phi(\hat{\Omega})$ characterized by the data of the transformation Φ, an approximation of the corresponding solution u_Φ is sought in the vectorial space spanned by the $\tilde{u}_1, \tilde{u}_2, \ldots, \tilde{u}_N$, where the $\tilde{u}_i$ are functions defined over Ω_Φ from the $\hat{u}_i$ through the same appropriate change of variables involving now the mapping Φ. In order for the approximated solution to be, up to a constant, as good as the analysis of the width might indicate, a Galerkin process is generally used, since from Céa's lemma, the error between the exact solution and the numerical one is upper bounded by some constant times the best fit error.

2.2. The multidomain case: the plumber's toolbox

Let us now add the domain decomposition argument. First, we assume that the domain Ω where the computation should be performed can be written as the *non-overlapping* union of subdomains Ω^k:

$$\overline{\Omega} = \bigcup_{k=1}^{K} \overline{\Omega}^k, \quad \Omega^k \cap \Omega^\ell = \emptyset, \text{ for } k \neq \ell. \tag{1}$$

Next, we assume that, as was said in the monodomain case, each subdomain Ω^k is the deformation of the "reference" domain $\hat{\Omega}$ through a regular enough, and one to one, mapping. Together with this geometric decomposition a functional decomposition is proposed since every Ω^k actually comes filled with the basis functions derived from the $\hat{u}_1, \hat{u}_2, \ldots, \hat{u}_N$. This allows us to define the finite-dimensional space

$$Y_N = \{v \in L^2(\Omega), v_{|\Omega^k} = \sum_{i=1}^{N} \alpha_i^k \mathcal{F}_k^{-1}[\hat{u}_i]\}, \tag{2}$$

which is a set of uncoupled, element by element, discrete functions. This is generally not yet adequate for the approximation of the problem of interest since some glue at the interfaces $\gamma_{k,\ell}$ between two adjacent domains $\overline{\Omega}^k \cap \overline{\Omega}^\ell$ has to be added to the elements of Y_N, the glue depending on the type of equations we are interested to solve (it will be relaxed C^0-continuity condition for a Laplace operator, or more generally relaxed C^1-continuity condition for a fourth-order operator[1]).

At this stage it should be noticed that, modulo an increase of complexity in the notations, there may exist not only one reference domain $\hat{\Omega}$ filled with its reduced basis functions but a few numbers so that the user can have more flexibilities in the design of the final global shape by assembling deformed basic shapes like a plumber would do for a central heating installation.

The reduced basis element method is then defined as a Galerkin approximation over the space X_N being defined from Y_N by imposing these relaxed continuity constraints. We refer to the next section for details concerning the way the relaxed continuity conditions are imposed.

3. The reduced basis element method in action

3.1. The Laplace problem

We synthesize here the experiments that have been done on the Laplace problem. The motivation comes from the design of an optimal thermal fin. Typical $K = 3$-stages and $K = 4$-stages thermal fins we consider are depicted and described in Figure 2.

We assume no heat generation within the thermal fin itself. A constant heat flux, q (generated from an electronic device, say) enters the fin at the fin root,

---

[1] A precise definition of the meaning of relaxed is proposed in the next section

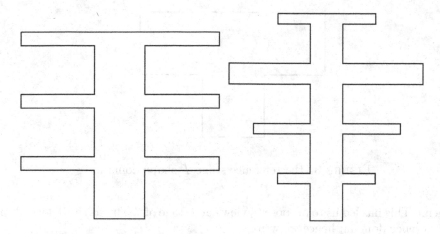

FIGURE 2. The left plot shows a typical $K = 3$-stages fin with similar wings, the right one shows a symmetric $K = 4$-stages fin with a variable wing-size. The root of the fin is at the bottom.

Γ_{root}, and leaves through the remaining surface of the fin. One motivation for this analysis can be to optimize the fin (the number of stages, the physical dimensions, and the thermal conductivities) so as to best remove this heat.

The heat loss from the fin surface due to convection, is modelled by prescribing Robin type boundary conditions

$$-\kappa \frac{\partial u}{\partial n} = \text{Bi } u \quad \text{on} \quad \partial\Omega \setminus \Gamma_{\text{root}} , \tag{3}$$

where κ denotes the piecewise constant conductivity in Ω.

The governing equation for the temperature u in the fin is the Laplace equation; more precisely, the thermal fin problem can be stated in variational form as : Find $u \in H^1(\Omega)$ such that

$$a_\Omega(u, v) = f_\Omega(v) \qquad \forall v \in H^1(\Omega) , \tag{4}$$

where

$$a_\Omega(u, v) = \int_\Omega \kappa \nabla u \cdot \nabla v \, dA + \text{Bi} \int_{\partial\Omega \setminus \Gamma_{\text{root}}} u v \, dS , \tag{5}$$

$$f_\Omega(v) = \int_{\Gamma_{\text{root}}} q v \, dS , \tag{6}$$

where we remind that κ is piecewise constant and is a (multi)parameter to be optimized. We note that $a_\Omega(\cdot, \cdot)$ is a symmetric, positive definite bilinear form, and $f_\Omega(\cdot)$ is a linear form. It is standard to show that this problem has a unique solution u.

When solving the problem on various fins, we observe that the temperature distribution is characterized by a certain amount of "repetitiveness" over the stages

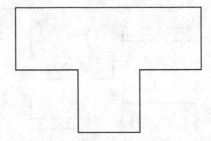

FIGURE 3. the reference stage T-shape domain.

of the fin. This has led us to propose to view each stage of the fin as the deformation of a reference domain; hence we write

$$\overline{\Omega} = \cup_{k=1}^{K} \overline{\Omega^k},$$

where each "building block" Ω^k is assumed to be the image of a reference one-stage fin $\hat{\Omega}$. The mapping Φ_k between $\hat{\Omega}$ and Ω^k is here chosen piecewise linear (and obviously continuous). We illustrate in Figure 3 the choice of reference domain $\hat{\Omega}$. The reduced basis element method assumes that $\hat{\Omega}$ is provided with basis functions $\hat{\zeta}_1, \hat{\zeta}_2, \ldots, \hat{\zeta}_N$, that are supposed to be linearly independent and mapped over each Ω^k through Φ_k. We thus introduce the space

$$Y_N(\Omega) = \{v_M \in L^2(\Omega)| \quad v_{M|\Omega^k} \circ \Phi_k \in \text{span}\{\hat{\zeta}_1, \hat{\zeta}_2, \ldots, \hat{\zeta}_N\}\}. \tag{7}$$

Note again that $Y_N(\Omega)$ is not an acceptable discretization space for $H^1(\Omega)$, which leads us to define a subspace $X_N(\Omega)$ by gluing the functions of $Y_N(\Omega)$ across the interfaces $\gamma_{k,\ell}$ between two adjacent stages

$$\overline{\gamma}_{k,\ell} = \overline{\Omega^k} \cap \overline{\Omega^\ell}. \tag{8}$$

Like in the mortar element method, the matching, expressing this "relaxed continuity" is done in a variational way by imposing

$$X_N(\Omega) = \{v_N \in Y_N(\Omega)| \quad \int_{\gamma_{k,\ell}} (v_{N|\Omega^k} - v_{N|\Omega^\ell})q = 0, \quad \forall q \in W_{k,\ell}\}, \tag{9}$$

where the space $W_{k,\ell}$ is defined in a proper way. An easy choice is the set of all polynomials on the interface $I\!\!P_n(\gamma_{k,\ell})$ with degree $\leq n$ but a smarter choice, based on the numerical analysis of nonconforming approximations, is to span $W_{k,\ell}$ with a few selected normal derivatives of the solutions that have been precomputed to construct the basis functions $\hat{\zeta}_1, \hat{\zeta}_2, \ldots, \hat{\zeta}_N$.

The discrete problem then reads: Find u_N in $X_N(\Omega)$ such that

$$\underline{a}_\Omega(u_N, v_N) + \text{Bi} \int_{\partial\Omega\backslash\Gamma_{\text{root}}} u\,v\,dS = f_\Omega(v_N), \quad \forall v_N \in X_N(\Omega), \tag{10}$$

where we have introduced the notation $\underline{a}_\Omega(u_N, v_N) = \sum_k \int_{\Omega^k} \kappa \nabla u \cdot \nabla v \, dA$ since the functions u_N are not in $H^1(\Omega)$ any more. It is standard to state that there exists a unique solution u_N to this discrete problem and that there is a constant $C > 0$, that is a function of the geometry of the problem, such that

$$\|u - u_N\|_{H^1(\Omega)} \leq C \inf_{v_N \in X_N(\Omega)} \|u - v_N\|_{H^1(\Omega)} + \text{consistency error.}$$

In the absence of a general theory about the status of the best fit that appears on the right-hand side (see however [16]), the feasibility experiment that we quoted in Subsection 2.1 allows us to get an idea of the size of the best fit. The consistency error involves the best fit of the fluxes at the interface $\gamma_{k,\ell}$ by elements of $W_{k,\ell}$.

Before closing this subsection, we should explain how the basis functions are actually chosen. As we have seen, these basis functions are to be used over varying fins with a different number of stages. The computations (here using the spectral element method, but this is not so important) have thus been done on a series of two-stage fins with various dimensions and conductivities. The corresponding solutions then give two candidates for a reduced basis over a single-stage geometry by domain reduction. After a simple change of variables, these solutions provide functions defined over the reference geometry $\hat{\Omega}$ and the basis functions $\hat{\zeta}_1, \hat{\zeta}_2, \ldots, \hat{\zeta}_N$ are selected within an ensemble of such functions (we make precise a selection procedure in Subsection 4.1).

Remark. It should be noticed here that the reduced basis element method, applied to the fin problem, has a lot a similarity with the plain reduced basis method that has been extensively used on this example for illustrating the power of the method (see [17], [25]). However, note that there is an additional dimension to the reduced basis element method due to the possibility of varying the number of stages. Even more, no precomputation (using your preferred method) on a four-stage problem has ever to be done to use the reduced element method. The precomputations are done on a two-stage fin, and the reduced basis element method can be applied on a fin with any number of stages.

3.2. The steady Stokes problem

A typical example of a hierarchical flow system Ω for which we are interested in simulating is shown in Figure 4 left. It has an inflow boundary Γ_{in}, an outflow boundary Γ_{out}, and wall boundaries Γ_w. As in the previous subsection this type of domain is composed of a non overlapping union of pipes (being obtained by deformations of a reference pipe $\hat{\mathcal{P}} = (-1, 1)^2$ as illustrated in Figure 5) and bifurcations (being obtained by deformations of the reference bifurcation $\hat{\mathcal{B}}$, represented in Figure 4 right, as is illustrated in Figure 6, see also Figure 7).

In what follows we thus assume that

$$\overline{\Omega} = \bigcup_{k=1}^K \overline{\Omega}^k, \text{ where each } \Omega^k = \Phi^k(\hat{\Lambda}) \,, \tag{11}$$

where Λ stands for $\mathcal{P}$ or $\mathcal{B}$ and Φ^k is a sufficiently regular and one to one mapping.

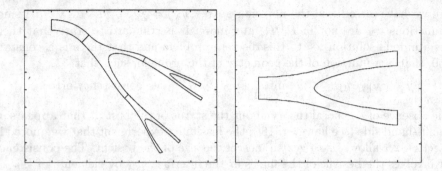

FIGURE 4. The left plot shows a typical domain for our flow problem. The inflow boundary is on the left and the 4 outflow boundaries are on the right. The domain has one pipe-block and three bifurcation-blocks. The plot on the right displays our reference bifurcation $\hat{B}$.

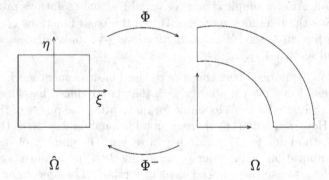

FIGURE 5. Mapping of the reference domain.

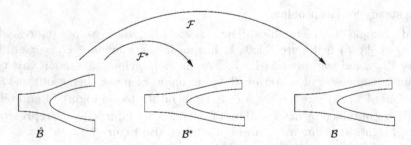

FIGURE 6. Different mappings for the bifurcation domains.

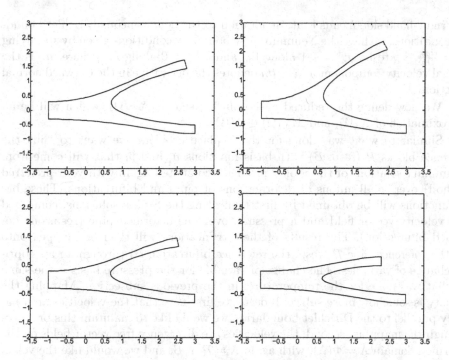

FIGURE 7. A few examples of deformations that are used to build the reduced basis on bifurcations.

On such domains we introduce the velocity space

$$X(\Omega) = \{ v \in (H^1(\Omega))^2, v_{|\Gamma_w} = 0, v_{t|\Gamma_{in}} = v_{t|\Gamma_{out}} = 0 \}, \tag{12}$$

where v_t is the tangential velocity component. We also define a pressure space

$$M(\Omega) = L^2(\Omega). \tag{13}$$

We assume that the flow is governed by the two-dimensional steady Stokes equations that, again written in variational formulation reads: find the velocity $u = (u_1, u_2) \in X(\Omega)$ and the pressure $p \in M(\Omega)$ such that

$$\begin{aligned} a_\Omega(u, v) + b_\Omega(v, p) &= l_\Omega(v) & \forall\, v \in X(\Omega) \\ b_\Omega(u, q) &= 0 & \forall\, q \in M(\Omega), \end{aligned} \tag{14}$$

where

$$a_\Omega(v, w) = \int_\Omega \nabla v \cdot \nabla w \, dA, \tag{15}$$

$$b_\Omega(v, q) = -\int_\Omega q \nabla \cdot v \, dA, \tag{16}$$

and

$$l_\Omega(v) = \int_{\Gamma_{in}} \sigma_n^{in} \, v \cdot \mathbf{n} \, dS + \int_{\Gamma_{out}} \sigma_n^{out} \, v \cdot \mathbf{n} \, dS. \tag{17}$$

The right-hand side means that, in addition to the homogeneous Dirichlet boundary conditions, we have the Neumann type boundary conditions given by specifying $\sigma_n = \frac{\partial u_n}{\partial n} - p$ to be $\sigma_n^{\mathrm{in}} = -1$ along Γ_{in} and $\sigma_n^{\mathrm{out}} = 0$ along Γ_{out}; here, u_n is the normal velocity component and $\partial/\partial n$ denotes the derivative in the outward normal direction.

We now define the reduced basis solution spaces: $X_N(\Omega)$, which will be an approximation of $X(\Omega)$, and $M_N(\Omega) \subset M(\Omega)$.

Similar to what was done for the Laplace problem, we want to "fill" the reference blocks $\hat{\mathcal{P}}$ (resp. $\hat{\mathcal{B}}$) with basis functions $\hat{u}_i$ and $\hat{p}_i$ that will come from preliminary solutions of the Stokes problem (computed again with your preferred method) over small unions of deformations of pipes and bifurcations. These basis functions will be obtained by first truncating the Stokes solutions, composed of a velocity vector field and a pressure, over one deformed pipe (resp. one deformed bifurcation). The results of these truncations will then be mapped onto the the the reference pipe $\hat{\mathcal{P}}$ (resp. the reference bifurcation $\hat{\mathcal{B}}$) through an appropriate change of variables. The change of variable for the pressure basis is the same as what was done for the temperature, in the previous subsection. Mapping the velocity is somehow more subtle. Indeed, we first note that the velocities are relatively parallel to the Dirichlet boundaries; we would like to maintain this property through the mapping, second, the velocities are divergence free vector fields on the deformed domain $\Lambda = \Phi(\hat{\Lambda})$, with again $\Lambda = \mathcal{P}$ or $\mathcal{B}$, and we would like the velocity basis to keep this property when mapped to the reference domain. The Piola transformation (see [27] and [3] for general properties) allows for this constraint:

$$\hat{u} = \Psi(u) = \mathcal{J}^{-1}(u \circ \Phi)|J|, \tag{18}$$

where $\mathcal{J}$ is the Jacobian matrix of Φ and $|J|$ its determinant. Let us insist on the fact that, would we map the velocities as scalar functions, the reduced basis method might work nevertheless, but the convergence rate would certainly not be as good as if we transform the velocities through the Piola transformation. Actually, it is well known that a divergence free field is, in 2D at least, the curl of a unique potential Ψ. The Piola transformation appears to be associated with the simplest mapping expressed on this potential, which is certainly a natural idea. In the reduced basis context, being smart pays off.

This allows us to define the reduced basis reference spaces as

$$\begin{aligned} M_N(\hat{\Lambda}) &= span\{\hat{p}_i, \quad i = 1, \ldots, N\}. \\ V_N(\hat{\Lambda}) &= span\{\hat{u}_i, \quad i = 1, \ldots, N\}, \end{aligned} \tag{19}$$

Note here that the basis functions $\hat{u}_i$ are divergence free (thus the standard notation "V" for the associated space, see, e.g., [9]).

The definition of the global spaces over Ω then proceeds from (11) by first setting

$$M_N(\Omega) = \{p \in L^2(\Omega), p_{|\Omega^k} = \hat{p} \circ (\Phi_k)^{-1}, \hat{p} \in M_N(\hat{\Lambda})\}, \tag{20}$$

and then define the velocity space to be

$$V_N(\Omega) = \left\{ \boldsymbol{u} \in L^2(\Omega)^2, \boldsymbol{u}_{|\Omega^k} = \Psi^{-1}(\hat{\boldsymbol{u}}), \hat{\boldsymbol{u}} \in V_N(\hat{\Lambda}), \right.$$
$$\left. \int_{\gamma_{k,\ell}} (\boldsymbol{u}_{|\Omega^k} - \boldsymbol{u}_{|\Omega^\ell})q = 0, \quad \forall q \in W_{k,\ell} \right\}. \tag{21}$$

Here $\gamma_{k,\ell}$ again denotes the interface between two adjacent subdomains Ω^k and Ω^ℓ and $W_{k,\ell}$ is a space of (vectorial) gluing functions.

It is interesting to note that, aside from the discontinuity across the interfaces, these discrete functions are perfect for the approximation of the velocity over Ω since they are divergence free. The approximation of the velocity solution of (14) can be obtained by solving only the "Laplace like" problem: find $\boldsymbol{u}_N \in V_N(\Omega)$ such that

$$\underline{a}_\Omega(\boldsymbol{u}_N, \boldsymbol{v}_N) = l_\Omega(\boldsymbol{v}_N) \quad \forall \boldsymbol{v}_N \in V_N(\Omega), \tag{22}$$

where we remind that the notation $\underline{a}_\Omega$ refers to the fact that the integral over Ω is split into a sum of integrals over Ω^k. In order to recover an approximation of the pressure, we proposed in [13] to first solve the problems: find $\hat{\boldsymbol{v}}_i = \arg\max_{\boldsymbol{w} \in (H_0^1(\hat{\Omega}))^2} \frac{\int_{\hat{\Omega}} \hat{p}_i \nabla \cdot \boldsymbol{w}}{\left(\int_{\hat{\Omega}} \nabla \boldsymbol{w}^2\right)^{1/2}}$, then define the reference space

$$Z_N(\hat{\Lambda}) = span\{\hat{\boldsymbol{v}}_i, i = 1, \dots, N\}, \tag{23}$$

and the global space

$$Z_N(\Omega) = \{\boldsymbol{v} \in L^2(\Omega)^2, \boldsymbol{v}_{|\Omega^k} = \Psi^{-1}(\hat{\boldsymbol{v}}), \hat{\boldsymbol{v}} \in Z_N(\hat{\Lambda})\}, \tag{24}$$

where we note that no interface condition should be imposed since these are locally H_0^1 functions. It is an easy matter to check that the inf sup condition is satisfied on the pair $Z_N \times M_N$

$$\inf_{q \in M_N} \sup_{v \in Z_N} \frac{\underline{b}_\Omega(v, q)}{\|v\|_{H^1}} = \beta,$$

(again the notation $\underline{b}_\Omega$ refers to the fact that the integral over Ω is split into a sum of integrals over Ω^k) where $\beta > 0$ may depend on N and Ω. This allows for recovering the discrete pressure by solving

$$\underline{b}_\Omega(p_N, \boldsymbol{v}_N) = l_\Omega(\boldsymbol{v}) - \underline{a}_\Omega(\boldsymbol{u}_N, \boldsymbol{v}_N), \quad \forall \boldsymbol{v}_N \in Z_N. \tag{25}$$

Note that by setting $X_N = V_N \oplus Z_N$ we have solved the Galerkin approximation of (14): find $\boldsymbol{u}_N \in X_N$ and $p_N \in M_N$ such that

$$\begin{aligned} \underline{a}_\Omega(\boldsymbol{u}_N, \boldsymbol{v}_N) + \underline{b}_\Omega(\boldsymbol{v}_N, p_N) &= l_\Omega(\boldsymbol{v}_N) & \forall \boldsymbol{v}_N \in X_N(\Omega), \\ \underline{b}_\Omega(\boldsymbol{u}_N, q_N) &= 0 & \forall q_N \in M_N(\Omega), \end{aligned} \tag{26}$$

Again, standard arguments in numerical analysis allow us to state that

$$\|\boldsymbol{u} - \boldsymbol{u}_N\|_{H^1} \le C \inf_{\boldsymbol{v}_N \in V_N} \|\boldsymbol{u} - \boldsymbol{v}_N\|_{H^1} + \text{consistency error}, \tag{27}$$

where an initial feasibility experiment may reveal how fast it goes to zero. Note that, due to the fact that the functions in the discrete space V_N are divergence free, the behavior of the inf sup parameter β with respect to N does not appear in this estimate. However, it does appear in the pressure approximation.

3.3. The steady Navier-Stokes problem

The extension to the steady Navier-Stokes equation is straightforward, at least as long as we stay at the level of the definition of the discrete problem and the numerical analysis. However, the implementation involves additional difficulties that we shall treat in the next section. First, let us recall a possible variational formulation of the problem: find the velocity $\boldsymbol{u} = (u_1, u_2) \in X(\Omega)$ and the pressure $p \in M(\Omega)$ such that

$$\begin{aligned} a_\Omega(\boldsymbol{u},\boldsymbol{v}) + b_\Omega(\boldsymbol{v},p) + c(\boldsymbol{u},\boldsymbol{v};\boldsymbol{u}) &= l_\Omega(\boldsymbol{v}) \qquad \forall\, \boldsymbol{v} \in X(\Omega)\,, \\ b_\Omega(\boldsymbol{u},q) &= 0 \qquad \forall\, q \in M(\Omega), \end{aligned} \qquad (28)$$

where the nonlinear term c take into account the convection contribution in the equations. This one is chosen here to be written as

$$2c(\boldsymbol{u},\boldsymbol{v};\boldsymbol{w}) = \int_\Omega \boldsymbol{w} \cdot \nabla \boldsymbol{u} \boldsymbol{v} dA - \int_\Omega \boldsymbol{w} \cdot \nabla \boldsymbol{v} \boldsymbol{u} dA\,,$$

to maintain stability of the discretization. Note that the inflow and outflow boundary conditions involves now the dynamical pressure $p + \frac{u^2}{2}$ instead of the pressure p.

The discretization space is again built from the computation of snapshots of the Navier-Stokes equations over the union of a few deformed references domains. These solutions are then restricted to one subdomain to provide, after the proper mapping, elements in $V_N(\hat\Lambda)$ and $M_N(\hat\Lambda)$. Finally, for a new instantiation of the geometry, the spaces $V_N(\Omega)$, $M_N(\Omega)$, $Z_N(\Omega)$ and $X_N(\Omega)$ are defined as in the linear situation.

Under standard hypothesis on the solution we are interested in for the Navier-Stokes problem, a convergence proof similar to (27) can again be obtained.

4. Numerical implementation

We start by emphasizing that that any reduced basis method necessarily involves the implementation of a more "classical" approximation method. Indeed – except for very particular and uninteresting problems – the knowledge of the solutions, that we named u_i, is impossible without referring to a discretization method (e.g., of finite element, spectral type...). This has some implications.

First of all, as explained in detail in [13], this blurs the statements on the reduced element method for the Stokes problem since the discrete reduced basis velocity functions are then not *exactly* divergence free any more. However, the divergence is very small, and is related to the discretization error. We have preferred to hide the difficulties that this involves since these are mostly technical issues, but when you want to be "less platonic" about the method and really try to implement it, you do have to deal with these technicalities.

The second difficulty is more general and comes from the fact that the solutions are only known through a preliminary basis, which, if we want the solution u_i to be well approximated, has to be *very* large. Knowing this, the rule of the game for the efficient implementation of any reduced basis method is to strictly prohibit any *online* reference to the extended basis. We allow *offline* precomputations of the solutions (that involves the extended basis) and some *offline* cross contribution of these solutions (based on their expression with respect to the extended basis) but this is forbidden *online*. We explain in the next section how this can be done.

4.1. Black box approach

From what we have just seen, both the Laplace problem and – with different notations – the Stokes problem, take the following form in the single domain case: find $u_N \in X_N(\Omega)$

$$a_\Omega(u_N, v_N) = f_\Omega(v_N), \quad \forall v_N \in X_N. \tag{29}$$

The solution u_N is sought as a linear combination of $\{u_j\}_{j=1,\ldots,\dim X_N}$ defined over Ω, locally, by a proper mapping of the $\{\hat{u}_i\}_{i=1,N}$ stored on a reference domain $\hat{\Lambda}$. First of all, it should be indicated that it is safer to define – through a Gram-Schmidt orthonormalization process – an orthonormal basis $\{\hat{\zeta}_i\}_{i=1,N}$ spanning the same space as the $\{\hat{u}_i\}_{i=1,N}$, that allows us to define ζ_j's over Ω by mapping and gluing as is explained in Subsection 2.2. This does not change the potential approximation properties of the reduced basis, but improves, to a large extent, the stability of the implementation. The solution procedure involves the evaluation of the elements of the stiffness matrix $a_\Omega(\zeta_i, \zeta_j)$. This computation involves some necessary differentiation and the evaluation of integrals over Ω, and this may be *very* costly. It should be stated here that the implementation of the reduced type method has to be much faster than the solution procedure that was used to compute the reduced basis, where much means many orders of magnitude. The $\mathcal{O}((\dim X_N)^2)$ entries of the stiffness matrix have thus to be evaluated through some smart way.

Let us begin by the easy case that is named *affine parametric dependence* where the entries $a_\Omega(\zeta_i, \zeta_j)$ appear to read

$$a_\Omega(\zeta_i, \zeta_j) = \sum_p g_p(\Omega) a_p(\hat{\zeta}_n, \hat{\zeta}_m), \tag{30}$$

where the bilinear forms a_p are domain independent. This is the case for the fin geometry where each subdomain, corresponding to one fin-stage, is composed of 4 rectangles that all map to a square through a simple affine mapping. Each a_p is the integral over a square of $\frac{\partial \hat{\zeta}_n}{\partial x} \frac{\partial \hat{\zeta}_m}{\partial x}$ or $\frac{\partial \hat{\zeta}_n}{\partial y} \frac{\partial \hat{\zeta}_m}{\partial y}$ while the $g_p(\Omega)$ take into account the dimension of the corresponding rectangle and the conductivity that, due to our hypothesis, is constant.

The expensive computation of the $a_{p,n,m} = a_p(\hat{\zeta}_n, \hat{\zeta}_m)$ can be done offline. Following the construction of the reduced basis these $a_{p,n,m}$ are stored, and for each new problem the evaluation of the stiffness matrix is done, online, in $P \times N^2$

operations, and solved in $\mathcal{O}((\dim X_N)^3)$ operations. These numbers are coherent with the rapid evaluation of the reduced basis method.

The hypothesis of *affine parametric dependency* is rather restrictive, and has to be generalized. In the case of quadratic or cubic dependency, the generalization is quite straightforward but even for linear problems such as Laplace or Stokes, when the geometry is the parameter, this is rarely the case and another approach has to be designed. In order to get a better understanding of the method, let us first assume that we want to compute $d_\Omega(\zeta_i, \zeta_j)$, defined as

$$d_\Omega(u, v) = \int_\Omega uv \, dA = \int_{\hat{\Omega}} uv J_\Phi \, d\hat{A} \,,$$

where J_Φ is the Jacobian determinant of the transformation that maps $\hat{\Omega}$ onto Ω (we assume momentarily that there is no domain decomposition in order to make the presentation less cumbersome). There is no reason in the general case that J_Φ will be affine, and thus the previous approach will not work. It is nevertheless likely that there exists a sequence of well-chosen transformations $\Phi_1^*, \ldots, \Phi_M^*, \ldots$ such that J_Φ may be well approximated by an expansion $J_\Phi \simeq \sum_{j=1}^M \beta_j J_{\Phi_j^*}$. An approximation of $d_\Omega(\zeta_i, \zeta_j)$ will then be given by

$$d_\Omega(\zeta_i, \zeta_j) \simeq \sum_{j=1}^M \beta_j \int_{\hat{\Omega}} \hat{\zeta}_i \hat{\zeta}_j J_{\Phi_j^*} \, d\hat{A} \,, \tag{31}$$

and again, the contributions $\int_{\hat{\Omega}} \hat{\zeta}_i \hat{\zeta}_j J_{\Phi_j^*} \, d\hat{A}$ will be precomputed offline. We do not elaborate here on how the Φ_j^* are selected, we shall discuss this in more generality latter. What we want to address is the evaluation of the coefficients $\beta_j = \beta_j(\Omega)$ in the approximation of J_Φ above. The idea is to use an interpolation procedure as is explained in [12]. Let $\mathbf{x}_1$ be the point where $|J_{\Phi_1^*}|$ achieves its maximum value. Assuming then that $\mathbf{x}_1, \ldots, \mathbf{x}_n$ have been defined, and are such that the $n \times n$ matrix with entries $J_{\Phi_k^*}(\mathbf{x}_\ell)$, $1 \le k, \ell \le n$ is invertible, we define $\mathbf{x}_{n+1}$ as being the point where $r_{n+1} = |J_{\Phi_{n+1}^*} - \sum_{k=1}^n \gamma_k J_{\Phi_k^*}|$ achieves it maximum value. Here the scalars γ_k are defined so that r_{n+1} vanishes at any $(\mathbf{x}_\ell)$ for $\ell = 1, \ldots, n$. This definition of the points $\mathbf{x}_k$ is possible as long the Φ_j are chosen such that the $J_{\Phi_k^*}$ are linearly independent (see [12]). The β_j are then evaluated also through the interpolation process

$$J_\Phi(\mathbf{x}_\ell) = \sum_{k=1}^M \beta_k J_{\Phi_k^*}(\mathbf{x}_\ell), \quad \forall 1 \le \ell \le M. \tag{32}$$

We have not much theory confirming the very good results that we obtain. An indicator that allows us to be quite confident in the interpolation process is the fact that the Lebesgue constant attached to the previously built points is, in all the examples we have encountered, rather limited. (We remind that the Lebesgue constant is the maximum of the ratio between the interpolation error and the best fit error.)

The same process is now used when implementing the Navier-Stokes problem, where we can decide to compute the solution to the discrete version of (28) through an iterative process. Given a current approximation $u_N^p \in V_N(\Omega)$, compute $u_N^{p+1} \in V_N(\Omega)$ as the solution of

$$a_\Omega(u_N^{p+1}, v) + c(u_N^{p+1}, v; u_N^p) = l_\Omega(v), \quad \forall v \in V_N(\Omega).$$

The evaluation of the stiffness matrix involves now the computation of $c(\zeta_i, \zeta_j; u_N^p)$, not only for each new geometry, but also at each iteration. It is an easy matter to realize that

$$c(\zeta_i, \zeta_j; u_N^p) = \int_{\hat{\Omega}} J_\Phi [u_N^p]^t \mathcal{J}_\Phi^{-1} \nabla \zeta_i \zeta_j d\hat{A}.$$

The online approximation of $J_\Phi [u_N^p]^t \mathcal{J}_\Phi^{-1}$ is done through interpolation on appropriate collocation points based on a set of functions $J_{\Phi_j^*} [u_N]^t (\Phi_j^*) \mathcal{J}_{\Phi_j^*}^{-1}, 1 \le j \le M$, where $u_N(\Phi_j^*)$ is the converged and previously offline computed solution associated with the geometry Φ_j^*. The construction of the updated part of the stiffness matrix is thus performed online in $\mathcal{O}(MN^2)$ operation, first by evaluating the β_k such that $J_\Phi [u_N^p]^t \mathcal{J}_\Phi^{-1} \simeq \sum_{k=1}^M \beta_k J_{\Phi_k^*} [u_N]^t (\Phi_k^*) \mathcal{J}_{\Phi_k^*}^{-1}$, then by approximating $c(\zeta_i, \zeta_j; u_N^p)$ by

$$c(\zeta_i, \zeta_j; u_N^p) \simeq \sum_{k=1}^M \beta_k \int_{\hat{\Omega}} J_{\Phi_k^*} [u_N]^t (\Phi_k^*) \mathcal{J}_{\Phi_k^*}^{-1} \nabla \zeta_i \zeta_j d\hat{A} .$$

4.2. Transfinite mappings

One major ingredient of the reduced basis element method is the design of the mapping between the reference domain and the current instantiation of the sub-domain. This design has to be efficient, and the resulting mapping has to be regular enough. There is a large flexibility in the definition of the different possible mappings, but they should all be one-to-one and map the boundaries of the computational domain onto the boundaries of the physical domain. When the domain of reference is a square, a standard way of defining a mapping is the Gordon-Hall transfinite interpolation approach; see [10].

The idea behind transfinite interpolation is to construct the image of the interior points of the physical domain as linear combinations of the image of the points on the boundaries. On the reference domain, $\hat{\Omega} = (-1,1)^2$, we construct one-dimensional weight functions $\phi_i(r)$, such that for $r_0 = -1$ and $r_1 = 1$ we get

$$\phi_i(r_j) = \delta_{ij}, 0 \le i, j \le 1. \tag{33}$$

The weight functions may be linear, but this is not a necessity. We may also use different weight functions in different spatial directions.

We assume that a representation of the boundaries of the physical domain is given with respect to the reference variables (ξ, η) by a bijective map. Each boundary will be the function of one variable, and we define the horizontal boundaries $x(\xi, -1)$ and $x(\xi, 1)$, and the vertical boundaries $x(-1, \eta)$ and $x(1, \eta)$, where $x = (x, y)$.

The transfinite mapping is then defined as

$$\Phi(\xi, \eta) = \phi_0(\xi)\boldsymbol{x}(-1, \eta) + \phi_1(\xi)\boldsymbol{x}(1, \eta) + \phi_0(\eta)\boldsymbol{x}(\xi, -1) + \phi_1(\eta)\boldsymbol{x}(\xi, 1)$$
$$- \sum_{i=0}^{1} \sum_{j=0}^{1} \phi_i(\xi)\phi_j(\eta)\boldsymbol{x}(r_i, r_j).$$

(34)

The mapping (34) will preserve the boundaries of the physical domain, and the interior points are determined via a linear transformation of the grid points defined on the reference domain.

When the reference domain $\hat{\Omega}$ is more complex, one way of working is to decompose it into a non-overlapping union of quadrilateral subdomains, do the same for the deformed domain Ω and define the mapping from $\hat{\Omega}$ onto Ω piecewise. The resulting mapping is generally continuous, piecewise regular but globally it is rarely $\mathcal{C}^1$. This is a redhibitory drawback for the use of the Piola transformation when dealing with the Stokes problem.

In order to improve the regularity of the mapping, we generalize (34) as follows. We assume $\hat{\Omega}$ and Ω are curved polygons with the same number of curved edges, say n. Let Γ_i (resp. $\hat{\Gamma}_i$) denote each edge of Ω (resp. $\hat{\Omega}$) ranked in a clockwise manner and such that $\Gamma_{n+1} = \Gamma_1$. Let $\boldsymbol{x}_i$ (resp. $\hat{\boldsymbol{x}}_i$) denote the vertex between Γ_i and Γ_{i+1} (resp. $\hat{\Gamma}_i$ and $\hat{\Gamma}_{i+1}$). We assume that each edge is parametrized by a one to one mapping ψ_i from $]0, 1[$ onto Γ_i so that $\psi_i(1) = \boldsymbol{x}_i$ (with obvious extension for the reference domain). We assume also that we are given projection operators π_i from $\hat{\Omega}$ onto $[0, 1]$ that associate with any point over $\hat{\Gamma}_{i+1}$ the value 1, with any point over $\hat{\Gamma}_{i-1}$ the value 0 and any point $\boldsymbol{x}$ over $\hat{\Gamma}_i$ the value $\hat{\psi}_i^{-1}(\boldsymbol{x})$. Finally, we introduce the weight functions φ_i with values in $[0, 1]$ that, similarly as in the original formulation satisfy $\varphi_i(\boldsymbol{x}) = 1$ over $\hat{\Gamma}_i$ and $\varphi_i(\boldsymbol{x}) = 0$ over any $\hat{\Gamma}_j$ with $j \neq i - 1, i, i + 1$. Then the mapping

$$\Phi(\xi, \eta) = \sum_{i=1}^{n} \{\varphi_i(\xi, \eta)\psi_i[\pi_i(\xi, \eta)] - \varphi_i[\pi_{i+1}(\xi, \eta)]\varphi_{i+1}[\pi_i(\xi, \eta)]\boldsymbol{x}_i\}, \quad (35)$$

preserves the boundary of the domains. Under mild assumptions over the ϕ's, it maps $\hat{\Omega}$ onto Ω. We refer to [14] for more about this strategy.

4.3. *A posteriori* error estimation

The reduced basis methods are known for rapid convergence rates. For application to realistic complex problem you are not interested in showing nice convergence plots; you are interested in getting the answer to your problem at a minimal cost. In addition, you want the result to be reliable. The number of elements in the reduced basis to be used for a given accuracy depends on the problem and the only way to get a hint whether you have used a rich enough basis set, is to refer to *a posteriori* error estimations.

Furthermore, in most cases it is not so much the solution of the PDE that is interesting; it is most often some outputs that can be computed from the knowledge of the solution. These outputs of interest are regular functionals evaluated over the solution. Let us consider the Stokes problem (14). For some specified output of interest, $s(\boldsymbol{u})$, we are thus interested in providing, after the solution $\boldsymbol{u}_N$ has been found, a computable lower bound $s^-(\boldsymbol{u}_N)$ and a computable upper bound $s^+(\boldsymbol{u}_N)$ such that

$$s^-(\boldsymbol{u}_N) \leq s(\boldsymbol{u}) \leq s^+(\boldsymbol{u}_N) .$$

In this work, we focus on compliant output, i.e.,

$$s(\boldsymbol{u}) = l_\Omega(\boldsymbol{u}). \tag{36}$$

and also, for the sake of simplification, on the mono-domain case.

We will follow the theory developed in [24] for operators which are continuous, coercive, symmetric and affine in terms of the parameter in a similar way as has been done in [28] for the steady Stokes problem with more standard parameter dependencies. The steady Stokes operator is symmetric and continuous, but not coercive, and due to the geometric dependency it is not affine either.

We introduce the diffusion operator

$$\hat{a}(\boldsymbol{v}, \boldsymbol{w}) = \int_{\hat{\Omega}} g(\Phi) \hat{\nabla}\hat{\boldsymbol{v}} \cdot \hat{\nabla}\hat{\boldsymbol{w}} d\hat{\Omega}, \tag{37}$$

on the reference domain, where $g(\Phi)$ is a geometry dependent positive function. The reconstructed error $\hat{\boldsymbol{e}} \in \hat{X}(\Omega)$ is then defined as the field that for some $g(\Phi)$ satisfies

$$\hat{a}(\boldsymbol{e}, \boldsymbol{v}) = l(\boldsymbol{v}) - a(\boldsymbol{u}_N, \boldsymbol{v}) - b(\boldsymbol{v}, p_N) \quad \forall \, \hat{\boldsymbol{v}} \in \hat{X}(\Omega). \tag{38}$$

The operator $g(\Phi)$ is chosen such that

$$\alpha_0 \|\boldsymbol{v}\|_X^2 \leq \hat{a}(\boldsymbol{v}, \boldsymbol{v}) \leq a(\boldsymbol{v}, \boldsymbol{v}) \quad \forall \, \boldsymbol{v} \in X(\Omega), \tag{39}$$

for some positive real constant α_0. For this reconstructed error we claim that

$$s^-(\boldsymbol{u}_N) = l(\boldsymbol{u}_N), \text{ and} \tag{40}$$

$$s^+(\boldsymbol{u}_N) = l(\boldsymbol{u}_N) + \hat{a}(\boldsymbol{e}, \boldsymbol{e}) , \tag{41}$$

are lower and upper bounds for $s(\boldsymbol{u})$.

Before we prove (40) and (41), we put $\boldsymbol{v} = \boldsymbol{u}_N$ in (14) and (26) to see that

$$a(\boldsymbol{u}_N, \boldsymbol{u} - \boldsymbol{u}_N) + b(\boldsymbol{u}_N, p) = 0. \tag{42}$$

The last term is zero and thus (42) reduces to

$$a(\boldsymbol{u}_N, \boldsymbol{u} - \boldsymbol{u}_N) = 0. \tag{43}$$

For the lower bound we get

$$\begin{aligned}
s^-(\boldsymbol{u}_N) &= s(\boldsymbol{u}) + l(\boldsymbol{u}_N - \boldsymbol{u}) \\
&= s(\boldsymbol{u}) + a(\boldsymbol{u}, \boldsymbol{u}_N - \boldsymbol{u}) + b(\boldsymbol{u}_N - \boldsymbol{u}, p) \\
&= s(\boldsymbol{u}) + a(\boldsymbol{u}, \boldsymbol{u}_N - \boldsymbol{u}) + a(\boldsymbol{u}_N, \boldsymbol{u} - \boldsymbol{u}_N) \\
&= s(\boldsymbol{u}) + a(\boldsymbol{u} - \boldsymbol{u}_N, \boldsymbol{u}_N - \boldsymbol{u}).
\end{aligned} \tag{44}$$

And we have the desired relationship

$$s^-(\boldsymbol{u}_N) \leq s(\boldsymbol{u}), \tag{45}$$

independent of $g(\Phi)$.

For the upper bound we denote the error on the deformed domain by $e^u = u - u_N$, and find that

$$
\begin{aligned}
2\hat{a}(e, e^u) &= l(u - u_N) + l(u - u_N) \\
&\quad -2a(u_N, u - u_N) \\
&= l(u - u_N) + a(u, u - u_N) \\
&\quad +b(u - u_N, p) - 2a(u_N, u - u_N) \\
&= l(u - u_N) + a(u - u_N, u - u_N) \\
&\quad -a(u_N, u - u_N) \\
&= l(u - u_N) + a(e^u, e^u).
\end{aligned}
\tag{46}
$$

To prove that (41) is an upper bound we now use (46) to get

$$
\begin{aligned}
s^+(u_N) &= l(u_N) + \hat{a}(e, e) \\
&= l(u_N) + \hat{a}(e, e) \\
&\quad -2\hat{a}(e, e^u) + l(u - u_N) + a(e^u, e^u) \\
&\quad +\hat{a}(e^u, e^u) - \hat{a}(e^u, e^u) \\
&= l(u) + \hat{a}(e - e^u, e - e^u) \\
&\quad +a(e^u, e^u) - \hat{a}(e^u, e^u) \\
&\geq s(u),
\end{aligned}
\tag{47}
$$

where the inequality is due to (39) and the coercivity of $\hat{a}(\cdot, \cdot)$.

It now remains to find a positive function $g(\Phi)$, such that

$$
a(v, v) \geq \hat{a}(v, v) \quad \forall\, v \in X(\Omega).
\tag{48}
$$

For a constant $g(\Phi) = \lambda$, we can use the theory of [24] to see that λ should be chosen as large as possible without violating (48). This largest constant may be found, as in [18], by computing the smallest eigenvalue of the generalized symmetric eigenvalue problem

$$
a(v, v) = \lambda \int_{\hat{\Omega}} \hat{\nabla}(v \circ \Phi) \cdot \hat{\nabla}(v \circ \Phi) d\hat{\Omega}.
\tag{49}
$$

We tried this approach also for the current problem, and used an inverse Rayleigh quotient iteration to estimate λ, but the resulting upper bound gap proved much too conservative.

To get a better estimate we consider the Jacobian, $\mathcal{J}(\Phi)$, of the mapping from $\hat{\Omega}$ to Ω. We start with the left-hand side of (48), and use the fact that

$$
\nabla = \mathcal{J}^{-T} \hat{\nabla},
\tag{50}
$$

to rewrite (15) mapped back on the reference domain

$$
\begin{aligned}
a(v, v) &= \int_{\hat{\Omega}} (\hat{\nabla}[v \circ \Phi])^T \mathcal{J}^{-1} \mathcal{J}^{-T} (\hat{\nabla}[v \circ \Phi]) |J| d\hat{\Omega} \\
&= \int_{\hat{\Omega}} w^T G w d\hat{\Omega},
\end{aligned}
\tag{51}
$$

where $w = \hat{\nabla}[v \circ \Phi]$, and $G = G(\Phi) = \mathcal{J}^T \mathcal{J} |J|$. At each point $\hat{x} \in \Omega$ we diagonalize the 2 symmetric positive-definite matrix G, that is, we write $G(\Phi(\hat{x})) = Q^T \Lambda Q$, where Q consists of the orthonormal eigenvectors of G. If we (at each point $\hat{\mathbf{x}} \in \hat{\Omega}$) replace the two diagonal elements of Λ with the smallest one $\Lambda_{\min}$, we get

$$\int_{\hat{\Omega}} \mathbf{w}^T G \mathbf{w} d\hat{\Omega} \geq \int_{\hat{\Omega}} \Lambda_{\min} (Q\mathbf{w})^T Q \mathbf{w} d\hat{\Omega}. \tag{52}$$

Since Q consists of the orthonormal eigenvectors, the last expression is equivalent to $\int_{\hat{\Omega}} \Lambda_{\min} \mathbf{w}^T \mathbf{w} d\hat{\Omega}$, and we end up with

$$a(\mathbf{v}, \mathbf{v}) \geq \int_{\hat{\Omega}} \Lambda_{\min} (\hat{\nabla}[\mathbf{v} \circ \Phi]) \cdot (\hat{\nabla}[\mathbf{v} \circ \Phi]) d\hat{\Omega} . \tag{53}$$

This is just (37) with $g(\Phi) = \Lambda_{\min}(\Phi)$, and thus (48) is satisfied.

If we replace $\Lambda_{\min}(\Phi)$ by $\overline{\Lambda}_{\min} = \min_{\hat{\mathbf{x}} \in \hat{\Omega}} \Lambda_{\min}(\Phi)$, we may put $g(\Phi)$ outside the integral and apply the theory of [24]. This will produce a more conservative upper bound, but the calculation of (41) can be split in an off-line/online procedure without applying the methodology of Subsection 4.1.

To illustrate numerically the results on the output bounds, we will use the domain $\Omega = \Phi(\hat{\Omega})$ defined in Figure 8.

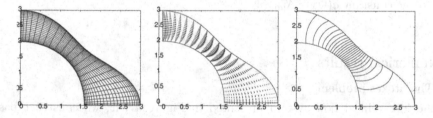

FIGURE 8. The domain Ω with velocity and pressure solution

The results for the previously defined deformed geometry are presented in Table 1. We see that the upper bound gap is relatively large compared to the lower bound gap. In the future, a different method to find an improved estimate of $g(\Phi)$ is desirable to reduce the upper bound gap.

We end this subsection by explaining how these a posteriori estimates can be used in order to select the basis solutions that are worth keeping to represent in an optimal way the set of all solutions. The strategy is based on a greedy algorithm in which the first parameter is selected at random, or at least so that the output is not zero. Then assuming that m basis functions are selected, the selection of the $m + 1$ parameter corresponds to the argmax of the a posteriori error on the output based on the approximation with the discrete space spanned with the m selected basis functions.

By choosing this argmax, we are indeed sure that the corresponding solution is quite far from the vectorial space spanned by the m first basis functions.

N	$s(\boldsymbol{u}) - s^-(\boldsymbol{u}_N)$	$s^+(\boldsymbol{u}_N) - s(\boldsymbol{u})$	$s_2^+(\boldsymbol{u}_N) - s(\boldsymbol{u})$
1	$2.82 \cdot 10^{-4}$	$5.28 \cdot 10^{-2}$	$1.01 \cdot 10^{-1}$
2	$1.87 \cdot 10^{-4}$	$9.86 \cdot 10^{-2}$	$1.85 \cdot 10^{-1}$
3	$1.35 \cdot 10^{-4}$	$9.74 \cdot 10^{-2}$	$1.90 \cdot 10^{-1}$
4	$1.32 \cdot 10^{-4}$	$9.70 \cdot 10^{-2}$	$1.86 \cdot 10^{-1}$
5	$7.67 \cdot 10^{-5}$	$3.02 \cdot 10^{-2}$	$4.52 \cdot 10^{-2}$
6	$7.44 \cdot 10^{-5}$	$6.86 \cdot 10^{-3}$	$1.20 \cdot 10^{-2}$
7	$1.04 \cdot 10^{-5}$	$1.56 \cdot 10^{-3}$	$2.60 \cdot 10^{-3}$
8	$7.03 \cdot 10^{-6}$	$2.62 \cdot 10^{-3}$	$4.32 \cdot 10^{-3}$
9	$7.02 \cdot 10^{-6}$	$1.61 \cdot 10^{-3}$	$2.39 \cdot 10^{-3}$
10	$4.24 \cdot 10^{-6}$	$6.10 \cdot 10^{-4}$	$1.02 \cdot 10^{-3}$
11	$4.16 \cdot 10^{-6}$	$6.21 \cdot 10^{-4}$	$1.05 \cdot 10^{-3}$
12	$3.15 \cdot 10^{-6}$	$6.16 \cdot 10^{-4}$	$9.68 \cdot 10^{-4}$
13	$2.82 \cdot 10^{-6}$	$4.63 \cdot 10^{-4}$	$7.50 \cdot 10^{-4}$
14	$1.94 \cdot 10^{-6}$	$4.32 \cdot 10^{-4}$	$6.97 \cdot 10^{-4}$
15	$1.94 \cdot 10^{-6}$	$3.82 \cdot 10^{-4}$	$6.71 \cdot 10^{-4}$

TABLE 1. Convergence of the lower and the upper bound gaps. Here, s^+ corresponds to the variable $g(\Phi) = \Lambda_{\min}(\Phi)$, while s_2^+ corresponds to the constant $g(\Phi) = \overline{\Lambda}_{\min}(\Phi)$.

5. Numerical results

5.1. The Stokes problem

For the hierarchical flow system presented in Figure 4 left, we construct basis functions for the pipe and bifurcation blocks separately. For the pipe we vary the deformation of the walled part of the boundary, and also the orientation of the outflow boundary relative to the inflow boundary. For the bifurcation we vary the opening angle of the two legs of the bifurcation, and the relative length of the two legs. The basis functions are computed as described above (truncating a multidomain solution, and applying the greedy algorithm). Using these basis functions for the pipe block and the three bifurcation blocks, we get the results presented in Table 2 for the steady Stokes problem.

We have used the same basis functions to approximate the solution of the bypass configuration shown in Figure 9. The results for the steady Stokes problem for this system are presented in Table 3.

5.2. The Navier-Stokes problem

The experiment on the steady Navier-Stokes problem is done on a monodomain pipe. The basis functions are found on a deformed quarter annulus by varying the deformation of the inner curved boundary. To solve the steady Navier-Stokes problem, we consider the corresponding time-dependent problem, and iterate in

N	N_1	N_2	$|u_N - u|_{H^1}$	$\|p_N - p\|_{L^2}$
36	9	9	$2.6 \cdot 10^{-3}$	$4.0 \cdot 10^{-1}$
44	11	11	$1.7 \cdot 10^{-3}$	$6.6 \cdot 10^{-2}$
52	13	13	$1.2 \cdot 10^{-3}$	$4.9 \cdot 10^{-2}$
65	15	15	$1.1 \cdot 10^{-3}$	$3.7 \cdot 10^{-2}$
105	15	30	$4.2 \cdot 10^{-4}$	$6.3 \cdot 10^{-3}$

TABLE 2. The error in the reduced basis steady Stokes solution on a multi-block system corresponding to Figure 4 left. Here, $N = N_1 + 3N_2$ is the total number of degrees-of-freedom in the reduced basis spaces X_N^0, X_N^e, and M_N, N_1 is the number of basis geometries used to generate the basis functions on the pipe block, and N_2 is the number of basis functions used on the bifurcation blocks.

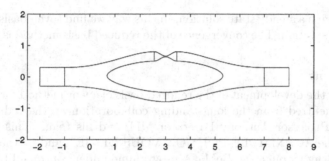

FIGURE 9. A bypass.

N	N_1	N_2	$|u_N - u|_{H^1}$	$\|p_N - p\|_{L^2}$
45	9	9	$9.3 \cdot 10^{-3}$	$3.3 \cdot 10$
55	11	11	$3.1 \cdot 10^{-3}$	$5.3 \cdot 10^{-1}$
65	13	13	$2.3 \cdot 10^{-3}$	$9.0 \cdot 10^{-2}$
75	15	15	$1.4 \cdot 10^{-3}$	$5.3 \cdot 10^{-2}$
105	15	30	$5.4 \cdot 10^{-4}$	$3.0 \cdot 10^{-2}$

TABLE 3. The error in the reduced basis steady Stokes solution on a multi-block bypass with three pipe blocks and two bifurcation blocks. Here, $N = 3N_1 + 2N_2$ is the total number of degrees-of-freedom in each of the reduced basis spaces X_N^0, X_N^e, and M_N, N_1 is the number of basis geometries used to generate the basis functions on the pipe block, and N_2 is the number of basis functions used on the bifurcation blocks.

$N/2$	$|u_N - u|_{H^1}$	$\|p_N - p\|_{L^2}$	$s_N^+ - s$	$s - s_N^-$
1	$1.5 \cdot 10^{-1}$	$1.2 \cdot 10^{-1}$	1.1	$3.9 \cdot 10^{-3}$
2	$1.7 \cdot 10^{-2}$	$1.1 \cdot 10^{-2}$	$1.5 \cdot 10^{-2}$	$1.0 \cdot 10^{-3}$
3	$2.0 \cdot 10^{-3}$	$8.8 \cdot 10^{-4}$	$3.8 \cdot 10^{-4}$	$2.6 \cdot 10^{-4}$
4	$1.1 \cdot 10^{-4}$	$2.5 \cdot 10^{-5}$	$1.0 \cdot 10^{-5}$	$5.8 \cdot 10^{-6}$
5	$5.3 \cdot 10^{-5}$	$3.6 \cdot 10^{-6}$	$2.5 \cdot 10^{-6}$	$1.5 \cdot 10^{-6}$
6	$5.1 \cdot 10^{-5}$	$3.4 \cdot 10^{-6}$	$1.3 \cdot 10^{-6}$	$1.4 \cdot 10^{-6}$
7	$3.9 \cdot 10^{-5}$	$2.1 \cdot 10^{-6}$	$2.2 \cdot 10^{-7}$	$4.2 \cdot 10^{-7}$

TABLE 4. Results when we use offline/online decoupling. The error in the reduced basis solution of the Navier-Stokes problem when the stopping criterion for the truth solution is 10^{-10}.

time until we reach a steady state solution. In this way we find seven basis function, and a reference solution. The convergence of the reduced basis method is presented in Table 4.

Acknowledgement

This project on the development of the reduced basis element method for fluid flow has greatly benefitted from the longstanding collaboration on the reduced basis methods with Professor Anthony Patera of MIT and his team. This work was supported by the ACI-NIM "LE-POUMON-VOUS-DIS-JE", and by the Research Council of Norway through the BeMatA programme under contract 147044/431.

References

[1] B.O. Almroth, P. Stern, F.A. Brogan – Automatic choice of global shape functions in structural analysis. AIAA Journal, 16 (1978) 525–528.

[2] L. Baffico, C. Grandmont, Y. Maday, and A. Osses – Homogenization of an elastic media with gaseous bubbles, in preparation.

[3] F. Brezzi and M. Fortin – *Mixed and Hybrid Finite Element Methods*. Springer Verlag, 1991.

[4] M. Briane, Y. Maday, and F. Madigou – Homogenization of a two-dimensional fractal conductivity, in preparation.

[5] C. Fetita, S.M., D. Perchet, F. Prêteux, M. Thiriet, and L. Vial – An image-based computational model of oscillatory flow in the proximal part of tracheobronchial trees, Computer Methods in Biomechanics and Biomedical Engineering, 8(4), 279–293, (2005).

[6] J.P. Fink and W.C. Rheinboldt – On the error behavior of the reduced basis technique for nonlinear finite element approximations. Zeitschrift für Angewandte Mathematik und Mechanik, 63(1), (1983) 21–28.

[7] R.L. Fox and H. Miura – An approximate analysis technique for design calculations. AIAA Journal, 9(1), (1971) 177–179.

[8] M.B. Giles and E. Süli – Adjoint methods for PDEs: a posteriori error analysis and postprocessing by duality. Acta Numerica, Vol. 11, 145–236, Cambridge University Press, 2002.

[9] V. Girault, P.A. Raviart – Finite element methods for Navier-Stokes equations: Theory and algorithms, (Springer Series in Computational Mathematics. Volume 5), Berlin and New York, Springer-Verlag (1986)

[10] W. Gordon and C. Hall – Transfinite element methods: blending-function interpolation over arbitrary curved element domains, Numer. Math. (21), 1973/74, pp. 109–129.

[11] C. Grandmont, Y. Maday, and B. Maury – A multiscale/multimodel approach of the respiration tree, in New Trends in Continuum Mechanics – M. Mihailescu-Suliciu, Ed. – Theta Foundation, Bucharest, Romania (2005).

[12] M.A. Grepl, Y. Maday, N.C. Nguyen, and A.T. Patera – Efficient reduced basis treatment of non-affine and nonlinear partial differential equations, submitted to M^2AN (2006).

[13] A.E. Løvgren, Y. Maday, and E.M. Rønquist – A reduced basis element method for the steady Stokes problem, to appear in M^2AN.

[14] A.E. Løvgren, Y. Maday, and E.M. Rønquist – in progress.

[15] L. Machiels, Y. Maday, I.B Oliveira, A.T. Patera, and D.V. Rovas — Output bounds for reduced-basis approximations of symmetric positive definite eigenvalue problems, CR Acad Sci Paris Series I 331:(2000) 153–158.

[16] Y. Maday, A.T. Patera, and G. Turinici – A Priori Convergence Theory for Reduced-Basis Approximations of Single-Parameter Elliptic Partial Differential Equations, J. Sci. Comput., 17 (2002), no. 1-4, 437–446.

[17] Y. Maday, L. Machiels, A.T. Patera, and D.V. Rovas – Blackbox reduced-basis output bound methods for shape optimization, in Proceedings 12th International Domain Decomposition Conference, Chiba, Japan (2000) 429–436.

[18] Y. Maday and E.M. Rønquist – The reduced-basis element method: Application to a thermal fin problem. SIAM Journal on Scientific Computing, 2004.

[19] A.K. Noor and J.M. Peters – Reduced basis technique for nonlinear analysis of structures. AIAA Journal, 18(4) (1980) 455–462.

[20] V. Milisic and A. Quarteroni. Analysis of lumped parameter models for blood flow simulations and their relation with 1D models, to appear in M^2AN, 2004.

[21] F. Murat and J. Simon, Sur le Contrôle par un Domaine Géometrique, Publication of the Laboratory of Numerical Analysis, University Paris VI, 1976.

[22] Pinkus, A. – n-Widths in Approximation Theory, Springer-Verlag, Berlin (1985).

[23] T.A. Porsching – Estimation of the error in the reduced basis method solution of nonlinear equations. Mathematics of Computation. 45(172) (1985) 487–496.

[24] C. Prud'homme, D.V. Rovas, K. Veroy, L. Machiels, Y. Maday, A.T. Patera, G. Turinici – Reliable real-time solution of parametrized partial differential equations: Reduced-basis output bound methods, *J. Fluids Engineering*, **124**, (2002) 70–80.

[25] C. Prud'homme, D.V. Rovas, K. Veroy, and A.T. Patera – A mathematical and computational framework for reliable real-time solution of parametrized partial differential equations. Programming. M^2AN Math. Model. Numer. Anal. 36 (2002), no. 5, 747–771.

[26] A. Quarteroni and A. Valli – Domain decomposition methods for partial differential equations, *Numerical Mathematics and Scientific Computation*, Oxford Science Publications, The Clarendon Press Oxford University Press, New York, (1999).

[27] P.A. Raviart and J.M. Thomas. – A mixed finite element method for 2nd order elliptic problems. In I. Galligani and E. Magenes, editors, Mathematical Aspects of Finite Element Methods, Lecture Notes in Mathematics, Vol. 606. Springer-Verlag, 1977.

[28] D.V. Rovas. – Reduced-Basis Output Bound Methods for Parametrized Partial Differential Equations. PhD thesis, Massachusetts Institute of Technology, Cambridge, MA, October 2002.

[29] M. Thiriet – http://www-rocq.inria.fr/who/Marc.Thiriet/Vitesv/

[30] A. Toselli and O. Widlund – *Domain decomposition methods – algorithms and theory*, Springer Series in Computational Mathematics, **34**, Springer-Verlag, Berlin, (2005).

[31] R. Verfurth – *A Review of A Posteriori Error Estimation and Adaptive Mesh-Refinement Techniques.* Wiley-Teubner, 1996.

Alf Emil Løvgren, Yvon Maday and Einar M. Rønquist

Analysis and Simulation of Fluid Dynamics
Advances in Mathematical Fluid Mechanics, 155–162
© 2006 Birkhäuser Verlag Basel/Switzerland

Asymptotic Stability of Steady-states for Saint-Venant Equations with Real Viscosity

Corrado Mascia and Frederic Rousset

1. Introduction

The Saint-Venant model, introduced in [4], gives raise to a simple and, nevertheless, rich hyperbolic system of conservation laws with a zero order reaction term. Apart from its interest as a model itself, we consider the analysis of the Saint-Venant system a useful preliminary step in the understanding of the rôle played by lower order term in qualitative properties/asymptotic behavior of solutions to conservation laws.

In what follows we deal with the viscous Saint-Venant model and analyze the stability of stationary steady states, introduced below. We skip all of the technical details of the analysis, stressing only the kind of result we are able to prove at the present and the main lines of the proof. Details will be given in a forthcoming paper.

The model

We consider the following *viscous Saint-Venant model* for shallow water

$$\begin{cases} \partial_t h + \partial_x(hv) = 0, \\ \partial_t(hv) + \partial_x \left(hv^2 + \frac{1}{2} gh^2 \right) = -g\, z'(x)\, h + \epsilon\, \partial_x(h\, \partial_x v) \end{cases} \tag{1}$$

The model is one dimensional in space, i.e., $x \in \mathbb{R}$. As usual, t is the time variable, $t \geq 0$. The constants g and ϵ represent, respectively, gravity and viscosity intensity. The function $z = z(x)$ represents the bottom topography and is considered as a given datum of the problem. The unknown functions are $h > 0$, height of the water, and v, velocity of the water.

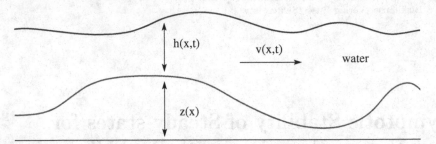

One of the main feature of this model is the presence of the viscosity term $\epsilon\,\partial_x(h\,\partial_x v)$, usually neglected in classical studies of Saint-Venant equation. Because of this term, the system is of hyperbolic-parabolic type and only partial smoothing effect should be expected.

In [1], the viscous Saint-Venant system (1) is derived from Navier-Stokes equations. The viscosity parameter ϵ essentially is the ratio between characteristic dimensions of height and length. The zero order approximation leads to the unviscous model, while a higher order approximation gives raise to the viscous model (1) we deal with.

The viscous Saint-Venant equations belong to the wider class of reaction–convection–diffusion system of the form

$$\partial_t U + \partial_x F(U) = G(x, U) + \partial_x(B(U)\,\partial_x U). \tag{2}$$

For specific models of this class, results on asymptotic stability of constant states/traveling waves have been proved. In what follow, we are interested in the asymptotic stability of a class of steady states, namely the *stationary steady states*, of (1). To this aim, we implement a strategy similar to the one used for stability of viscous shock waves for systems of conservation laws (i.e., $G \equiv 0$) in [5, 2, 3] based on linearization and detailed estimates on the Green function, through Laplace transform and detailed analysis of the resolvent equation.

Stationary steady states

Steady states of the viscous Saint-Venant system are solution of the form

$$(h, v)(x, t) = (H, V)(x).$$

Inserting in the equations, we get

$$HV = C \in \mathbb{R}, \qquad \left(HV^2 + \frac{1}{2}gH^2\right)' = -g\,z'(x)\,H + \epsilon\,(H\,V')'$$

or, equivalently

$$HV = C \in \mathbb{R}, \qquad CV' + g\left(H + z(x)\right)' H = \epsilon\,(H\,V')'.$$

A STATIONARY STEADY STATE (SSS) is a steady state with $V \equiv 0$, i.e., $C = 0$. Hence a SSS satisfies $H' = -z'(x)$, or, in other words,

$$H(x) = C - z(x) \qquad\qquad C \in \mathbb{R}.$$

A SSS is such that $H + z =$ constant, corresponding to a flat surface for the water. In all of the following, we concentrate our attention on SSS.

Asymptotic stability

Next, let us consider the Cauchy problem for (1) with an initial datum of the form

$$(\tilde{h}, \tilde{v})\Big|_{t=0} = (H, 0) + (h_0, v_0)$$

where $(H, 0)$ is a fixed SSS and (h_0, v_0) is a small perturbation of the steady state. Here $(\tilde{h}, \tilde{v})$ denote the solution to the system (1). The classical asymptotic stability question for the SSS is: *does (h, v), evolution of the initial perturbation (h_0, v_0), decay to zero as $t \to +\infty$ with respect to some metric?* Together with the stability question, the determination of the rate of decay is the natural companion problem: *is it possible to determine $F = F(t)$ with $F(t) \to 0$ as $t \to +\infty$ such that $(h, v) \sim F(t)$ as $t \to +\infty$ in an appropriate sense?* Finding the optimal rate of decay is fundamental to understand the main effects dominating the dynamic for large-times. In the case of the SSS solutions stability is "physically evident" thinking back to the phenomena described by the model, but the rate of decay is far from being trivial.

Let us present the assumptions on the function H (hence analogous assumptions holds for the bottom topography function z) guaranteeing stability. We assume that there exists $H_{\pm}, c_0 \in (0, +\infty)$ such that

$$H \quad \text{smooth}, \qquad \lim_{x \to \pm\infty} H(x) = H_{\pm} \in \mathbb{R}, \qquad H(x) \geq c_0 > 0.$$

The convergence to the asymptotic states is assumed to be exponential.

Under the above assumption, asymptotic stability can be proved.

Theorem 1. *If (h_0, v_0) is sufficiently small in $W^{1,p} \cap L^1$ for $p > 1$, then*

$$\|(h, v)\|_{W^{1,p}} \leq C(1 + t)^{-\alpha_p} \qquad \alpha_p := \frac{1}{2}\left(1 - \frac{1}{p}\right)$$

for some $C > 0$, independent of t.

The decay rate α_p is the same decay rate of the one-dimensional heat kernel. This shows that diffusion effects dominate in both component for large time even if the system is not uniformly parabolic. Similar results have been found in the analysis of stability of viscous shock waves, due to the presence of diffusion waves. In the case of shock waves, the solution are never isolated since they belong (at least) to the manifold determined by their own spatial translation. For SSS this is not the case: steady stationary solution are isolated in L^p. Indeed, in this case, these solutions are the analogous of the constant states for viscous conservation laws (they ARE constant in the case of the flat bottom topography $z \equiv$ constant). This reflects in the convergence of the perturbed solution to the SSS itself.

2. Method of the proof

Next, we give a short and schematic overview on the proof of the theorem. The basic steps are the following

1. *Linearization*: the equation for the perturbation can be rewritten as

$$\partial_t u = \mathcal{L}u + \text{H.O.T.},$$

where $\mathcal{L}$ is an appropriate linear differential operator. Coefficients of operator $\mathcal{L}$ are x-dependent, but, thanks to the assumptions on H, it is asymptotically constant.

2. *Resolvent equation*: by applying the Laplace transform, the problem of determining the solution of the linearized equation turns into solving the resolvent equation

$$(\mathcal{L} - \lambda I)u = F.$$

The solution of this equation can be represented through an appropriate resolvent kernel G_λ

$$u(x) = \int_{\mathbb{R}} G_\lambda(x, y)F(y)\, dy.$$

One of the original feature of the present problem is that the resolvent kernel is not a function, but a distribution, due to the presence of a (decaying) delta function term.

Once the delta function term is singled out, the smooth part of the resolvent kernel can be suitable approximated for small/large λ, using the property of $\mathcal{L}$ of being asymptotically constant.

3. *Linear stability*: using anti-laplace transform formula and the above approximation of the resolvent kernel G_λ, it is possible to give approximate expression for the Green function G of the original linearized problem. From

$$\partial_t u = \mathcal{L}u, \quad u\big|_{t=0} = u_0, \qquad \Longleftrightarrow \qquad u(x,t) = \int_{\mathbb{R}} G(x, y; t)u_0(y)\, dy,$$

we obtain linearized stability results with rate decay (same of the Green function).

4. *Nonlinear stability*: the solution of the complete nonlinear equation can be represented by means of the usual Duhamel formula. Thanks to decay properties of the Green function, it is possible to prove the nonlinear stability results together with decay estimates by application of an iterative argument to control the evolution of the $W^{1,p}$ norm.

The above strategy closely follows the one applied in the analysis of stability of shock waves for systems of conservation laws with viscosity. The main novelty is in the analysis of the resolvent equation where a delta function appears.

Linearization

The viscous Saint-Venant model fits in the general structure

$$\partial_t F^0(\tilde{w}) + \partial_x F^1(\tilde{w}) + Q(x)\tilde{w} = \partial_x(B(\tilde{w})\,\partial_x\tilde{w}),$$

where, in our case,

$$\tilde{w} = \begin{pmatrix} \tilde{h} \\ \tilde{v} \end{pmatrix}, \qquad F^0(\tilde{w}) = \begin{pmatrix} \tilde{h} \\ \tilde{h}\,\tilde{v} \end{pmatrix}, \qquad F^1(\tilde{w}) = \begin{pmatrix} \tilde{h}\,\tilde{v} \\ \tilde{h}\tilde{v}^2 + \frac{1}{2}\,g\,\tilde{h}^2 \end{pmatrix},$$

$$Q(x) = \begin{pmatrix} 0 & 0 \\ gz'(x) & 0 \end{pmatrix}, \qquad\qquad B(\tilde{w}) = \begin{pmatrix} 0 & 0 \\ 0 & \epsilon\tilde{h} \end{pmatrix}.$$

Let $W = W(x)$ be a given solution of the previous equation and consider the equation for the perturbation $w := \tilde{w} - W$. Substituting and using the equation satisfied by W, we get the nonlinear equation for w

$$\partial_t \left(F^0(W + w) - F^0(W) \right) + \partial_x \left(F^1(W + w) - F^1(W) \right) + Q(x)\,w$$

$$= \partial_x \left[(B(W + w) - B(W))\,W' \right] + \partial_x (B(W + w)\,\partial_x w). \tag{3}$$

Therefore the linearized system is

$$dF^0(W)\,\partial_t w = \partial_x (B(W)\,\partial_x w) + \partial_x \left[dB(W)\,w\,W - dF^1(W)\,w \right] - Q(x)\,w$$

that is $\partial_t w = \mathcal{L}\,w$, where

$$\mathcal{L}w := dF^0(W)^{-1} \left\{ \partial_x (B(W)\,\partial_x w) + \partial_x \left[dB(W)\,w\,W - dF^1(W)\,w \right] - Q\,w \right\}.$$

In the case of the viscous Saint-Venant system, setting $W = (H, 0)$, we get the following linear system

$$\begin{cases} \partial_t h + \partial_x (Hv) = 0 \\ H\partial_t v + gH\partial_x h = \epsilon\,\partial_x (H\partial_x v). \end{cases}$$

Multiplying the first equation by gh and the second by v, we get

$$\frac{1}{2}\,\partial_t (g\,h^2 + H\,v^2) + \partial_x (g\,H\,h\,v - \epsilon\,H\,v\,\partial_x v) + \epsilon\,H\,|\partial_x v|^2 = 0$$

from which the (linear) energy estimate follows

$$|(\sqrt{g}\,h, \sqrt{H}\,v)|_{L^2}^2 + 2\epsilon \int_0^t |\sqrt{H}\,\partial_x v|_{L^2}^2\,(s)\,ds = |(\sqrt{g}\,h_0, \sqrt{H}\,v_0)|_{L^2}^2$$

showing linear L^2-stability of the equilibrium solution.

Resolvent equation

Setting $w := Hv$ and applying Laplace transform, $\partial_t \mapsto \lambda\cdot$, we get the resolvent system

$$\begin{cases} \lambda h + \partial_x w = F_1 \\ \lambda w + gH\partial_x h + \epsilon\,\partial_x (\partial_x (hH)\,w) - \epsilon\partial_x^2 w = F_2, \end{cases}$$

to be solved for any given F_1, F_2. Formally, we expect to find a function $G_\lambda = G_\lambda(x, y)$ such that the solution operator of the resolvent system is

$$F = (F_1, F_2) \qquad \mapsto \qquad (h, w) := \int_{\mathbb{R}} G_\lambda(x, y)\,F(y)\,dy.$$

But using the first equation $\partial_x w = F_1 - \lambda\,h$, it is possible to reduce the order of the system! This phenomenon is due to the presence of a delta-function term in G_λ. Let us consider an easy example showing the same property: disregarding the term with the function H in the resolvent system, we get the simplified version

$$\begin{cases} \lambda h + \partial_x w = F_1 \\ \lambda w - \epsilon \partial_x^2 w = F_2. \end{cases}$$

Then, by using the relation $\partial_x w = F_1 - \lambda h$, we get

$$\begin{cases} \partial_x w + \lambda h = F_1 \\ \epsilon \lambda \partial_x h + \lambda w = \epsilon \partial_x F_1 + F_2. \end{cases}$$

The presence of the term $\partial_x F_1$ shows that the solution operator is not smoothing in some directions. Setting $h := \check{h} + \lambda^{-1} F_1$, we get the system for the unknown $(\check{h}, v)$,

$$\begin{cases} \partial_x w + \lambda \check{h} = F_1 \\ \epsilon \lambda \partial_x \check{h} + \lambda w = F_2, \end{cases}$$

whose solution operator is smoothing. This means that the resolvent kernel G_λ is the sum of two terms: a smooth one and a delta function.

In the case of viscous Saint-Venant system, skipping all of the calculations, the following representation for the resolvent kernel holds

$$G_\lambda(x,y) = \check{G}_\lambda(x,y) + \frac{\delta_y}{\lambda + gH(x)/\epsilon} \begin{pmatrix} 1 & 0 \\ 0 & 0 \end{pmatrix}.$$

It is important to stress that the delta function term appears as a singularity of the resolvent kernel in the stable half-plane $\{\lambda : \operatorname{Re}\lambda < 0\}$, hence it will give raise to an exponential decaying term.

The regular part $\check{G}_\lambda$ of the resolvent kernel G_λ can be approximated using an approach similar to the one developed in the analysis of stability of viscous shock waves. Roughly speaking, $\check{G}_\lambda$ is given by linear combination of decaying/growing modes approximated by considering the linearized operator $\mathcal{L}$ as a perturbation of the piecewise constant operator $\mathcal{L}_\infty$

$$\mathcal{L}_\infty := \begin{cases} \mathcal{L}_- & x < 0 \\ \mathcal{L}_+ & x > 0. \end{cases} \qquad \text{where} \quad \mathcal{L}_\pm := \lim_{x \to \pm\infty} \mathcal{L}.$$

Linear stability

By applying anti-Laplace transform and spectral information of the operator $\mathcal{L}$ (namely, no point spectrum is in the unstable half-plane $\{\lambda : \operatorname{Re}\lambda \geq 0\}$), we can derive pointwise estimates on the Green function G of the linearized operator $\mathcal{L}$.

The singular term gives to the Green function G the contribution

$$\delta_y(x) \exp\left(-g\,H(x)\,t/\epsilon\right)$$

that is uniformly exponentially decaying in time, thanks to the assumption $H(x) \geq c_0 > 0$ for some c_0.

For the regular part $\check{G}$, decay integral estimates can be obtained as in the viscous shock wave case, with the simplification that, since 0 is not an eigenvalue of $\mathcal{L}$ (SSS is an isolated steady state), no excited term is present. The prototype estimate is

$$\left\| \int_{\mathbb{R}} \check{G}(x, y, t)\, f(y)\, dy \right\|_{L^p} \le C(1+t)^{-\alpha_p} \|f\|_{L^p}$$

where $\alpha_p := (1 - 1/p)/2$. Analogous estimates hold for the derivatives of $\check{G}$.

Linearized stability is now evident: if W denote the solution of the linear Cauchy problem

$$\partial_t W = \mathcal{L} W, \qquad\qquad W(0) = W_0,$$

then it can be represented as

$$W(x, t) = \int_{\mathbb{R}} G(x, y; t)\, W_0(y)\, dy = \int_{\mathbb{R}} \check{G}(x, y; t)\, W_0(y)\, dy + W_0(x) e^{-g\, H(x)\, t/\epsilon},$$

showing the linearized decay of the perturbations with same rate of one-dimensional heat kernel.

Nonlinear stability

The nonlinear stability argument is based on Duhamel formula and on an appropriate iteration argument. Let us consider the conservative variables in the perturbation equation (3)

$$z = z(w) := F^0(W + w) - F^0(W).$$

The equation for z can be rewritten as

$$z_t = \mathcal{M} z + \partial_x \mathcal{R},$$

where

$$\mathcal{M} z := dF^0(W)^{-1} \left\{ \partial_x(B(W)\, \partial_x w) + \partial_x \left[dB(W)\, w\, W - dF^1(W)\, w \right] - Q\, w \right\}.$$

and $\mathcal{R} = O(|z|^2 + |\partial_x z|^2)$ is the nonlinear part.

If $G = G(x, y; t)$ is the Green function of the linear equation $z_t = \mathcal{M} z$, then the solution of the nonlinear problem with initial datum $z(x, 0) = z_0(x)$ satisfies

$$z(x, t) = \int_{\mathbb{R}} G(x, y; t)\, z_0(y)\, dy - \int_0^t \int_{\mathbb{R}} G_y(x, y; t)\, \mathcal{R}(y, s)\, dy\, ds.$$

Using the decomposition of the Green function previously described

$$z(x, t) = \int_{\mathbb{R}} \check{G}(x, y; t)\, z_0(y)\, dy - \int_0^t \int_{\mathbb{R}} \check{G}_y(x, y; t)\, \mathcal{R}(y, s)\, dy\, ds + e^{-g\, H(x)\, t/\epsilon} \Pi z_0(x),$$

where Π denotes the projection over the second variable (here we have used the property $\Pi \mathcal{R} = 0$). Setting

$$\zeta(t) := \sup\{|z(\cdot, s)|_{W^{1,p}}^p\, (1+s)^{\alpha_p}\ :\ s \in [0, t],\ p \in [1, +\infty]\},$$

it is possible to show that $\zeta(t) \le C(C_0 + \zeta^2(t))$ where C_0 is an appropriate constant depending on the initial data and C is a universal constant. From this relation the decay estimate follows.

Note that, with respect to the case of the asymptotic stability of shock waves, the situation is simplified by the fact that the SSS solution is isolated and hence there is no need to determine an appropriate nonlinear projection operator controlling the possible movement of the solution along the manifold of steady state.

References

[1] Gerbeau J.-F., Perthame B., *Derivation of viscous Saint-Venant system for laminar shallow water; numerical validation*, Discrete Contin. Dyn. Syst. Ser. B 1 (2001), no. 1, 89–102.

[2] Mascia C., Zumbrun K., *Pointwise Green function bounds for shock profiles of systems with real viscosity*, Arch. Ration. Mech. Anal. 169 (2003), no. 3, 177–263.

[3] Mascia C., Zumbrun K., *Stability of large-amplitude viscous shock profiles of hyperbolic-parabolic systems*, Arch. Ration. Mech. Anal. 172 (2004), no. 1, 93–131.

[4] de Saint-Venant A.J.C., *Théorie du mouvement non-permanent des eaux, avec application aux crues des rivière at à l'introduction des marées dans leur lit*, C. R. Acad. Sci. Paris 73 (1871), 147–154.

[5] Zumbrun K., Howard P., *Pointwise semigroup methods and stability of viscous shock waves* Indiana Univ. Math. J. 47 (1998), no. 3, 741–871.

Corrado Mascia
Dipartimento di Matematica "G. Castelnuovo"
University of Rome "La Sapienza"

Frederic Rousset
CNRS, University of Nice – Sophia Antipolis

Analysis and Simulation of Fluid Dynamics
Advances in Mathematical Fluid Mechanics, 163–181
© 2006 Birkhäuser Verlag Basel/Switzerland

Numerical Simulations of the Inviscid Primitive Equations in a Limited Domain

A. Rousseau, R. Temam and J. Tribbia

Abstract. This work is dedicated to the numerical computations of the primitive equations (PEs) of the ocean without viscosity with the nonlocal (mode by mode) boundary conditions introduced in [RTT05b]. We consider the 2D nonlinear PEs, and firstly compute the solutions in a "large" rectangular domain Ω_0 with periodic boundary conditions in the horizontal direction. Then we consider a subdomain Ω_1, in which we compute a second numerical solution with transparent boundary conditions. Two objectives are achieved. On the one hand the absence of blow-up in these computations indicates that the PEs without viscosity are well posed when supplemented with the boundary conditions introduced in [RTT05b]. On the other hand they show a very good coincidence on the subdomain Ω_1 of the two solutions, thus showing also the computational relevance of these new boundary conditions. We end this study with some numerical simulations of the linearized primitive equations, which correspond to the theoretical results established in [RTT05b], and evidence the transparent properties of the boundary conditions.

1. Introduction: Motivations and objectives

In this work, we intend to present our numerical simulations of the 2D primitive equations (PEs) without viscosity supplemented with the nonreflective and nonlocal boundary conditions introduced in [RTT05b]. To this aim, we first compute the PEs in a "large" (x, z) domain $\Omega_0 = (0, L) \times (-H, 0)$, with no-flux boundary conditions at top and bottom, and periodic boundary conditions in the horizontal direction (see Figure 1).

We then consider a sub-domain $\Omega_1 = (a, b) \times (-H, 0)$, where $0 < a < b < L$, so that $\Omega_1 \subset \Omega_0$. We perform numerical simulations of the PEs on Ω_1, with the no-flux boundary condition at top and bottom, but we use the transparent boundary conditions at $x = a$ and $x = b$ (the actual values are taken from the calculations in Ω_0). The initial condition is the same as on Ω_0 (restriction to Ω_1, see Figure 2).

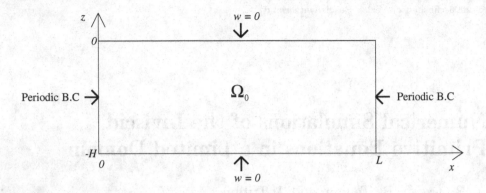

FIGURE 1. Domain Ω_0

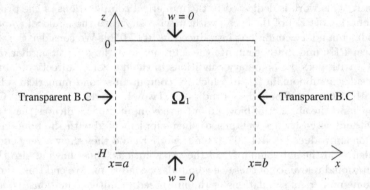

FIGURE 2. Subdomain Ω_1

Our objectives are twofold: firstly in view of extending the theoretical results of [RTT05b] to the *nonlinear* PEs, we test boundary conditions similar to those in [RTT05b]; secondly to show that the proposed boundary conditions are well suited for the problem of numerical simulations in a limited domain. This is done by observing that the solutions computed on Ω_1 only (with the nonreflecting boundary conditions) match well, on Ω_1, with the solutions computed on the whole domain Ω_0.

This article is organized as follows. In Section 2, we present the equations and the so-called normal mode expansion. Then we introduce in Section 3 the numerical schemes that we use in order to solve the primitive equations (2.1). The spectral method is employed for the vertical direction, whereas the x and t derivatives are discretized using finite differences. In Section 4, we present the numerical computations of the nonlinear PEs. The first results are dedicated to the periodic boundary conditions for Ω_0; then we implement the nonreflective boundary conditions introduced in [RTT05b] for Ω_1, and we end this section with a comparison between the two different numerical solutions.

2. The equations and the normal expansion

In this study we consider the nonlinear primitive equations without viscosity, independent of y, see, $e.g.$, [RTT05b]:

$$\frac{\partial u}{\partial t} + (\overline{U}_0 + u)\frac{\partial u}{\partial x} + w\frac{\partial u}{\partial z} - fv + \frac{\partial \phi}{\partial x} = F_u, \qquad (2.1a)$$

$$\frac{\partial v}{\partial t} + (\overline{U}_0 + u)\frac{\partial v}{\partial x} + w\frac{\partial v}{\partial z} + fu = F_v - f\overline{U}_0, \qquad (2.1b)$$

$$\frac{\partial \psi}{\partial t} + (\overline{U}_0 + u)\frac{\partial \psi}{\partial x} + (N^2 + \frac{\partial \psi}{\partial z})w = F_\psi, \qquad (2.1c)$$

$$\frac{\partial \phi}{\partial z} = -\frac{\rho}{\rho_0}g = \psi, \qquad (2.1d)$$

$$\frac{\partial u}{\partial x} + \frac{\partial w}{\partial z} = 0. \qquad (2.1e)$$

We consider the equations in the bidimensional domain $\Omega_0 = (0, L) \times (-H, 0)$, and supplement them with an initial data u_0, v_0, ψ_0. The boundary condition taken at $z = -H$ and $z = 0$ is classically $w = 0$, and we will consider in this study two different sets of boundary conditions in the horizontal direction: the periodic ones and the non-reflective ones.

2.1. The linearized primitive equations

We first start with the linearized version of (2.1) on which the study of (2.1) is based. The linearization of the PEs (2.1) around the steady flow $(\overline{U}_0, 0, 0)$ reads

$$\frac{\partial u}{\partial t} + \overline{U}_0\frac{\partial u}{\partial x} - fv + \frac{\partial \phi}{\partial x} = F_u, \qquad (2.2a)$$

$$\frac{\partial v}{\partial t} + \overline{U}_0\frac{\partial v}{\partial x} + fu = F_v - f\overline{U}_0, \qquad (2.2b)$$

$$\frac{\partial \psi}{\partial t} + \overline{U}_0\frac{\partial \psi}{\partial x} + N^2 w = F_\psi, \qquad (2.2c)$$

$$\frac{\partial \phi}{\partial z} = -\frac{\rho}{\rho_0}g = \psi, \qquad (2.2d)$$

$$\frac{\partial u}{\partial x} + \frac{\partial w}{\partial z} = 0, \qquad (2.2e)$$

where N is the so-called buoyancy frequency, assumed to be constant. In this study, we assume that the constant $\overline{U}_0$ is positive. Naturally, this hypothesis is not restrictive and the study could easily be extended to the case where $\overline{U}_0$ is negative.

Classically we proceed by separation of variables and actually look for the unknown functions (u, v, w, ϕ, ψ) under the form (see [RTT05b] for more details):

$$(u, v, \phi) = \mathcal{U}(z)(\hat{u}, \hat{v}, \hat{\phi})(x, t), \quad (w, \psi) = \mathcal{W}(z)(\hat{w}, \hat{\psi})(x, t), \qquad (2.3)$$

where the functions $\hat{u}, \hat{v}, \hat{w}$, and $\hat{\phi}$ only depend on x and t. Introducing the decomposition (2.3) into equations (2.2) shows that $\mathcal{W}$ (and then $\mathcal{U}$) solves a two-point eigenvalue problem with the boundary condition $\mathcal{W} = 0$ at top and bottom ($z = 0, -H$). We thus obtain the normal modes $\mathcal{W}_m(z)$ and $\mathcal{U}_m(z)$ such that

$$\mathcal{U}_m(z) = \sqrt{\frac{2}{H}} \cos(N \lambda_m z), \quad \mathcal{U}_0(z) = \frac{1}{\sqrt{H}}, \tag{2.4}$$

$$\mathcal{W}_m(z) = \sqrt{\frac{2}{H}} \sin(N \lambda_m z), \tag{2.5}$$

$$\lambda_m = \frac{m \pi}{N H}. \tag{2.6}$$

We then look for the general solution in the form a series

$$(u, v, \phi) = \sum_{m \geq 0} \mathcal{U}_m(z) \, (\hat{u}_m, \hat{v}_m, \hat{\phi}_m) \, (x, t), \tag{2.7}$$

$$(w, \psi) = \sum_{m \geq 1} \mathcal{W}_m(z) \, (\hat{w}_m, \hat{\psi}_m) \, (x, t). \tag{2.8}$$

We notice that $\forall \, m' \geq 0, m \geq 1$, we have the usual orthogonality properties:

$$\begin{cases} \displaystyle\int_{-L_3}^{0} \mathcal{U}_{m'}(z) \mathcal{U}_m(z) \, dz = \delta_{m',m}, \\ \displaystyle\int_{-L_3}^{0} \mathcal{U}_{m'}(z) \mathcal{W}_m(z) \, dz = 0, \\ \mathcal{U}_m'(z) = -N \lambda_m \mathcal{W}_m(z), \\ \mathcal{W}_m'(z) = N \lambda_m \mathcal{U}_m(z). \end{cases} \tag{2.9}$$

In the numerical simulations, we will truncate the series after M terms. Naturally, the larger M is, the more accurate the method is expected to be, but the heavier the computations are. Typically, $M = 10$ is satisfying from the physical point of view.

The case of the steady mode $m = 0$ is very simple, and is explained in [RTT05b]. From now on we only consider the modes $m \geq 1$.

Writing the linear PEs mode by mode, and writing $(u_m, v_m, w_m, \psi_m, \phi_m)$ instead of $(\hat{u}_m, \hat{v}_m, \hat{w}_m, \hat{\psi}_m, \hat{\phi}_m)$, we obtain the following system of integrodifferential equations ($1 \leq m \leq M$):

$$\frac{\partial u_m}{\partial t} + \overline{U}_0 \frac{\partial u_m}{\partial x} - f \, v_m + \frac{\partial \phi_m}{\partial x} = F_{u,m}, \tag{2.10a}$$

$$\frac{\partial v_m}{\partial t} + \overline{U}_0 \frac{\partial v_m}{\partial x} + f \, u_m = F_{v,m}, \tag{2.10b}$$

$$\frac{\partial \psi_m}{\partial t} + \overline{U}_0 \frac{\partial \psi_m}{\partial x} + N^2 \, w_m = F_{\psi,m}. \tag{2.10c}$$

$$\phi_m = -\frac{\psi_m}{N\,\lambda_m}, \tag{2.10d}$$

$$w_m = -\frac{1}{N\,\lambda_m}\frac{\partial u_m}{\partial x}; \tag{2.10e}$$

Taking equations (2.10d) and (2.10e) into account, equations (2.10a)–(2.10c) become:

$$\frac{\partial u_m}{\partial t} + \overline{U}_0\frac{\partial u_m}{\partial x} - f\,v_m - \frac{1}{N\,\lambda_m}\frac{\partial \psi_m}{\partial x} = F_{u,m}, \tag{2.11a}$$

$$\frac{\partial v_m}{\partial t} + \overline{U}_0\frac{\partial v_m}{\partial x} + f\,u_m = F_{v,m}, \tag{2.11b}$$

$$\frac{\partial \psi_m}{\partial t} + \overline{U}_0\frac{\partial \psi_m}{\partial x} - \frac{N}{\lambda_m}\frac{\partial u_m}{\partial x} = F_{\psi,m}. \tag{2.11c}$$

We have M systems of three coupled integro-differential equations (time-dependent with one space variable). We will discretize this system in Section 3 below.

For every $m \le M$, we introduce $\xi_m = u_m - \psi_m/N$, $\eta_m = u_m + \psi_m/N$, and let v_m unchanged. In terms of these variables, the system (2.10) becomes:

$$\frac{\partial \xi_m}{\partial t} + \left(\overline{U}_0 + \frac{1}{\lambda_m}\right)\frac{\partial \xi_m}{\partial x} - f\,v_m = F_{\xi,m}, \tag{2.12a}$$

$$\frac{\partial v_m}{\partial t} + \overline{U}_0\frac{\partial v_m}{\partial x} + f\,\frac{\xi_m + \eta_m}{2} = F_{v,m}, \tag{2.12b}$$

$$\frac{\partial \eta_m}{\partial t} + \left(\overline{U}_0 - \frac{1}{\lambda_m}\right)\frac{\partial \eta_m}{\partial x} - f\,v_m = F_{\eta,m}. \tag{2.12c}$$

The physical quantities can be obtained from ξ_m, η_m and v_m with:

$$u_m(x,t) = \frac{\xi_m + \eta_m}{2}(x,t), \tag{2.13a}$$

$$w_m(x,t) = -\frac{u_{m_x}}{N\,\lambda_m}(x,t), \tag{2.13b}$$

$$\psi_m(x,t) = \frac{N\,(\eta_m - \xi_m)}{2}(x,t), \tag{2.13c}$$

$$\phi_m(x,t) = -\frac{\psi_m}{N\,\lambda_m}(x,t). \tag{2.13d}$$

It is crucial to notice that $\overline{U}_0$, $\overline{U}_0 + 1/\lambda_m$ are always positive, whereas $\overline{U}_0 - 1/\lambda_m$ can either be positive or negative[1], depending on the value of $m \le M$; actually, the sign of these three characteristic values will determine the way we discretize the equations (2.12) in the horizontal direction.

---

[1] This is actually why the PEs in a bounded domain are ill-posed with any set of local boundary conditions, see [OS78, TT03, PR05, RTT05a, RTT05b].

Thanks to (2.6), there exists a critical value m_c such that $\overline{U}_0 - 1/\lambda_m$ is negative (resp. positive) if $m \leq m_c$ (resp. $m > m_c$). The corresponding modes are then called subcritical (resp. supercritical).

2.2. Boundary conditions for the linear case

We now recall the boundary conditions first introduced in [RTT05b]. The subcritical modes ($m \leq m_c$) and the supercritical ones ($m > m_c$) have to be handled separately. Namely, we prescribe for $m \leq m_c$:

$$\xi_m(0,t) = \xi_{m,g}^l(t), \tag{2.14a}$$

$$v_m(0,t) = v_{m,g}^l(t), \tag{2.14b}$$

$$\eta_m(L,t) = \eta_{m,g}^r(t), \tag{2.14c}$$

where the quantities $\xi_{m,g}^l$, $v_{m,g}^l$ and $\eta_{m,g}^r$ are some given functions that depend on time. The boundary conditions for the supercritical modes ($m > m_c$) are

$$\xi_m(0,t) = \xi_{m,g}^l(t), \tag{2.15a}$$

$$v_m(0,t) = v_{m,g}^l(t), \tag{2.15b}$$

$$\eta_m(0,t) = \eta_{m,g}^0(t), \tag{2.15c}$$

where $\xi_{m,g}^l$, $v_{m,g}^l$ and $\eta_{m,g}^l$ are also given. We notice that (2.14) differ from (2.15), since two characteristics (resp. three) enter the domain at $x = 0$ (resp. $x = L$) for subcritical modes (resp. the supercritical ones).

From the continuous viewpoint, the boundary conditions (2.14) can be written in the form of integral equations:

$$\int_{-L_3}^0 u(0,z)\,\mathcal{U}_m(z)\,dz - \frac{1}{N}\int_{-L_3}^0 \psi(0,z)\,\mathcal{W}_m(z)\,dz = 0, \tag{2.16a}$$

$$\int_{-L_3}^0 v(0,z)\,\mathcal{U}_m(z)\,dz = 0, \tag{2.16b}$$

$$\int_{-L_3}^0 u(L_1,z)\,\mathcal{U}_m(z)\,dz + \frac{1}{N}\int_{-L_3}^0 \psi(L_1,z)\,\mathcal{W}_m(z)\,dz = 0. \tag{2.16c}$$

The supercritical boundary conditions (2.15) can similarly be rewritten in the form of an infinite sequence of integral equations:

$$\int_{-L_3}^0 u(0,z)\,\mathcal{U}_m(z)\,dz = 0, \tag{2.17a}$$

$$\int_{-L_3}^0 v(0,z)\,\mathcal{U}_m(z)\,dz = 0, \tag{2.17b}$$

$$\int_{-L_3}^0 \psi(0,z)\,\mathcal{W}_m(z)\,dz = 0, \tag{2.17c}$$

We recall that the linearized PEs (2.2) supplemented with the boundary conditions (2.16)–(2.16) lead to a well-posed problem. See [RTT05b] for more details.

2.3. The modal form of the nonlinear primitive equations

We return to equations (2.1), and perform the same normal mode decomposition. We obtain the nonlinear form of equations (2.10a)–(2.10c), (2.10d) and (2.10e) being unchanged. For $1 \leq m \leq M$ we have:

$$\frac{\partial u_m}{\partial t} + \overline{U}_0 \frac{\partial u_m}{\partial x} - f\, v_m + \frac{\partial \phi_m}{\partial x} + B_{u,m}(U) = F_{u,m}, \qquad (2.18a)$$

$$\frac{\partial v_m}{\partial t} + \overline{U}_0 \frac{\partial v_m}{\partial x} + f\, u_m + B_{v,m}(U) = F_{v,m}, \qquad (2.18b)$$

$$\frac{\partial \psi_m}{\partial t} + \overline{U}_0 \frac{\partial \psi_m}{\partial x} + N^2\, w_m + B_{\psi,m}(U) = F_{\psi,m}, \qquad (2.18c)$$

where $B_{u,m}$, $B_{v,m}$ and $B_{\psi,m}$ are the following modal parts of the nonlinearities:

$$B_{u,m} = \int_{-H}^{0} (u\, \frac{\partial u}{\partial x} + w\, \frac{\partial u}{\partial z})\, \mathcal{U}_m\, dz, \qquad (2.19a)$$

$$B_{v,m} = \int_{-H}^{0} (u\, \frac{\partial v}{\partial x} + w\, \frac{\partial v}{\partial z})\, \mathcal{U}_m\, dz, \qquad (2.19b)$$

$$B_{\psi,m} = \int_{-H}^{0} (u\, \frac{\partial \psi}{\partial x} + w\, \frac{\partial \psi}{\partial z})\, \mathcal{W}_m\, dz. \qquad (2.19c)$$

with u, v, ψ, w truncated to M modes.

2.4. Boundary conditions for the nonlinear case

We make the same change of variables $\xi_m = u_m - \psi_m/N$, $\eta_m = u_m + \psi_m/N$, and obtain the nonlinear version of (2.12), namely:

$$\frac{\partial \xi_m}{\partial t} + (\overline{U}_0 + \frac{1}{\lambda_m}) \frac{\partial \xi_m}{\partial x} - f\, v_m + B_{\xi,m}(U) = F_{\xi,m}, \qquad (2.20a)$$

$$\frac{\partial v_m}{\partial t} + \overline{U}_0 \frac{\partial v_m}{\partial x} + f\, \frac{\xi_m + \eta_m}{2} + B_{v,m}(U) = F_{v,m}, \qquad (2.20b)$$

$$\frac{\partial \eta_m}{\partial t} + (\overline{U}_0 - \frac{1}{\lambda_m}) \frac{\partial \eta_m}{\partial x} - f\, v_m + B_{\eta,m}(U) = F_{\eta,m}, \qquad (2.20c)$$

where $B_{\xi,m} = B_{u,m} - B_{\psi,m}/N$ and $B_{\eta,m} = B_{u,m} + B_{\psi,m}/N$.

We assume in the following that the initial data is such that the nonlinear part is small compared to the stratified flow $(\overline{U}_0, 0, 0)$, so that the characteristic values do not change sign, at least during a certain period of time. Assuming so, we conjecture that the boundary conditions provided for the linearized system will give a well-posed problem for the nonlinear equations, at least for some time. We leave the theoretical analysis to subsequent studies, and perform here the corresponding numerical simulations based on this hypothesis, which is comforted by the lack of numerical blow-up. Hence the boundary conditions that we consider for the nonlinear case are (2.16)–(2.17).

3. Numerical scheme

3.1. Vertical discretization by spectral method

In the vertical direction, we proceed by normal modes decomposition as in (2.7), (2.8). From the numerical point of view, we will need to transform some grid-data into modal coefficients in the $\mathcal{U}_m$ or $\mathcal{W}_m$ bases of $L^2(-H,0)$, and *vice versa*.

Given a function f represented by its values f_l on a grid $z_l = -H + l\Delta z$, $0 \leq l \leq l_{\max}$, $\Delta z = H / l_{\max}$, we want to transform it into coefficients f_m, $0 \leq m \leq M$. To this aim we use the second order central point integration method, with the z_l as collocation points. For the functions u, v and ϕ, we decompose them in the $\mathcal{U}_m$ basis of $L^2(-H,0)$. For $0 \leq m \leq M$:

$$\{u_m, v_m, \phi_m\} = \Delta z \sum_{l=0}^{l_{\max}-1} \frac{\mathcal{U}_m(z_l) \cdot \{u,v,\phi\}(z_l) + \mathcal{U}_m(z_{l+1}) \cdot \{u,v,\phi\}(z_{l+1})}{2}, \tag{3.1}$$

and for w and ψ, $1 \leq m \leq M$:

$$\{w_m, \psi_m\} = \Delta z \sum_{l=0}^{l_{\max}-1} \frac{\mathcal{W}_m(z_l) \cdot \{w,\psi\}(z_l) + \mathcal{W}_m(z_{l+1}) \cdot \{w,\psi\}(z_{l+1})}{2}. \tag{3.2}$$

This approach which is that proposed by the physicists is different from the more mathematical approach to spectral and pseudo-spectral methods (as in, *e.g.*, [BM97, GH01]). The advantage of such a choice is that the orthogonality relations (2.9) are satisfied from the numerical point of view. Further studies and comparisons of the two approaches will be needed in the future.

On the contrary, if the function is given by its modal coefficients, the values on the z-grid z_l, $0 \leq l \leq l_{\max}$ is simply given by:

$$(u, v, \phi)(z_l) = \sum_{m=0}^{M} (u_m, v_m, \phi_m) \mathcal{U}_m(z_l), \tag{3.3}$$

$$(w, \psi)(z_l) = \sum_{m=0}^{M} (w_m, \psi_m) \mathcal{W}_m(z_l). \tag{3.4}$$

In the numerical simulations, we are given some initial data on the physical grid $(z_l)_{0 \leq l \leq l_{\max}}$. We transform them into modal coefficients thanks to formulas (3.1) or (3.2), and if the problem is linear, we keep them all along the computations, except for graphic purposes, for which we use inverse formulas (3.3)–(3.4) to return to physical space. Naturally, in the nonlinear case, we need to operate (3.1)–(3.4) once at every time step, in order to avoid the computation of a convolution product, that would cost too much in term of CPU time and is not considered an appropriate numerical procedure. We compute the nonlinear terms of the equations in the physical space (x, z) thanks to Fourier and inverse Fourier transforms.

3.2. Finite differences in time and space (horizontal direction)

Looking at the form of (2.20), we choose to discretize these equations in the horizontal direction with the finite differences method. Naturally, care has to be taken to the sign of the characteristic values, in order to take an upwind (hence stable) spatial discretization of the x-derivative. Whereas $\overline{U}_0$ and $\overline{U}_0 + 1/\lambda_m$ are always positive, the third characteristic value of the mth mode – in the linear case – is $\overline{U}_0 - 1/\lambda_m$. and can either be positive or negative for the actual physical values that we consider.

In the nonlinear case, since the initial data is small compared to $\overline{U}_0$ ([TT03]), we implicitly assume that m_c remains unchanged for a certain period of time. Until a full nonlinear theory is performed, a first step in the verification of this hypothesis would be to linearize Equation (2.18) (or (2.11)–(2.20)) around the current state which may amount to replacing $\overline{U}_0$ by $\overline{U}_0 + u$, but may also involve a more complex analysis already in the linearized case. These involved issues are investigated in a work in progress.

That is, for every subcritical mode $m \leq m_c$, we discretize (2.20) as follows:

$$\frac{\xi_{m,j}^{n+1} - \xi_{m,j}^n}{\Delta t^n} + (\overline{U}_0 + \frac{1}{\lambda_m}) \frac{\xi_{m,j}^n - \xi_{m,j-1}^n}{\Delta x} - f\, v_{m,j}^n = F_{\xi,m,j}^n - B_{\xi,m,j}^n, \quad (3.5a)$$

$$\frac{v_{m,j}^{n+1} - v_{m,j}^n}{\Delta t^n} + \overline{U}_0 \frac{v_{m,j}^n - v_{m,j-1}^n}{\Delta x} + f\, \frac{\xi_{m,j}^n + \eta_{m,j}^n}{2} = F_{v,m,j}^n - B_{v,m,j}^n, \quad (3.5b)$$

$$\frac{\eta_{m,j}^{n+1} - \eta_{m,j}^n}{\Delta t^n} + (\overline{U}_0 - \frac{1}{\lambda_m}) \frac{\eta_{m,j+1}^n - \eta_{m,j}^n}{\Delta x} - f\, v_{m,j}^n = F_{\eta,m,j}^n - B_{\eta,m,j}^n. \quad (3.5c)$$

where the right-hand side of (3.5) contains the nonlinear terms, computed explicitly thanks to an Adams-Bashforth scheme.

Equations (3.5a) and (3.5b) hold for $1 \leq j \leq J$, whereas (3.5c) is written for $0 \leq j \leq J - 1$. There are no equations for $\xi_{m,0}^{n+1}$, $v_{m,0}^{n+1}$ and $\eta_{m,J}^{n+1}$, these quantities being given by the boundary conditions as required in [RTT05b]. On the contrary, if $m > m_c$ (supercritical case), we choose for $1 \leq j \leq J$:

$$\frac{\xi_{m,j}^{n+1} - \xi_{m,j}^n}{\Delta t^n} + (\overline{U}_0 + \frac{1}{\lambda_m}) \frac{\xi_{m,j}^n - \xi_{m,j-1}^n}{\Delta x} - f\, v_{m,j}^n = F_{\xi,m,j}^n - B_{\xi,m,j}^n, \quad (3.6a)$$

$$\frac{v_{m,j}^{n+1} - v_{m,j}^n}{\Delta t^n} + \overline{U}_0 \frac{v_{m,j}^n - v_{m,j-1}^n}{\Delta x} + f\, \frac{\xi_{m,j}^n + \eta_{m,j}^n}{2} = F_{v,m,j}^n - B_{v,m,j}^n, \quad (3.6b)$$

$$\frac{\eta_{m,j}^{n+1} - \eta_{m,j}^n}{\Delta t^n} + (\overline{U}_0 - \frac{1}{\lambda_m}) \frac{\eta_{m,j}^n - \eta_{m,j-1}^n}{\Delta x} - f\, v_{m,j}^n = F_{\eta,m,j}^n - B_{\eta,m,j}^n. \quad (3.6c)$$

Either $\xi_{m,0}^{n+1}$, $v_{m,0}^{n+1}$ and $\eta_{m,0}^{n+1}$ are given by the boundary conditions defined as in [RTT05b] (transparent boundary conditions case), or they satisfy the periodicity conditions (4.2) below (periodical case).

For every function $f(x, z, t)$, $f_{m,j}^n$ represents $f_m(x_j, t_n)$ for $0 \leq j \leq J$, $0 \leq n \leq n_{\max}$, with

$$0 = x_0 < x_1 < \cdots < x_j < \cdots < x_J = L, \tag{3.7}$$

$$0 = t_0 < t_1 < \cdots < t_n < \cdots < t_{n_{\max}} = T, \tag{3.8}$$

$$\Delta x = x_{j+1} - x_j = \frac{L}{J}, \tag{3.9}$$

$$\Delta t^n = t_{n+1} - t_n. \tag{3.10}$$

In the numerical experiments, we choose an homogeneous space discretization ($\Delta x = \text{const} = L/J$). For the sake of simplicity, we choose an explicit time-scheme, with a constant time-step Δt, which will be restricted by the well-known CFL condition to guarantee stability in the linear case:

$$\Delta t \leq \frac{\Delta x}{\max_{1 \leq m \leq M}(\overline{U}_0, \overline{U}_0 + \frac{1}{\lambda_m}, |\overline{U}_0 - \frac{1}{\lambda_m}|)} = \frac{\Delta x}{\overline{U}_0 + \frac{1}{\lambda_1}}. \tag{3.11}$$

Naturally, when the equations are nonlinear, the CFL condition is not enough to guarantee stability. Actually, the characteristic values depend on time since $\overline{U}_0$ has to be replaced by $u + \overline{U}_0$, but we assume that the initial data is such that $|u_0| \ll \overline{U}_0$, which is physically relevant, [TT03]. We actually base our computations on those of the quoted article [TT03]: in this article the initial data is such that the ratio between the perturbation and the reference flow $\overline{U}_0 \mathbf{e_x}$ is less than 10%, which is physically relevant. In the case of numerical simulations with periodic boundary conditions, we multiply the initial data of [TT03] by $\sin(\pi x/L)$ to make it periodic and avoid any boundary layer at $t = 0$.

4. Numerical simulations

We present hereafter two different sets of numerical results. In Section 4.1 we re-solve the nonlinear PEs in a domain $\Omega_0 = (0, L) \times (-H, 0)$ with periodic boundary conditions in the horizontal (x) direction, and $w = 0$ at $z = -H, 0$. These numerical results will provide the boundary conditions needed for the computations of Section 4.2 below, and we evidence in Section 4.3 the transparent properties of the boundary conditions introduced, thanks to a comparison between the solutions computed in Section 4.1 and those of Section 4.2.

The computations are done as follows. We fix M (the number of modes) and compute $(u_m^0, v_m^0, \psi_m^0)_{0 < m \leq M}$ from the given data u^0, v^0, ψ^0 thanks to (3.1)–(3.2).

Then, for every mode $m \leq M$, we consider the modal equations (2.20) and their discretization (3.5)–(3.6), and supplement them with the appropriate boundary conditions, either (4.3) for the periodical case or (4.7)–(4.8) for the case of transparent boundary conditions. We recall here that for every m, $(\xi_m, \eta_m) = (u_m - \psi_m/\lambda_m, u_m + \psi_m/\lambda_m)$ will be the numerical unknowns to be computed, so that

$(u_m, w_m, \psi_m, \phi_m)$ can be obtained with

$$u_m(x,t) = \frac{\xi_m + \eta_m}{2}(x,t), \tag{4.1a}$$

$$w_m(x,t) = -\frac{u_{m_x}}{N\lambda_m}(x,t), \tag{4.1b}$$

$$\psi_m(x,t) = \frac{N(\eta_m - \xi_m)}{2}(x,t), \tag{4.1c}$$

$$\phi_m(x,t) = -\frac{\psi_m}{N\lambda_m}(x,t). \tag{4.1d}$$

As a consequence, we will only consider the quantities (ξ_m, v_m, η_m) in the sequel, the other physical quantities being easily computed thanks to (4.1).

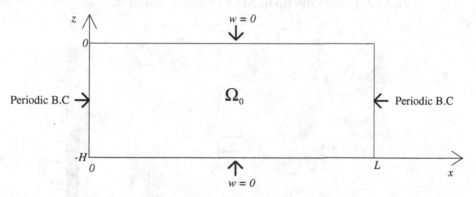

FIGURE 3. Domain Ω_0

4.1. Periodic boundary conditions for the large domain Ω_0

In the periodical case, we consider the following modal boundary conditions:

$$\xi_m(0,t) = \xi_m(L,t), \tag{4.2a}$$

$$v_m(0,t) = v_m(L,t), \tag{4.2b}$$

$$\eta_m(0,t) = \eta_m(L,t). \tag{4.2c}$$

For each time step $\Delta t^n = \Delta t$ satisfying (3.11) we compute the unknown functions $(\xi_m^{n+1}, v_m^{n+1}, \eta_m^{n+1})$ thanks to (3.5) and (3.6), with the numerical boundary conditions:

$$\xi_{m,0}^{n+1} = \xi_{m,J}^{n+1}, \tag{4.3a}$$

$$v_{m,0}^{n+1} = v_{m,J}^{n+1}, \tag{4.3b}$$

$$\eta_{m,0}^{n+1} = \eta_{m,J}^{n+1}. \tag{4.3c}$$

The following figures plot u, v and ψ in the domain Ω_0 at two different times. Figures 4, 5 and 6 represent the initial data $(t = 0)$ for these three quantities, whereas Figures 7, 8 and 9 represent u, v and ψ at $t = t_1 > 0$.

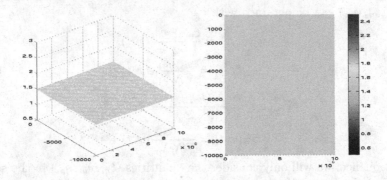

FIGURE 4. Periodic Boundary Condition. Initial data u_0.

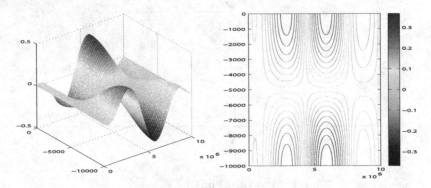

FIGURE 5. Periodic Boundary Condition. Initial data v_0.

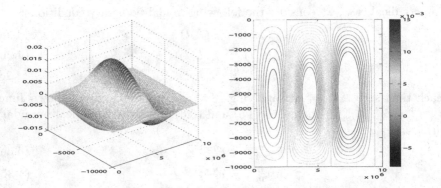

FIGURE 6. Periodic Boundary Condition. Initial data ψ_0.

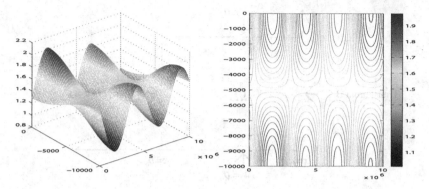

FIGURE 7. Periodic Boundary Condition. Values of u at $t = t_1$.

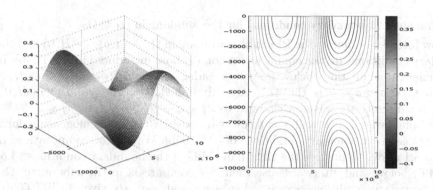

FIGURE 8. Periodic Boundary Condition. Values of v at $t = t_1$.

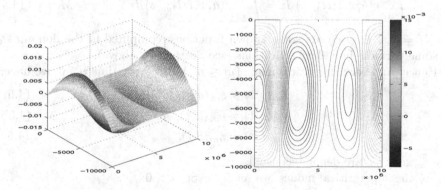

FIGURE 9. Periodic Boundary Condition. Values of ψ at $t = t_1$.

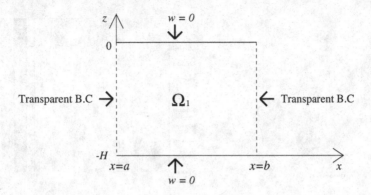

FIGURE 10. Subdomain Ω_1

4.2. Transparent boundary conditions for the subdomain $\Omega_1 \subset \Omega_0$

We now intend to simulate the PEs in the subdomain $\Omega_1 = (a, b) \times (-H, 0)$ and the boundary conditions are the nonlinear version of those introduced in [RTT05b]. In the numerical simulations below we will consider a domain $\Omega_1 = (a, b) \times (-H, 0)$ such that $0 < a < b < L$, so that Ω_1 is actually imbedded in $\Omega_0 = (0, L) \times (-H, 0)$. The space discretization is now changed to $x_j = a + j(b - a)/J$, $0 \leq j \leq J$.

At the boundaries $x = a$ and $x = b$, we will consider the nonhomogeneous form of the transparent boundary conditions of [RTT05b]. We use the computations of Section 3.1 above to provide the right-hand side of the boundary conditions (4.5) and (4.6) below, and afterwards use them for comparison in the subdomain Ω_1. These boundary conditions, expressed in a general way, are given in [RTT05b]. They consist in *an infinite set of integral boundary conditions*. For example:

$$\int_{-H}^{0} v(a, z, t)\, \mathcal{U}_m(z)\, dz = \int_{-L_3}^{0} \tilde{v}(a, z, t)\, \mathcal{U}_m(z)\, dz, \quad \forall m \leq M, \qquad (4.4)$$

where $\tilde{U} = (\tilde{u}, \tilde{v}, \tilde{w}, \tilde{\psi}, \tilde{\phi})$ are some known functions, computed in the domain Ω_0 with some periodic boundary conditions (see Section 3.1 above).

Hence, for every subcritical mode ($m \leq m_c$) and every time $t > 0$, we have:

$$\xi_m(a, t) = \tilde{\xi}_m(a, t), \qquad (4.5a)$$

$$v_m(a, t) = \tilde{v}_m(a, t), \qquad (4.5b)$$

$$\eta_m(b, t) = \tilde{\eta}_m(b, t), \qquad (4.5c)$$

where $\tilde{\xi}_m$ and $\tilde{\eta}_m$ are defined as usual.

For the supercritical modes, we set for every $t > 0$:

$$\xi_m(a, t) = \tilde{\xi}_m(a, t), \qquad (4.6a)$$

$$v_m(a, t) = \tilde{v}_m(a, t), \qquad (4.6b)$$

$$\eta_m(a, t) = \tilde{\eta}_m(a, t). \qquad (4.6c)$$

To implement these boundary conditions, we discretize equations (2.20) with the finite differences method, taking into account the sign of $\overline{U}_0 - 1/\lambda_m$ for the discretization of the first x-derivative of η_m in equation (2.20c) (see equations (3.5) and (3.6) of Section 3 above).

For each time step $\Delta t^n = \Delta t$ satisfying (3.11) we compute the unknown functions $(\xi_m^{n+1}, v_m^{n+1}, \eta_m^{n+1})$ thanks to (3.5) and (3.6), with the numerical boundary conditions:

$$\xi_{m,0}^{n+1} = \tilde{\xi}_m(a, t_{n+1}), \tag{4.7a}$$

$$v_{m,0}^{n+1} = \tilde{v}_m(a, t_{n+1}), \tag{4.7b}$$

$$\eta_{m,J}^{n+1} = \tilde{\eta}_m(b, t_{n+1}), \tag{4.7c}$$

if m is subcritical ($m \leq m_c$). If m is supercritical ($m > m_c$), we set

$$\xi_{m,0}^{n+1} = \tilde{\xi}_m(a, t_{n+1}), \tag{4.8a}$$

$$v_{m,0}^{n+1} = \tilde{v}_m(a, t_{n+1}), \tag{4.8b}$$

$$\eta_{m,0}^{n+1} = \tilde{\eta}_m(a, t_{n+1}). \tag{4.8c}$$

The following figures plot u, v and ψ in the domain Ω_1 at two different times. Figures 11, 12 and 13 represent the initial data ($t = 0$) for these three quantities, whereas Figures 14, 15 and 16 represent u, v and ψ at $t = t_1 > 0$.

Here, one can see that Figures 14, 15 and 16 respectively match with Figures 7, 8 and 9 in the domain Ω_1.

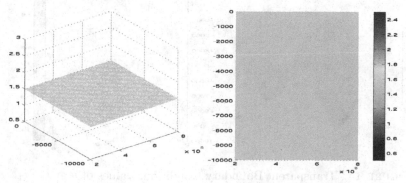

FIGURE 11. Transparent Boundary Condition. Initial data u_0.

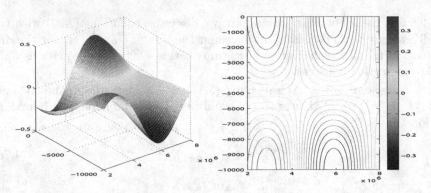

FIGURE 12. Transparent Boundary Condition. Initial data v_0.

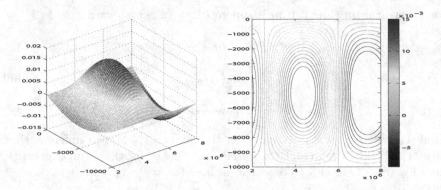

FIGURE 13. Transparent Boundary Condition. Initial data ψ_0.

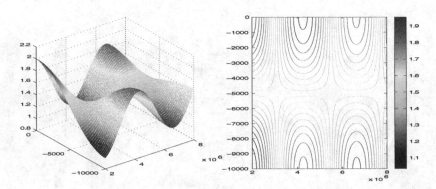

FIGURE 14. Transparent Boundary Condition. Values of u at $t = t_1$.

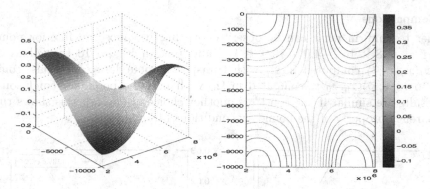

FIGURE 15. Transparent Boundary Condition. Values of v at $t = t_1$.

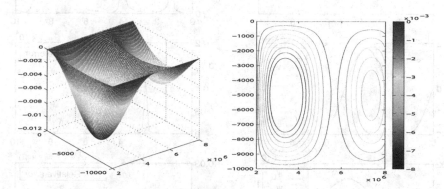

FIGURE 16. Transparent Boundary Condition. Values of ψ at $t = t_1$.

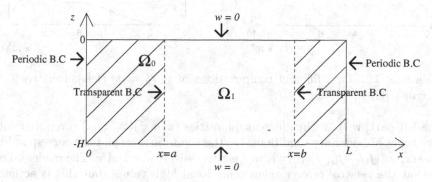

FIGURE 17. Subdomains Ω_0 and Ω_1

4.3. Comparisons

In order to confirm what can be observed, we finally choose an interior point
$(x_0, z_0) = (5.8 \times 10^6, -4.0 \times 10^3) \in \Omega_1$, and plot in Figure 18 the values of
$(u, v, \psi)(x_0, z_0, t)$ computed in Ω_1 with transparent boundary conditions, compared to $(u, v, \psi)(x_0, z_0, t)$ computed in Ω_0 with periodic boundary conditions.
The results are similar if one considers another choice of (x_0, z_0); this shows the
transparency property of the boundary conditions (4.5)–(4.6).

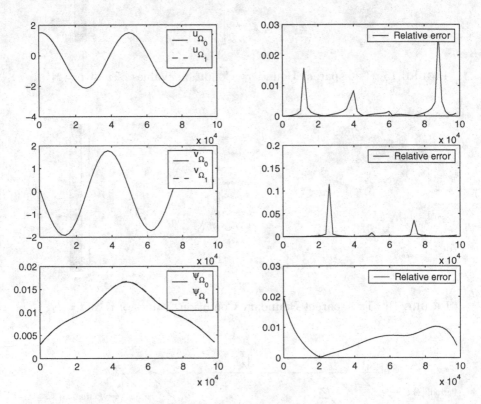

FIGURE 18. Two different computations of $\psi(x_0, z_0, t)$ (left). Relative
error (right).

On the left part, we plot the different quantities $(u, v, \psi)(x_0, z_0, t)$ computed with
the two types of boundary conditions. On the right part, we plot the corresponding
relative errors $|f_{\Omega_0} - f_{\Omega_1}|/|f_{\Omega_0}|$ where f is successively u, v and ψ. The reader might
think that the relative error reaches some local high values, but this is actually
due to the fact that the quantity u_{Ω_0} (or v_{Ω_0}, ψ_{Ω_0}) vanishes; these local maximum
are not meaningful.

5. Conclusion

In this article, the primitive equations of the ocean are considered. After recalling the study of the linearized version of these equations, we present some numerical results on the nonlinear system of equations. The boundary conditions that are implemented are those of the linear problem, which is physically relevant, at least for some time. We achieve here two goals: firstly we started here the extension of the theoretical results of [RTT05b] to the nonlinear primitive equations. Secondly the proposed boundary conditions are well suited for the problem of numerical simulations in a limited domain, as shown by the very good matching of the two different numerical solutions.

References

[BM97]   C. Bernardi and Y. Maday. Spectral methods. In *Handbook of numerical analysis, Vol. V*, Handb. Numer. Anal., V, pages 209–485. North-Holland, Amsterdam, 1997.

[GH01]   D. Gottlieb and J.S. Hesthaven. Spectral methods for hyperbolic problems. *J. Comput. Appl. Math.*, 128(1-2):83–131, 2001. Numerical analysis 2000, Vol. VII, Partial differential equations.

[OS78]   J. Oliger and A. Sundström. Theoretical and practical aspects of some initial boundary value problems in fluid dynamics. *SIAM J. Appl. Math.*, 35(3):419–446, 1978.

[PR05]   M. Petcu and A. Rousseau. On the δ-primitive and Boussinesq type equations. Advances in Differential Equations, to appear, 2005.

[RTT05a] A. Rousseau, R. Temam, and J. Tribbia. Boundary conditions for an ocean related system with a small parameter. In *Nonlinear PDEs and Related Analysis*, volume 371, pages 231–263. Gui-Qiang Chen, George Gasper and Joseph J. Jerome Eds., Contemporary Mathematics, AMS, Providence, 2005.

[RTT05b] A. Rousseau, R. Temam, and J. Tribbia. Boundary conditions for the 2D linearized PEs of the ocean in the absence of viscosity. Discrete and Continuous Dynamical Systems, to appear, 2005.

[TT03]   R. Temam and J. Tribbia. Open boundary conditions for the primitive and Boussinesq equations. *J. Atmospheric Sci.*, 60(21):2647–2660, 2003.

A. Rousseau
Laboratoire d'Analyse Numérique, Université Paris–Sud
Orsay, France

R. Temam
The Institute for Scientific Computing and Applied Mathematics
Indiana University, Bloomington, IN, USA

J. Tribbia
National Center for Atmospheric Research
Boulder, Colorado, USA

Analysis and Simulation of Fluid Dynamics
Advances in Mathematical Fluid Mechanics, 183–199
© 2006 Birkhäuser Verlag Basel/Switzerland

Some Recent Results about the Sixth Problem of Hilbert

Laure Saint-Raymond

Abstract. The sixth problem proposed by Hilbert, in the occasion of the International Congress of Mathematicians held in Paris in 1900, asks for a global understanding of the gas dynamics. For a perfect gas, the kinetic equation of Boltzmann provides a suitable model of evolution for the statistical distribution of particles. Hydrodynamic models are obtained as first approximations when collisions are frequent. In incompressible regime, rigorous convergence results are now established by describing precisely the corrections to the hydrodynamic approximation, namely physical phenomena such as relaxation or oscillations on small spatio-temporal scales, and checking that they do not disturb the mean motion.

Quelques résultats récents sur le sixième problème de Hilbert: Résumé

Le sixième problème pos(e par Hilbert au Congrès International des Mathématiciens en 1900 appelait une compréhension globale de la dynamique des gaz. Pour un gaz parfait, l'équation cinétique de Boltzmann fournit un bon modèle d'évolution de la distribution statistique des particules. Les modèles hydrodynamiques sont obtenus en première approximation quand les collisions sont très nombreuses. En régime incompressible, on sait maintenant démontrer des résultats de convergence rigoureux, en décrivant précisément les corrections à l'approximation hydrodynamique, en particulier les phénomènes de type relaxation ou oscillations sur des peties échelles spatio-temporelles, et en vérifiant qu'ils ne modifient pas le mouvement moyen.

1. The problem of fluid limits

The problem of fluid limits has been asked by David Hilbert in the occasion of the International Congress of Mathematicians, held in Paris in 1900:

"Le livre de Monsieur Boltzmann nous incite à établir [...] de manière complète et rigoureuse ls méthodes basées sur l'idée de passage à la limite, et qui, de la conception atomique, nous conduisent aux lois du mouvement des continua."

The heart of the question is then to get a unified theory of the gas dynamics, including the various levels of description.

1.1. Modelling perfect gases

From a macroscopic point of view, meaning on sufficiently large scales, a gas can be considered as a continuous medium. Thus it can be modelled by the classical fluid mechanics.

The state of the gas is therefore described by a small number of variables such as the pressure, temperature, bulk velocity... And its evolution is governed by a system of partial differential equations, constituted of the local conservation laws as well as some state relation expressing the inherent features of the gas. Such models are obtained by a phenomenological approach which proves itself for more than three centuries: it goes back to Euler for inviscid gases, and to Navier and Stokes for viscous gases.

From a microscopic point of view, the gas is however constituted of a large number of molecules, which are themselves constituted of a large number of atoms and electrons... Modelling the gas dynamics seems then to be a very complicated job based on the solid mechanics.

In such an approach the state of the gas is indeed described by the positions and the velocities of all elementary particles. And its evolution is governed by a complex system of ordinary differential equations, coming from Newton's principle of dynamics and depending on the nature of the microscopic interactions. Of course, at the macroscopic scale, the individual motion of any elementary particle is not observable: only the statistical effect of the microscopic interactions is interesting.

In the particular case of perfect gases, this statistical effect can be modelled directly, so that an intermediate level of description can be introduced. Note that the validity of such statistical models, the so-called kinetic models, requires a strong assumption of rarefaction satisfied only by perfect gases. In all the sequel, we will restrict our attention to these gases (meaning in particular that the results exhibited here bring a very partial answer to Hilbert's question).

In kinetic theory, the state of the gas is described through the distributions of positions and velocities of all species of elementary particles. Assume for the sake of simplicity that there is only one species: the evolution of the gas is then governed by a partial differential equation, taking into account both the transport of particles and the statistical effect of their microscopic interactions. The hypothesis of rarefaction is necessary in order that the probability that two particles interact depends only on the distribution of positions and velocities: indeed the statistical independence of the particles is proved to propagate only under such an assumption.

The problem of Hilbert in our particular framework consists then in understanding how these three types of models can be linked together, or in other words in giving some sense to the scheme shown in Figure 1.

Air considered as a fluid
described by its pressure, temperature
and velocity $p(t,x)$, $T(t,x)$, $v(t,x)$.

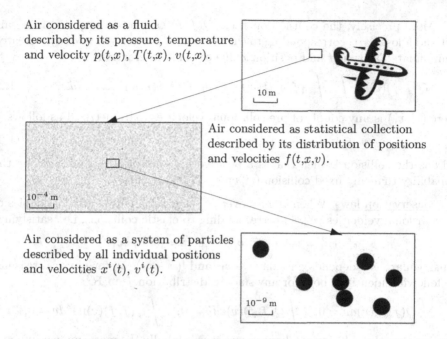

10 m

Air considered as statistical collection
described by its distribution of positions
and velocities $f(t,x,v)$.

10^{-4} m

Air considered as a system of particles
described by all individual positions
and velocities $x^i(t)$, $v^i(t)$.

10^{-9} m

FIGURE 1. Levels of description for a perfect gas

1.2. The Boltzmann equation

Our starting point here is the Boltzmann equation, which is the most commonly used kinetic equation for monoatomic perfect gases (constituted of neutral particles).

Let us just recall that it is formally obtained from the microscopic system of ordinary differential equations, using the associated Liouville equation and the BBGKY hierarchy in the thermodynamic limit, that is when the number of elementary particles tends to infinity whereas their individual size tends to zero.

The mathematical justification of such an asymptotics is however extremely complicated, and it is completely achieved only for very small times (about half of the occurrence time of the first collision) and for smooth distribution functions.

The mathematical formulation. In the Boltzmann model, the state of the gas is described by its number density $f \equiv f(t, x, v)$, where $f(t, x, v)dtdxdv$ is the number of elementary particles with position x and velocity v at time t.

The evolution of the density f is then given by a partial differential equation, taking into account both the free transport, and the statistical effect of collisions:

$$\partial_t f + v \cdot \nabla_x f = Q(f, f), \qquad (1.1)$$

where the collision operator Q is quadratic and acts only on the v variable, meaning that the microscopic interactions are assumed to be binary and localized in time and space (which is another formulation of the perfect gas assumption).

More precisely, the collision operator $Q(f, f)$ can be decomposed in a gain part and a loss part, corresponding to the account of particles of velocity v created or annihilated at time t and position x due to some interaction:

$$Q(f, f)(v) = \int_{\mathbf{R}^3} \int_{S^2} \{f(v_1')f(v') - f(v_1)f(v)\}b(v - v_1, \omega)dv_1 d\omega, \qquad (1.2)$$

where (v', v_1') is any couple of pre-collisional velocities, parametrized as follows

$$v' = v - (v - v_1) \cdot \omega\omega, \quad v_1' = v_1 + (v - v_1) \cdot \omega\omega, \qquad (1.3)$$

and b is the collision kernel, that is a positive function of v, v_1 and ω giving the probability that any fixed collision $(v', v_1') \mapsto (v, v_1)$ occurs.

The conservation laws. When ω goes over S^2, one obtains by (1.3) all couples of pre-collisional velocities (v', v_1') corresponding to elastic collisions, i.e., satisfying

$$v' + v_1' = v + v_1, \quad |v'|^2 + |v_1'|^2 = |v|^2 + |v_1|^2.$$

In particular, as b depends only on $|v - v_1|$ and $|(v - v_1) \cdot \omega|$ for physical reasons, the following identities hold for any smooth distribution f on $\mathbf{R}^3$:

$$\int Q(f, f)(v)dv = 0, \quad \int Q(f, f)(v)v dv = 0, \quad \int Q(f, f)(v)|v|^2 dv = 0,$$

meaning that the collision mechanism preserves globally the mass, momentum and kinetic energy.

As a consequence, the thermodynamic fields associated with any solution of the Boltzmann equation satisfy formally the conservation laws

$$\partial_t \int f(t, x, v)dv + \nabla_x \cdot \int f(t, x, v)v dv = 0$$

$$\partial_t \int f(t, x, v)v dv + \nabla_x \cdot \int f(t, x, v)v \otimes v dv = 0 \qquad (1.4)$$

$$\partial_t \int f(t, x, v)|v|^2 dv + \nabla_x \cdot \int f(t, x, v)v|v|^2 dv = 0$$

The statistical effect of microscopic interactions. Using the same symmetry argument, it is easy to check that for any smooth distribution f on $\mathbf{R}^3$:

$$\int Q(f, f) \log f(v)dv$$

$$= \frac{1}{4} \iiint \{f(v_1')f(v') - f(v_1)f(v)\} \log \left(\frac{f(v)f(v_1)}{f(v')f(v_1')} \right) b(v - v_1, \omega)dv_1 d\omega dv \leq 0$$

with equality if and only if

$$\text{for all } (v, v_1, \omega) \in \mathbf{R}^3 \times \mathbf{R}^3 \times S^2, \quad f(v)f(v_1) = f(v')f(v_1'),$$

or in other words if and only if f is a Gaussian distribution. This means that statistically the collisions induce a relaxation mechanism which brings the gas on local thermodynamic equilibrium as predicted by Maxwell.

As a consequence, the entropy associated with any solution of the Boltzmann equation formally decreases

$$\partial_t \int f \log f(t,x,v)dv + \nabla_x \cdot \int f \log f(t,x,v)vdv \le 0 \qquad (1.5)$$

Nevertheless properties (1.4) and (1.5) are not known to be satisfied by the solutions built by DiPerna and Lions [3], which are the only solutions of the Boltzmann equation known to exist globally in time without any restriction of size or smoothness on the initial data.

In this framework the assumptions on the initial data are indeed only physical estimates, namely controls on the energy and entropy of the gas. In the case when the density is a fluctuation around a global thermodynamic equilibrium different from vacuum, say for instance around

$$M(v) = \frac{1}{(2\pi)^{3/2}} \exp\left(-\frac{|v|^2}{2}\right),$$

the convenient quantity to be controlled is the relative entropy defined by

$$H(f/M) = \iint \left(f \log \frac{f}{M} - f + M\right) dvdx. \qquad (1.6)$$

Note that the integrand is always nonnegative.

Theorem 1 (DiPerna-Lions). *Let f^0 be a nonnegative function of $L^1_{loc}(\mathbf{R}^3 \times \mathbf{R}^3)$ such that*

$$H(f^0/M) < +\infty \qquad (1.7)$$

Then there exists a nonnegative function $f \in C(\mathbf{R}^+, L^1_{loc}(\mathbf{R}^3 \times \mathbf{R}^3))$ such that $f_{|t=0} = f^0$ satisfying

- *the renormalized Boltzmann equation*

$$\partial_t \Gamma(f) + v \cdot \nabla_x \Gamma(f) = \Gamma'(f)Q(f,f) \qquad (1.8)$$

   for all $\Gamma \in C^1(\mathbf{R}^+)$ such that $\Gamma(0) = 0$ and $\sup_{z\in\mathbf{R}^+}(1+z)|\Gamma'(z)| < +\infty$;

- *the local conservation of mass*

$$\partial_t \int f(t,x,v)dv + \nabla_x \cdot \int f(t,x,v)vdv = 0; \qquad (1.9)$$

- *the global entropy inequality*

$$H(f(t)/M) + \int_0^t D(f(s))ds \le H(f^0/M), \qquad (1.10)$$

   with

$$D(f) = \frac{1}{4} \iiint \{f(v_1)f(v) - f(v_1')f(v')\} \log\left(\frac{f(v)f(v_1)}{f(v')f(v_1')}\right) b(v-v_1,\omega)dv_1 d\omega dv.$$

1.3. The various hydrodynamic regimes

Fluid models are then obtained in the fast relaxation limit, that is when the elementary particles undergo many collisions on the observation time scale. Indeed the collision process is therefore expected to constrain the gas to be locally at thermodynamic equilibrium, meaning that its distribution is completely defined by macroscopic parameters, namely the density $R(t,x)$, bulk velocity $U(t,x)$ and temperature $T(t,x)$:

$$f(t,x,v) \sim \frac{R(t,x)}{(2\pi T(t,x))^{3/2}} \exp\left(-\frac{|v - U(t,x)|^2}{2T(t,x)}\right).$$

In order to measure the sharpness of such an approximation, one introduces a dimensionless number, the so-called Knudsen number Kn, which is defined as the ratio between the mean free-path, that is the typical length a particle goes over between two collisions, and the typical observation length. The hydrodynamic limit is then the macroscopic model which governs asymptotically the evolution of the density, bulk velocity and temperature in the limit when the Knudsen number vanishes.

This macroscopic model depends on another feature of the gas, namely its compressibility. In order to measure the ability of the gas to reflect back pressure variations, one introduces another dimensionless number, the so-called Mach number Ma, which is defined as the ratio between the bulk velocity and the sound speed, that is the typical velocity of the elementary particles. When the Mach number is very small, pressure variations are rapidly compensated by the thermic agitation and the gas is almost incompressible:

$$\nabla_x \cdot (RU(t,x)) = 0.$$

Introducing the various physical parameters in the Boltzmann equation leads to the following

$$Ma\partial_t f + v \cdot \nabla_x f = \frac{1}{Kn}Q(f,f)$$

where t, x and v denote from now on dimensionless variables. Then, based on the previous formal analysis, we expect the scheme in Figure 2 to govern the fast relaxation limits.

Note that compressible viscous models are not obtained as a fluid limit of the Boltzmann equation. The Boltzmann equation holds indeed for perfect gases, meaning that the size of elementary particles is negligible at the macroscopic scale and that there is consequently no excluded volume in the state relation. In this framework the viscosity comes only from the fact that correlations are induced by a strong thermic agitation, which implies in particular that the gas is incompressible. This is expressed in the von Karman relation, giving the Reynolds number $\Re$ (which is inversely proportional to the kinematic viscosity) in terms of the Knudsen and Mach numbers:

$$\Re = \frac{Ma}{Kn}$$

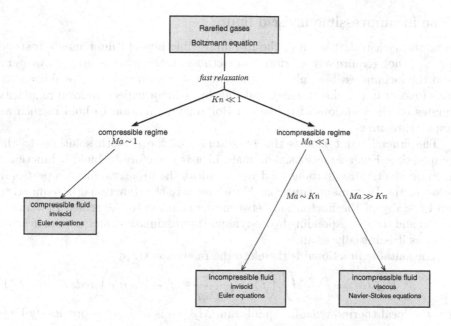

FIGURE 2. Hydrodynamic approximations of the Boltzmann equation

1.4. Aims of the mathematical study

The formal derivation of the fluid limits presented in the previous paragraph goes back to Hilbert for the compressible limit, and to Chapman and Enskog for the viscous correction. Nevertheless the mathematical justification of such asymptotics presents an interest both from the modelling point of view and from the theoretical point of view.

First of all, it allows to identify the physical phenomena which are neglected in the hydrodynamic approximations: fast relaxation which introduces a small time layer, fast oscillations which do not occur in general in the mean motion but could disturb it in some particular framework... One gets therefore the validity conditions of each fluid approximation. It is then possible to split the space in many subdomains, to use different levels of modelling depending on the characteristics of each subdomain, and to determine the suitable boundary conditions in order to ensure a convenient connection.

The rigorous derivation of the fluid limits requires furthermore an extensive understanding of the mathematical theory of the Boltzmann equation. The partial results obtained on the Hilbert problem and presented in the sequel of this paper have allowed in particular to establish refined properties of the renormalized solutions to the Boltzmann equation: a priori estimates coming from the entropy dissipation, control on the possible loss of energy by a defect measure... This study has also asked for new results of functional analysis, namely a compactness criterion in L^1 based both on a velocity averaging result and on a dispersion property.

2. The incompressible inviscid limit

The mathematical derivation of the incompressible inviscid limit uses a method which does not require any a priori compactness on the macroscopic parameters: indeed the incompressible Euler equations are not known to have global weak solutions (except in two dimensions), and thus one cannot expect uniform regularity estimates on the solutions of the scaled Boltzmann equation to hold in such an asymptotic regime.

The crucial point here is the L^2 stability of the smooth solutions to the incompressible Euler equations. The main idea is therefore to build a functional defined on kinetic distributions, and which enjoys the same stability properties in relation to the Boltzmann equation. More precisely this functional is required to control the size of the fluctuation between any solution to the scaled Boltzmann equation and the corresponding hydrodynamic approximation, and to remain small as soon as it is initially small.

The suitable functional is therefore the relative entropy

$$H(f/M_{R,U,T}) = \iint \left(f \log \frac{f}{M_{R,U,T}} - f + M_{R,U,T} \right) dv dx \qquad (2.11)$$

where the local thermodynamic equilibrium $M_{R,U,T}$ is defined as previously by

$$M_{R,U,T}(t,x,v) = \frac{R(t,x)}{(2\pi T(t,x))^{3/2}} \exp\left(-\frac{|v - U(t,x)|^2}{2T(t,x)} \right),$$

and which is just the modulated form of the Lyapunov functional $H(f/M)$ defined by (1.6).

2.1. The convergence result

Theorem 2. *Let (f_ε^0) be a family of well-prepared initial data, meaning that*

$$H(f_\varepsilon^0/M) \le C_0 \varepsilon^2 \text{ for some } C_0 > 0,$$
$$\frac{1}{\varepsilon^2} H(f_\varepsilon^0/M_{1,\varepsilon u^0,1}) \to 0 \text{ as } \varepsilon \to 0, \qquad (2.12)$$

where u^0 denotes some divergence-free vector field in $H^s(\mathbf{T}^3)$ ($s > \frac{5}{2}$).

Denote by (f_ε) any family of renormalized solutions to the scaled Boltzmann equation

$$\varepsilon \partial_t f_\varepsilon + v \cdot \nabla_x f_\varepsilon = \frac{1}{\varepsilon^q} Q(f_\varepsilon, f_\varepsilon) \text{ with } q > 1, \qquad f_{\varepsilon|t=0} = f_\varepsilon^0. \qquad (2.13)$$

Then the following asymptotic results hold:

- *The family of fluctuations $\left(\frac{1}{\varepsilon}(f_\varepsilon - M) \right)$ is relatively weakly compact in $L^1([0,T] \times \mathbf{T}^3 \times \mathbf{R}^3)$ for any $T > 0$;*
- *Any limiting fluctuation is an infinitesimal Maxwellian*

$$M \left(\rho + u \cdot v + \theta \frac{|v|^2 - 3}{2} \right)$$

where ρ, u and θ belongs to $L^2([0,T] \times \mathbf{T}^3)$ for any $T > 0$;

- *The moments of the limiting fluctuation satisfy*

$$\rho = \theta = 0,$$

$$\partial_t u + \nabla_x \cdot (u \otimes u) + \nabla_x p = 0, \quad \nabla_x \cdot u = 0, \qquad (2.14)$$

$$u_{|t=0} = u^0,$$

as long as the solution to the incompressible Euler equations is smooth (or in other words satisfies the energy equality).

Before giving the sketch of the proof of Theorem 2, let us comment a little on this convergence result, particularly on the restrictions (2.12) given on the initial data.

The macroscopic parameters are required to satisfy initially two types of assumptions: the incompressibility and Boussinesq constraints

$$\nabla_x \cdot u^0 = 0, \quad \rho^0 + \theta^0 = 0,$$

and the <u>lack of temperature fluctuation</u>

$$\theta^0 = 0.$$

The first one ensures that there is no acoustic wave in the system (i.e., no fast oscillation of the macroscopic parameters), and the second one implies that the mean temperature fluctuation is identically zero. Therefore, under both assumptions, the temperature fluctuation is expected to converge strongly to 0 as $\varepsilon \to 0$, and the energy flux is more easily controlled.

In addition to these assumptions on the thermodynamic fields, there is an hypothesis on the initial <u>velocity profile</u>. Indeed the initial distribution is supposed to satisfy

$$f_\varepsilon^0 \sim M_{1, \varepsilon u^0, 1},$$

meaning that the gas is almost at local thermodynamic equilibrium. This ensures that there is no kinetic layer, and thus no relaxation to control.

Thus the present result provides only a partial answer to Hilbert's question in inviscid incompressible regime. Even for simplified kinetic models as the BGK equation for which the relaxation is easy to control, the relative entropy method does not allow to conclude since the energy flux (which depends on the third moments of the distribution) cannot be controlled by the relative entropy and entropy dissipation (which behave as second moments). An alternative would be to consider relativistic models for which velocities are bounded.

2.2. The relative entropy method

As seen in the introduction of this section, the modulated entropy introduced by Yau in the framework of the Ginzburg-Landau equation [14] is also an interesting tool to deal with the hydrodynamic limits of kinetic equations. It has been used for the first time in the framework of the Boltzmann equation by Golse [2], who has obtained a convergence result in incompressible inviscid regime on condition of the following assumptions:

(i) the solutions f_ε of the scaled Boltzmann equation are supposed to satisfy the local conservation of momentum, which is not known to hold for renormalized solutions built by DiPerna and Lions;

(ii) the family of fluctuations $\left(\frac{1}{\varepsilon}(f_\varepsilon - M)\right)$ is supposed to satisfy a compactness property, giving both a control on large velocities and on fast spatial oscillations.

The first restriction has been removed by Lions and Masmoudi [8] introducing a defect measure, and the second one by the author [12] using a dissipation and a loop estimates. But the scheme of the proof is not modified, it consists in establishing a Gronwall estimate for the modulated entropy.

The modulated entropy inequality. A straightforward computation using the local conservation of mass

$$\varepsilon \partial_t \int f_\varepsilon dv + \nabla_x \cdot \int f_\varepsilon v dv = 0,$$

the local conservation of momentum with defect measure

$$\varepsilon \partial_t \int f_\varepsilon v dv + \nabla_x \cdot \int f_\varepsilon v^{\otimes 2} dv + \nabla_x \cdot \int m_\varepsilon = 0,$$

and the entropy inequality with defect measure

$$H(f_\varepsilon(t)/M) + \int tr(m_\varepsilon)(t) + \frac{1}{\varepsilon^{q+1}} \int_0^t D(f_\varepsilon)(s)ds \leq H(f_\varepsilon^0)$$

leads to

$$H(f_\varepsilon/M_{1,\varepsilon w,1})(t) + \int tr(m_\varepsilon)(t) + \frac{1}{\varepsilon^{q+1}} \int_0^t D(f_\varepsilon)(s)ds$$

$$\leq H(f_\varepsilon^0/M_{1,\varepsilon w^0,1}) - \int_0^t \int Dw : m_\varepsilon(s)ds$$

$$+\varepsilon \int_0^t \iint f_\varepsilon(\partial_t w + w \cdot \nabla_x w) \cdot (\varepsilon w - v)(s,x,v)dvdxds \qquad (2.15)$$

$$- \int_0^t \iint f_\varepsilon Dw : (v - \varepsilon w)^{\otimes 2}(s,x,v)dvdxds$$

where $m_\varepsilon \in L_{loc}^\infty(\mathbf{R}^+, \mathcal{M}(\mathbf{T}^3, M_3(\mathbf{R})))$ is a symmetric positive matrix-valued measure, and w is any smooth divergence-free vector field.

The defect measure m_ε introduced by Lions and Masmoudi corresponds to a possible loss of compactness (with respect to velocities) in the approximation scheme of the Boltzmann equation. Indeed, for fixed ε, the global entropy inequality (which is uniformly satisfied by the approximate solutions of the Boltzmann equation) provides a uniform bound on the second moment but no equiintegrability for large velocities.

The next step is then to estimate the right-hand side of (2.15), and more precisely to determine the limit of the acceleration term and to prove that the flux term can be controlled by the entropy dissipation and the modulated entropy so as to get a Gronwall inequality.

The acceleration term. Using the entropy bound

$$\frac{1}{\varepsilon^2} H(f_\varepsilon(t)/M) \le C_0,$$

and Young's inequality, one can prove that any subsequence of

$$\left(\frac{1}{\varepsilon}(f_\varepsilon - M)(1 + |v|^2) \right)$$

is uniformly bounded in $L^1([0,T] \times \mathbf{T}^3 \times \mathbf{R}^3)$ and equiintegrable (with respect to ε).

Then, passing to the limit (up to extraction of a subsequence) by a weak compactness argument leads to

$$\frac{1}{\varepsilon} \int f_\varepsilon(v - \varepsilon w) dv \rightharpoonup \bar{u} - w \text{ in } L^1([0,T] \times \mathbf{T}^3).$$

Note that the local conservation of mass implies furthermore that

$$\nabla_x \cdot \bar{u} = 0.$$

The flux term. In order to estimate the flux term, the idea is to split it into two terms, introducing the local thermodynamic equilibrium associated with f_ε. Indeed the Chapman-Enskog expansion

$$f_\varepsilon = M_{f_\varepsilon}(1 + \varepsilon^{\frac{q+3}{2}} q_\varepsilon)$$

coupled with the estimate on q_ε coming from the entropy dissipation is expected to provide a sharper approximation than the Hilbert expansion

$$f_\varepsilon = M_{1,\varepsilon w,1}(1 + \varepsilon g_\varepsilon)$$

coupled with the estimate on the fluctuation g_ε coming from the modulated entropy.

In the case of a simple relaxation model, namely for the BGK equation, this strategy is very easy to implement, whereas in the case of the Boltzmann equation the entropy dissipation does not control exactly the distance between the distribution and the corresponding Maxwellian. A more suitable decomposition is obtained by projection on the kernel of the linearized collision operator

$$\mathcal{L}_w : g \mapsto \frac{1}{M_{1,\varepsilon w,1}} Q(M_{1,\varepsilon w,1}, M_{1,\varepsilon w,1}g).$$

The fundamental property is that $\mathcal{L}_w$ is an unbounded self-adjoint operator on $L^2(M_{1,\varepsilon w,1}(v)dv)$ satisfying the coercivity estimate

$$\|g - \Pi_w g\|^2_{L^2(M_{1,\varepsilon w,1}(v)dv)} \le C \int g \mathcal{L}_w g M_{1,\varepsilon w,1}(v)dv$$

where Π_w denotes the orthogonal projection on the kernel of $\mathcal{L}_w$, i.e., on the subspace spanned by 1, v and $|v|^2$. Therefore using a convenient squareroot renormalization of the fluctuation allows to obtain the expected Gronwall estimate:

$$\iint f_\varepsilon Dw : (v - \varepsilon w)^{\otimes 2} dv dx \leq CD(f_\varepsilon) + CH(f_\varepsilon/M_{1,\varepsilon w,1}),$$

where C denotes some nonnegative constant depending only on w and its derivatives.

<u>Passing to the limit.</u> From Gronwall's lemma and a classical density argument we deduce that the inequality

$$
\frac{1}{\varepsilon^2} H(f_\varepsilon/M_{1,\varepsilon w,1})(t) \leq \frac{1}{\varepsilon^2} H(f_\varepsilon^0/M_{1,\varepsilon w^0,1}) \exp\left(2\int_0^t \|Dw\|_\infty(s) ds\right)
$$
$$
+ \int_0^t \iint f_\varepsilon(\partial_t w + w \cdot \nabla_x w) \cdot (w - \frac{1}{\varepsilon}v) dv dx \exp\left(2\int_s^t \|Dw\|_\infty(\sigma) d\sigma\right)
\tag{2.16}
$$

holds for all divergence-free vector field $w \in W^{1,\infty}([0,T] \times \mathbf{T}^3)$.

In particular, if u is the smooth solution of the incompressible Euler equations with initial data u^0, the acceleration term tends to zero as $\varepsilon \to 0$ since

$$\int (\partial_t u + u \cdot \nabla_x u) \cdot (u - \bar{u}) dx = -\int \nabla_x p \cdot (u - \bar{u}) = 0$$

by the incompressibility condition coming from the local conservation of mass. From the assumption

$$\frac{1}{\varepsilon^2} H(f_\varepsilon^0/M_{1,\varepsilon u^0,1}) \to 0,$$

we then deduce that

$$\frac{1}{\varepsilon^2} H(f_\varepsilon/M_{1,\varepsilon u,1}) \to 0 \text{ in } L^\infty([0,T]).$$

In particular any limiting bulk velocity $\bar{u}$ is necessarily the solution u of the incompressible Euler equations.

3. The incompressible viscous limit

The mathematical derivation of the incompressible viscous limit uses a method which is much more flexible, for instance which does not require any assumption on the initial data (except the scaling assumption). Indeed the theory of the incompressible Navier-Stokes equations is understood in a very general framework, since the viscosity (and thermal conductivity) prevent singularities to occur.

The crucial point here is the weak L^2 stability of the Leray solutions to the incompressible Navier-Stokes equations, which comes from the regularity estimates given by the energy dissipation. The main idea to deal with the viscous asymptotics is therefore to get uniform regularity estimates on the solutions to the scaled Boltzmann equation from the entropy dissipation.

Such regularity estimates are actually established not for the kinetic distribution but only for the corresponding macroscopic parameters, using some velocity averaging lemma. The convergence result is then obtained by passing to the limit in the moment equations associated to the renormalized Boltzmann equation.

3.1. The convergence result

Theorem 3. *Let (f_ε^0) be a family of initial data such that*

$$H(f_\varepsilon^0/M) \leq C_0\varepsilon^2 \text{ for some } C_0 > 0, \tag{3.17}$$

meaning only that the initial Mach number is of order ε, which is relevant with the scaling of the Boltzmann equation.

Denote by (f_ε) any family of renormalized solutions to the scaled Boltzmann equation

$$\varepsilon\partial_t f_\varepsilon + v \cdot \nabla_x f_\varepsilon = \frac{1}{\varepsilon}Q(f_\varepsilon, f_\varepsilon),$$
$$f_{\varepsilon|t=0} = f_\varepsilon^0. \tag{3.18}$$

Then the following asymptotic results hold:

- *The family of fluctuations $\left(\frac{1}{\varepsilon}(f_\varepsilon - M)\right)$ is relatively weakly compact in $L^1([0,T] \times \mathbf{T}^3 \times \mathbf{R}^3)$ for any $T > 0$;*

- *Any limiting fluctuation is an infinitesimal Maxwellian*

$$M\left(\rho + u \cdot v + \theta\frac{|v|^2 - 3}{2}\right)$$

where ρ, u and θ belongs to $L^2([0,T] \times \mathbf{T}^3)$ for any $T > 0$;

- *The moments of the limiting fluctuation satisfy*

$$\partial_t u + \nabla_x \cdot (u \otimes u) + \nabla_x p = \nu\Delta_x u, \quad \nabla_x \cdot u = 0,$$
$$\partial_t \theta + \nabla_x \cdot (u\theta) = \kappa\Delta_x\theta, \quad \rho + \theta = 0,$$

$$u_{|t=0} = \lim_{\varepsilon\to 0} P\left(\frac{1}{\varepsilon}\int f_\varepsilon^0 v dv\right), \tag{3.19}$$

$$\theta_{|t=0} = \lim_{\varepsilon\to 0}\frac{1}{\varepsilon}\int (f_\varepsilon^0 - M)\frac{|v|^2 - 5}{2}dv,$$

where P denotes the Leray projection on divergence-free vector fields, and κ and ν are nonnegative constants depending only on the cross-section b defining Boltzmann's collision operator.

The convergence result stated in Theorem 3 indicates that it is possible to have a global overview of the various physical phenomena occurring in viscous regime, since no assumption on the initial data is required.

Indeed the macroscopic parameters can actually be decomposed in two parts, the first one oscillating rapidly according to the equations of acoustic waves, and the second one converging strongly to the solution of the incompressible Navier-Stokes equations:

$$\begin{pmatrix} \rho_\varepsilon \\ u_\varepsilon \\ \theta_\varepsilon \end{pmatrix}(t,x) = \begin{pmatrix} \rho_{\mathrm{osc}} \\ u_{\mathrm{osc}} \\ \theta_{\mathrm{osc}} \end{pmatrix}\left(\frac{t}{\varepsilon},x\right) + \begin{pmatrix} \bar{\rho} \\ \bar{u} \\ \bar{\theta} \end{pmatrix}(t,x) + O(\varepsilon)$$

In particular, if the incompressibility and Boussinesq constraints are satisfied initially (as in Theorem 2), there is no acoustic wave and, up to extraction of a subsequence, the renormalized moments converge strongly in $L^2([0,T] \times \mathbf{T}^3)$. If moreover the initial moments are sufficiently smooth, the solution of the limiting system is unique for small times and the whole sequence of moments converges strongly.

In such a viscous regime the convergence result can also be extended to more realistic spatial domains Ω, supplementing the Boltzmann equation with some boundary conditions [11]. Indeed boundary layers such as the Prandtl layer are not expected to appear since the viscosity and thermal conductivity smooth the profiles near the boundary.

If the kinetic boundary condition is a specular reflection, the limiting system includes a slipping condition, the so-called Navier condition:

$$\bar{u}(t,x) \cdot n(x) = 0, \quad \frac{\partial \bar{u}}{\partial n} \wedge n = 0, \quad \frac{\partial \bar{\theta}}{\partial n} = 0 \text{ on } \delta\Omega.$$

If the kinetic boundary condition is a diffuse reflection, the limiting system includes a stopping condition, known as the Dirichlet condition:

$$\bar{u} = 0, \quad \bar{\theta} = \bar{\rho} = 0 \text{ on } \delta\Omega.$$

Note that the main difficulty to deal with in this framework is to give sense to the renormalized Boltzmann equation with boundary conditions [10].

3.2. The weak compactness method

The weak convergence result stated in Theorem 3 does not give any information neither on the convergence rate, even nor on the asymptotic behavior of the kinetic distribution. In order to characterize the limiting macroscopic parameters, the idea of Grad is therefore to take limits only in the moment equations [6]. It has been implemented by Bardos, Golse and Levermore [1], who have established the Navier-Stokes limit of the Boltzmann equation in the following framework:

(i)   the scaled Boltzmann equation (3.18) is replaced by its stationary (or time-discretized) form, in order to get rid of all difficulties linked with a possible lack of compactness with respect to time;

(ii) the renormalized solutions f_ε of the scaled Boltzmann equation are assumed to satisfy the local conservation of momentum;

(iii) the family of fluctuations $\left(\frac{1}{\varepsilon}(f_\varepsilon - M)\right)$ is supposed to satisfy a compactness property giving both a control on large velocities and on fast spatial oscillations.

The first restriction has been removed by Lions and Masmoudi, describing precisely the fast temporal oscillations, known as acoustic waves [9]. Then works by Golse and Levermore [4] and by Levermore and Masmoudi [7] have proved that the local conservation of momentum (and energy) can be obtained asymptotically in viscous regime without any assumption. The last restriction has finally been removed first by the author in the BGK case [13], then by Golse and the author for the Boltzmann equation [5], using the entropy dissipation to control large velocities and a new velocity averaging result to control the fast spatial oscillations.

The renormalized moment equations. Integrating the renormalized form of the Boltzmann equation (1.8) with respect to velocities leads to

$$\partial_t \int \frac{1}{\varepsilon}(f_\varepsilon - M)\gamma\left(\frac{f_\varepsilon}{M}\right)\zeta(v)\mathbb{1}_{|v|\leq K_\varepsilon}dv + \nabla_x \cdot \int \frac{1}{\varepsilon}(f_\varepsilon - M)\gamma\left(\frac{f_\varepsilon}{M}\right)v\zeta(v)\mathbb{1}_{|v|\leq K_\varepsilon}dv$$

$$= \int \frac{1}{\varepsilon^3}Q(f_\varepsilon, f_\varepsilon)\hat{\gamma}\left(\frac{f_\varepsilon}{M}\right)\zeta(v)\mathbb{1}_{|v|\leq K_\varepsilon}dv = D_\varepsilon(\zeta) \qquad (3.20)$$

where γ is some C^∞ truncation function with compact support, and $\hat{\gamma}$ is defined by $\hat{\gamma}(z) = \gamma(z) + (z-1)\gamma'(z)$.

Then, if ζ denotes any collision invariant, i.e., any element of the space spanned by 1, v and $|v|^2$, the right side member is expected to tends to zero as $K_\varepsilon \to +\infty$ since $f_\varepsilon/M = 1 + O(\varepsilon)$.

The proof consists then to establish that the conservation defects vanish asymptotically for a suitable choice of K_ε, and to determine the limit of the flux terms. Such an asymptotic study requires beforehand to establish refined a priori estimates (replacing assumption (iii)).

<u>Refined a priori estimates.</u> As in the previous section, the main idea to obtain some control on large velocities is to compare any solution of the scaled Boltzmann equation to the corresponding local thermodynamic equilibrium, using the projection on the kernel of the linearized collision operator

$$\mathcal{L} : g \mapsto \frac{1}{M}Q(M, Mg)$$

(which satisfies a coercivity estimate in $L^2(Mdv)$), and a squareroot renormalization of the fluctuation (which belongs to the convenient functional space).

This provides the expected control on large velocities

$$\frac{1}{\varepsilon^2}\left(\sqrt{\frac{f_\varepsilon}{M}} - 1\right)^2 M(1 + |v|^2) \text{ is uniformly bounded in } L^1([0,T] \times \mathbf{T}^3 \times \mathbf{R}^3),$$

as well as some equiintegrability with respect to the v-variable.

Then, in order to control the fast spatial oscillations and more precisely to establish some weak spatial compactness, the fundamental idea is to transfer some equiintegrability from the v-variable to the x-variable by means of the free transport operator, or in other words to use some dispersion property.

Combining this weak compactness on the family of kinetic fluctuations

$$\frac{1}{\varepsilon^2}\left(\sqrt{\frac{f_\varepsilon}{M}}-1\right)^2$$

with the bound on the advection

$$(\varepsilon\partial_t+v\cdot\nabla_x)\frac{1}{\varepsilon}\left(\sqrt{\frac{f_\varepsilon}{M}}-1\right)$$

provides strong compactness on the macroscopic parameters by means of a velocity averaging lemma.

The conservation defects. Equipped with these preliminary results, one can prove that

$$D_\varepsilon(\zeta)\to 0 \text{ in } L^1([0,T]\times\mathbf{T}^3)$$

for any collision invariant ζ, using the symmetries of the collision operator.

The flux terms. The kinetic fluxes $v\otimes v-\frac{1}{3}|v|^2 Id$ and $\frac{1}{2}v(|v|^2-5)$ (denoted generically by $\phi(v)$) belong to $(\mathrm{Ker}\mathcal{L})^\perp$. Then, as $\mathcal{L}$ is an unbounded self-adjoint operator satisfying the Fredholm alternative, the flux terms can be rewritten in terms of $\tilde{\phi}(v)$ (which is the element of $(\mathrm{Ker}\mathcal{L})^\perp$ such that $\mathcal{L}\tilde{\phi}=\phi$) and of

$$\frac{1}{\varepsilon}\mathcal{L}\left(\frac{1}{\varepsilon}\left(\sqrt{\frac{f_\varepsilon}{M}}-1\right)\right).$$

Using the bilinearity of the collision operator Q, this last term can be split into two parts, the first one giving the diffusion term and the second one leading to the convection term.

The limiting diffusion term is obtained by a weak compactness argument, using the L^2 bound on $\frac{1}{\varepsilon^2}Q(\sqrt{Mf_\varepsilon},\sqrt{Mf_\varepsilon})$ coming from the entropy dissipation, and identifying its weak limit as a spatial derivative of the limiting fluctuation.

$$-\frac{2}{\varepsilon^2}\int Q(\sqrt{Mf_\varepsilon},\sqrt{Mf_\varepsilon})\tilde{\phi}(v)dv \rightharpoonup -\int Mv\cdot\nabla_x\left(\rho+u\cdot v+\theta\frac{|v|^2-3}{2}\right)\tilde{\phi}(v)dv.$$

The convection term can be computed explicitly as a nonlinear function of the macroscopic parameters, using the projection Π on the kernel of linearized collision operator. Taking limits in such a term requires therefore some spatial regularity on the moments, which is obtained as explained in the previous paragraph combining dispersion and velocity averaging results. Furthermore one has to check that fast temporal oscillations, namely acoustic waves, do not produce any constructive interference and thus do not occur in the limiting equation, which is obtained by a compensated compactness argument.

$$\frac{2}{\varepsilon^2}\int\left(\Pi\left(\sqrt{\frac{f_\varepsilon}{M}}-1\right)\right)^2(v\otimes v-\frac{1}{3}|v|^2 Id)dv \rightharpoonup (u\otimes u-\frac{1}{3}|u|^2 Id).$$

Any limiting bulk velocity then satisfies the incompressible Navier-Stokes equations, whereas the corresponding limiting temperature is the unique solution of the advection-diffusion equation, known as the Fourier equation.

References

[1] C. Bardos, F. Golse & C.D. Levermore. Fluid Dynamic Limits of Kinetic Equations II: Convergence Proofs for the Boltzmann Equation, *Comm. Pure Appl. Math.*, **46**(1993), 667–753.

[2] F. Bouchut, F. Golse & M. Pulvirenti. Kinetic Equations and Asymptotic Theory, *B. Perthame and L. Desvillettes eds., Series in Applied Mathematics,* **4** (2000), Gauthier-Villars, Paris.

[3] R.J. DiPerna & P.L. Lions. On the Cauchy Problem for the Boltzmann Equation: Global Existence and Weak Stability Results. *Annals of Math.*, **130** (1990), 321–366.

[4] F. Golse & C.D. Levermore. The Stokes-Fourier and Acoustic Limits for the Boltzmann Equation, *Comm. Pure Appl. Math.*,

[5] F. Golse & L. Saint-Raymond. The Navier-Stokes Limit of the Boltzmann Equation for Bounded Collision Kernels, *Invent. Math.*, **55** (2004), 81–161.

[6] H. Grad. Asymptotic theory of the Boltzmann equation II, *Proc. 3rd Internat. Sympos., Palais de l'Unesco* **1** (1963), Paris.

[7] C.D. Levermore & N. Masmoudi. From the Boltzmann Equation to an Incompressible Navier-Stokes-Fourier System. *Preprint.*

[8] P.L. Lions & N. Masmoudi. From Boltzmann Equations to the Stokes and Euler Equations II, *Arch. Ration. Mech. Anal.*, **158** (2001), 195–211.

[9] P.L. Lions & N. Masmoudi. From Boltzmann Equations to Navier-Stokes Equations I, *Arch. Ration. Mech. Anal.*, **158** (2001), 173–193.

[10] S. Mischler. On weak-weak convergences and applications to the initial boundary value problem for kinetic equations. *Preprint.*

[11] N. Masmoudi & L. Saint-Raymond. From the Boltzmann equation to the Stokes-Fourier system in a bounded domain, *Comm. Pure Appl. Math.*, **56**(2003), 1263–1293.

[12] L. Saint-Raymond. Convergence of solutions to the Boltzmann equation in the incompressible Euler limit, *Arch. Ration. Mech. Anal.*, **166** (2003), 47–80.

[13] L. Saint-Raymond. From the BGK model to the Navier-Stokes equations, *Ann. Sci. École Norm. Sup.*, **36** (2003), 271–317.

[14] H.T. Yau. Relative entropy and hydrodynamics of Ginzburg-Landau models. *Letters in Math. Phys.*, **22** (1991), 63–80.

Laure Saint-Raymond
Laboratoire J.-L. Lions UMR 7598
Université Paris VI
175, rue du Chevaleret
F-75013 Paris, France
e-mail: saintray@ann.jussieu.fr

Analysis and Simulation of Fluid Dynamics

Advances in Mathematical Fluid Mechanics, 201–228

© 2006 Birkhäuser Verlag Basel/Switzerland

On Compressible and Incompressible Vortex Sheets

Paolo Secchi

Abstract. We introduce the main known results of the theory of incompressible and compressible vortex sheets. Moreover, we present recent results obtained by the author with J.F. Coulombel about compressible vortex sheets in two space dimensions, under a supersonic condition that precludes violent instabilities. The problem is a nonlinear free boundary hyperbolic problem with two difficulties: the free boundary is characteristic and the Lopatinski condition holds only in a weak sense, yielding losses of derivatives. In [18, 20] we prove the existence of such piecewise smooth solutions to the Euler equations close enough to stationary vortex sheets. Since the a priori estimates for the linearized equations exhibit a loss of regularity, our existence result is proved by using a suitable modification of the Nash-Moser iteration scheme.

1. Introduction

Let us consider Euler equations of isentropic gas dynamics in the whole space $\mathbb{R}^n, n \geq 2$. Denoting by $\mathbf{u}$ the velocity of the fluid and ρ the density, the equations read:

$$\begin{cases} \partial_t \rho + \nabla \cdot (\rho \mathbf{u}) = 0, \\ \partial_t (\rho \mathbf{u}) + \nabla \cdot (\rho \mathbf{u} \otimes \mathbf{u}) + \nabla p = 0, \end{cases} \tag{1}$$

where $p = p(\rho)$ is the pressure law. In all this paper, p is a $\mathcal{C}^\infty$ function of ρ, defined on $]0, +\infty[$, with $p'(\rho) > 0$ for all ρ. The speed of sound $c(\rho)$ in the fluid is then defined by the relation

$$c(\rho) := \sqrt{p'(\rho)}.$$

The Euler equations for inviscid compressible flow are the prototypical example of symmetric hyperbolic system. It is well known that, given smooth initial data, then there exists a smooth solution to (1) locally in time. The first proof of this result is due to T. Kato [30] for the Cauchy problem and to H. Beirão da

Veiga [5] for the initial-boundary value problem, under the usual non-slip boundary condition. In the same period, in an independent paper, R. Agemi [1] also proved the local existence of solutions to the initial-boundary value problem. As already observed, all these results are local in time.

For the three-dimensional Euler equations for polytropic gases, Serre and Grassin [26, 27, 51] studied the existence of global smooth solutions under appropriate assumptions on the initial data for both isentropic and non-isentropic cases. It was proved in [26] that the three-dimensional Euler equations for a polytropic gas have global smooth solutions, provided that the initial entropy and the initial density are small enough and the initial velocity forces particles to spread out.

In general, however, the global existence of regular solutions can't be expected because the formation of singularities in finite time may occur. The breakdown of the smoothness property may come from the appearance of discontinuities in the solution, i.e., shock waves which develop no matter how smooth the initial data are, see [54]. Another possibility occurs when we consider initial data containing vacuum. In this case the singularity develops near the interface separating the gas and the vacuum and can appear in the solution before the shock waves, see [38], [58]. New achievements about the free boundary problem for perfect compressible and incompressible fluids in vacuum may be found in [36, 37].

While the theory in the one-dimensional case is essentially complete, no general existence theorem is known in the multi-dimensional case for solutions which present discontinuities. A fundamental part in the study of quasi-linear hyperbolic equations is the Riemann problem, i.e., the initial value problem where initial data are piecewise constant with a jump in between. This initial discontinuity generates elementary waves of three kinds: centered rarefaction waves, shock waves and contact discontinuities. In general, the solution of the Riemann problem is expected to develop singularities or fronts of the above kind for all the characteristic fields. Since the general case is too difficult, we will restrict the problem to the case with only one single wave front separating two smooth states.

The first attempt to extend the theory to several space variables is due to A. Majda [39, 40], who showed the short-time existence and stability of a single shock wave, see also [7] and the references therein for a different approach. A general presentation of Majda's result with some improvements may be found in Métivier [46]. See [25] for the uniform stability of weak shocks when the shock strength tends to zero. The existence of rarefaction waves was then showed by S. Alinhac [2]. The present paper deals with the contact discontinuities. While rarefaction waves are continuous solutions with only a singularity at the initial time given by the initial jump, shocks and contact discontinuities are solutions with a discontinuity which persists in time; it is therefore useful to point out the differences between these two cases. Let us briefly recall the main definitions.

Consider a general $N \times N$ system of conservation laws in $\mathbb{R}^n$

$$\partial_t U + \sum_{j=1}^{n} \partial_{x_j} f_j(U) = 0,$$

where $f_j \in \mathcal{C}^\infty(\mathbb{R}^N; \mathbb{R}^N)$. Given $U \in \mathbb{R}^N$, denote

$$A_j(U) := f_j'(U), \quad A(U, \nu) := \sum_{j=1}^{n} \nu_j A_j(U) \quad \forall \nu \in \mathbb{R}^n;$$

let $\lambda_k(U, \nu)$ be the (real) eigenvalues (characteristic field) of the matrix $A(U, \nu)$,

$$\lambda_1(U, \nu) \leq \cdots \leq \lambda_N(U, \nu).$$

Let us denote by $r_k(U, \nu)$ the right eigenvectors of $A(U, \nu)$.

Consider a planar discontinuity at $(\underline{t}, \underline{x})$ with front

$$\Sigma := \{\nu \cdot (x - \underline{x}) = \sigma(t - \underline{t})\},$$

where σ is the velocity of the front. Denote by $U^\pm$ the values of U at $(\underline{t}, \underline{x})$ from each side of Σ. We have the following definition introduced by P. Lax [34].

Definition. $U^\pm$ is a *shock* if there exists $k \in \{1, \ldots, N\}$ such that $\nabla \lambda_k \cdot r_k \neq 0 \ \forall U$, $\forall \nu$ (λ_k is said *genuinely nonlinear*) and

$$\lambda_{k-1}(U^-, \nu) < \sigma < \lambda_k(U^-, \nu),$$
$$\lambda_k(U^+, \nu) < \sigma < \lambda_{k+1}(U^+, \nu).$$

The first inequality on the left (resp. the last on the right) is ignored when $k = 1$ (resp. $k = N$).

$U^\pm$ is a *contact discontinuity* if there exists $k \in \{1, \ldots, N\}$ such that $\nabla \lambda_k \cdot r_k \equiv 0 \ \forall U$, $\forall \nu$ (λ_k is said *linearly degenerate*) and

$$\lambda_k(U^+, \nu) \leq \sigma = \lambda_k(U^-, \nu) \qquad \text{or} \qquad \lambda_k(U^+, \nu) = \sigma \leq \lambda_k(U^-, \nu).$$

In case of shocks, the definition shows that the velocity of the front is always different from the characteristic fields. It follows that the shock front is a *noncharacteristic* interface. On the contrary, the contact discontinuity is a *characteristic* interface, because of the possible equalities. Another crucial difference between shocks and contact discontinuities is that in the first case one has the *uniform stability*, which is the extension of the uniform Kreiss-Lopatinskii condition for standard mixed problems. In case of contact discontinuities the Kreiss-Lopatinskii condition holds only weakly, and not uniformly. This fact has consequences for the a priori energy estimate of solutions. We will analyze this point more precisely in the sequel.

2. Incompressible vortex sheets

Before going on, we briefly recall the main points in the theory of incompressible vortex sheets. In this context the problem is also known as the Kelvin-Helmholtz problem. For more details, we refer the reader to the books [11, 42, 44].

A velocity discontinuity across a streamline in an inviscid flow is called a *vortex sheet*. To simplify, in the theory it is usually assumed that the flow is irrotational on either sides of the front of discontinuity. Even more, sometimes it is assumed that the velocity field is piecewise constant. Thus a vortex sheet has vorticity concentrated as a measure (delta function) along a surface of codimension one. In three-space dimensions, the vorticity is concentrated along a surface in the space; in two-space dimensions, along a curve in the plane. If the solution is piecewise constant on the two sides of the interface of discontinuity, one has planar vortex sheets in the three-dimensional case and rectilinear vortex sheets in the two-dimensional case, respectively. Most of the theory has been developed for the 2D case.

Vortex sheets are known to be extremely unstable structures since a long time. The first qualitative observation of such instability is due to Helmholtz (1868). Some years later, Kelvin (1894) and Rayleigh (1896) provided the first mathematical analysis of this phenomenon. The instability may be seen by considering how the position of the sheet evolves in time. Below we describe the main ideas (details of the proof may be found in [42, 44]).

Assume that the sheet Σ is a curve dividing the plane into two regions Ω^+ and Ω^-. We look for a parametrization $z(\alpha, t)$ of the curve, where the position z is determined by a parameter α at time t. Some calculations show that for the parameter α one may choose the circulation Γ across the sheet, because it may be used as a local Lagrangian coordinate. Accordingly, the sheet may be described by a parametrization $z(\Gamma, t)$ at time t. Combining suitably the information coming from the Euler equations of motion, the incompressibility condition and the Biot-Savart law, it follows that the position of the vortex sheet is described by the evolution equation

$$\frac{\partial}{\partial t} z^*(\Gamma, t) = \frac{1}{2\pi i} PV \int \frac{d\Gamma'}{z(\Gamma, t) - z(\Gamma', t)},$$

where z^* denotes the complex conjugate. This is the so-called *Birkhoff-Rott* (B-R) equation. Let us consider a perturbation of a flat, constant solution $z(\Gamma, t) = \Gamma$; let this perturbed sheet have the form

$$z(\Gamma, t) = \Gamma + \zeta(\Gamma, t).$$

The computation of the linearized equations gives to leading order the following equation for the perturbation ζ:

$$\frac{\partial}{\partial t} \zeta^*(\Gamma, t) = \frac{1}{2\pi i} PV \int \frac{\zeta'(\Gamma', t)}{\Gamma' - \Gamma} d\Gamma' = \frac{1}{2} \mathcal{H} \zeta'(\Gamma, t), \tag{2}$$

where ζ' denotes the derivative with respect to Γ and where $\mathcal{H}$ is the Hilbert transform

$$\mathcal{H} f(x) = \frac{1}{\pi i} PV \int \frac{f(y)}{y - x} dy.$$

We look for solutions of the form

$$\zeta(\Gamma, t) = A_k(t) e^{ik\Gamma} + B_k(t) e^{-ik\Gamma}, \quad k > 0.$$

Plugging this into (2) gives differential equations

$$A'_k = -\frac{1}{2}ik\overline{B}_k, \quad B'_k = -\frac{1}{2}ik\overline{A}_k,$$

with solutions

$$A_k(t) = A^+_k e^{kt/2} + A^-_k e^{-kt/2}, \quad B_k(t) = B^+_k e^{kt/2} + B^-_k e^{-kt/2},$$

where $A^+_k = -iB^{+*}_k$, $A^-_k = iB^{-*}_k$. As a consequence the kth Fourier mode has a component that grows like $e^{|k|t/2}$. The growth rate implies that the linear evolution problem is *ill posed* in the sense of Hadamard. In fact, given an initial condition, the solution at later times does not depend continuously on the initial data in any Sobolev norm. This instability is called *Kelvin-Helmholtz* instability.

Birkhoff [6] conjectured that the nonlinear B-R equation with analytic initial data is well posed, at least locally in time. The existence local in time of a unique analytic solution was proved by Sulem et al. [55], by application of a Nirenberg-Nishida version of the Cauchy-Kowalewskaya theorem. Duchon and Robert [22] proved that for a special class of initial conditions there exists a global analytic solution for all time.

Numerical experiments show the appearance of singularities in the shape of a cusp or a logarithmic spiral and that, for periodic initial data of suitable size ϵ, the critical time when the first singularity occurs has order $\mathcal{O}(\log \epsilon)$, see [49, 45, 31].

Caflish and Orellana [9] proved existence almost up to the time of expected singularity formation for analytic data close to the flat sheet. Duchon and Robert [22] and Caflish and Orellana [10] constructed examples of solutions of the B-R equation where a curvature singularity develops in finite time from analytic data. These examples show that the initial value problem for the B-R equation is ill-posed in $C^{1+\alpha}(\mathbb{R}), \alpha > 0$, and in $H^s(\mathbb{R}), s > 3/2$, in the Hadamard sense. The Kelvin-Helmholtz problem is formally reversible in time. In order to analyze the appearance of singularities one can consider problems with singular data for $t = 0$ and regular solution for $t > 0$. The construction of Caflish and Orellana corresponds to the above situation. The solution is singular at $t = 0$ and the singularity is a cusp as observed in numerical computations.

The ill-posedness of the Kelvin-Helmholtz problem is also considered by Lebeau [35]. Under the hypothesis that the curve where the vorticity is concentrated satisfies a Hölder type condition, for any locally defined in time solution u of the Euler equations, the curve of discontinuity of u and the density of vorticity on this curve are analytic. Thus, the only solutions are those described by the result of Sulem et al. [55].

Because analytic sheets quickly develop singularities, the analytic setting is too restrictive for practical applications and excludes any analysis at and beyond the critical time of singularity formation. Wu [57] looks for solutions of the B-R equation with low regularity, such that for each fixed time t both sides of the equation are functions locally in L^2, and proves results of existence and analiticity of solutions on chord-arc curves. Both Lebeau's and Wu's results use the recent observation that the B-R equation is of "elliptic" type.

Instead of considering the B-R equation for the evolution of the vortex sheet, one may directly consider the 2D Euler equation in a suitable weak formulation expressed in terms of the velocity, so to be compatible with a vorticity which is a measure, see [42, 11]. The global existence result of such general weak solutions to the Euler equations is not available.

Numerical experiments show the role of the sign of the vorticity. Krasny [32] considered an initial flat finite sheet with vortex sheet strength having different signs; the evolution in time shows that the sheet rolls up generating vortices with great complexity at a small scale. In the light of this, Delort [21] considered the case with as initial vorticity a measure with fixed sign and proved the global existence of a weak solution for the Euler equations. Refinements of this result can be found in [23, 41].

The problem of uniqueness of the weak solution is still unsolved. Examples of weak solutions with the velocity field $u \in L^2(\mathbb{R}^2 \times (-T, T))$ that is compactly supported in space-time were constructed by Scheffer [50] and Shnirelman [53]. This gives non-uniqueness of weak solutions in $L^2(\mathbb{R}^2 \times (-T, T))$. However non-uniqueness in the physically relevant class of conserved energy $u \in L^\infty([0, +\infty); L^2_{loc}(\mathbb{R}^2))$ remains open. There are some numerical evidences of non-uniqueness of weak solutions for vortex sheet data. Also, it is not known which is the relationship between Delort's weak solution and the B-R solution when the initial data are analytic, even before the critical time is reached.

3. Compressible vortex sheets

We consider again vortex sheets in a two-dimensional isentropic compressible flow. Let $\Sigma := \{x_2 = \varphi(t, x_1)\}$ be a smooth interface and $(\rho, \mathbf{u})$ a smooth function on either side of Σ:

$$(\rho, \mathbf{u}) := \begin{cases} (\rho^+, \mathbf{u}^+) & \text{if } x_2 > \varphi(t, x_1) \\ (\rho^-, \mathbf{u}^-) & \text{if } x_2 < \varphi(t, x_1). \end{cases}$$

Definition. $(\rho, \mathbf{u})$ is a *weak solution* of (1) if and only if it is a classical solution on both sides of Σ and it satisfies the *Rankine-Hugoniot conditions* at Σ:

$$\partial_t \varphi \, [\rho] - [\rho \mathbf{u} \cdot \nu] = 0,$$
$$\partial_t \varphi \, [\rho \mathbf{u}] - [(\rho \mathbf{u} \cdot \nu)\mathbf{u}] - [p]\nu = 0, \tag{3}$$

where $\nu := (-\partial_{x_1} \varphi, 1)$ is a (space) normal vector to Σ. As usual, $[q] = q^+ - q^-$ denotes the jump of a quantity q across the interface Σ.

Following Lax [34], we shall say that $(\rho, \mathbf{u})$ is a contact discontinuity if (3) is satisfied in the following way:

$$\partial_t \varphi = \mathbf{u}^+ \cdot \nu = \mathbf{u}^- \cdot \nu,$$
$$p^+ = p^-.$$

Because p is monotone, the previous equalities read

$$\partial_t \varphi = \mathbf{u}^+ \cdot \nu = \mathbf{u}^- \cdot \nu,$$
$$\rho^+ = \rho^-. \tag{4}$$

Since the density and the normal velocity are continuous across the interface Σ, the *only jump* experimented by the solution is on the *tangential velocity*. (Here, normal and tangential mean normal and tangential with respect to Σ.) For this reason, a contact discontinuity is a *vortex sheet* and we shall make no distinction in the terminology we use.

Therefore the problem is to show the existence of a weak solution to (1), (4). Observe that the interface Σ, or equivalently the function φ, is part of the unknowns of the problem; we thus deal with a *free boundary problem*. Due to (4), the free boundary is characteristic with respect to both left and right sides.

Some information comes from the study of the *linearized* equations near a *piecewise constant* vortex sheet.

The stability of linearized equations for planar and rectilinear compressible vortex sheets around a piecewise constant solution has been analyzed some time ago by Miles [48, 24], using tools of complex analysis. For a vortex sheet in $\mathbb{R}^n$, the situation may be summarized as follows (see [52]):

for $n \geq 3$, the problem is always *violently unstable*;

for $n = 2$, there exists a critical value for the jump of the tangential velocity such that:

if $|[\mathbf{u} \cdot \tau]| < 2\sqrt{2}c(\rho)$ the problem is *violently unstable* (subsonic case), (5)

if $|[\mathbf{u} \cdot \tau]| > 2\sqrt{2}c(\rho)$ the problem is *weakly stable* (supersonic case),

where $c(\rho) := \sqrt{p'(\rho)}$ is the sound speed and τ is a tangential unit vector to Σ.

As will be seen below, the problem given by the linearized equations obtained from (1) and the transmission boundary conditions at the interface (4) may be formulated as a nonstandard boundary value problem, which is well posed if the analogue of the Kreiss-Lopatinskii condition is satisfied (see, e.g., [52]). In (5), violent instability means that the Kreiss-Lopatinskii condition is violated, so that there exist exponentially exploding modes of instability. This instability corresponds to ill-posedness in the sense of Hadamard. Weak stability means that the Kreiss-Lopatinskii condition is satisfied (there are no growing modes) but not uniformly. In this case the solution can become unstable, but the instability is much slower to develop than in the case of violent instability.

In the instability case no a priori energy estimate for the solution is possible, because of the ill-posedness. In the weak stability case, an $L^2 - L^2$ energy estimate (for the solution with respect to the data) is not expectable, because the Kreiss-Lopatinskii condition doesn't hold uniformly. However, it is reasonable to look for an energy estimate with loss of derivatives with respect to the data. This is different from the case of shocks, where the Kreiss-Lopatinskii condition holds uniformly, so that an energy estimate without loss of derivatives may be proved (see Majda [39, 40]).

The above result in 2D formally agrees with the theory of incompressible vortex sheets outlined before. In fact, in the incompressible limit the speed of sound tends to infinity, and the above result yields that two-dimensional incompressible vortex sheets are always unstable (the Kelvin-Helmhotz instability).

Recalling that in the theory for incompressible vortex sheets, solutions are shown to exist in the class of analytic functions, one may look for analytic solutions also in the compressible instability case; here the existence of a local in time analytic solution for the nonlinear problem may be obtained by applying Harabetian's result [28].

In the transition case $|[\mathbf{u} \cdot \tau]| = 2\sqrt{2}c(\rho)$ the problem is also weakly stable, as shown in [19], in a weaker sense than in the supersonic case. The complete analysis of linear stability of contact discontinuities for the nonisentropic Euler equations is carried out in [17], for both cases $n = 2$ and $n = 3$. Moreover, we refer to [29] for the study of the instability of vortex sheets when heat conduction is taken into account. The one-dimensional stability of contact discontinuities has received a general treatment in [12] and [13]. Unfortunately, the isentropic Euler equations do not admit contact discontinuities in one space dimension.

From now on we consider the 2D supersonic *weakly stable* regime.

Motivated by the interest for the motion and structure of galactic jets in astrophysics, Woodward [56] has performed numerical simulations of supersonic vortex sheets. In contrast to the predictions of linearized stability in the supersonic case, Woodward has shown that solutions to the nonlinear vortex sheet problem are also unstable, giving insight about how the instability occurs. First nonlinear kinks appear on the slip-stream surface with corresponding paired shock and rarefaction waves emanating into respective sides of the interface; these kink modes grow self-similarly in time, then collide and interact. When enough vorticity is generated through nonlinear interaction of the kink modes, the compressible vortex sheet rolls up, in a similar fashion to the subsonic case of instability, where however roll-up occurs on a much more rapid time-scale, as for classical incompressible vortex sheets.

To explain this apparent paradox, Artola and Majda [4] have studied the problem by methods of geometric optics. They consider an expansion of the form

$$u \sim \epsilon^2 U(x, t, \varphi/\epsilon)$$

as a small amplitude rectilinear wave incident on the vortex sheet. The plane wave strikes the vortex sheet yielding both a distortion of the slip-stream as well as perturbed transmitted and reflected sound waves. For most angles of incidence, the nonlinear response has leading order asymptotic expansion of the same form as the incident wave. In particular, they have the same order of amplitude $\mathcal{O}(\epsilon^2)$.

However, there are three special angles of incidence associated with the boundary phase functions, where the above expansion ceases to be valid. Associated with each of the three angles of incidence, there is a set of simplified asymptotic equations which predict the nonlinear dynamic response to the incident wave and lead to the nonlinear development of propagating kink modes of instability along the vortex sheet. Due to nonlinear resonance, the nonlinear response at the three angles of incidence has different asymptotic expansion from the incident one. In particular, the response has much larger order $\mathcal{O}(\epsilon)$.

Thus, the analysis shows a loss of one power of ϵ between the incident waves of order $\mathcal{O}(\epsilon^2)$ and the reflected waves with order $\mathcal{O}(\epsilon)$. The result suggests the loss of one derivative in the energy estimate of the solution.

In a recent paper with J.F. Coulombel [18], we show that supersonic constant vortex sheets are linearly stable, in the sense that the linearized system (around a piecewise constant solution) obeys an L^2-energy estimate. The failure of the uniform Kreiss-Lopatinskii condition yields an energy estimate with the loss of one tangential derivative from the source terms to the solution. Moreover, since the problem is characteristic, the estimate we prove exhibits a loss of control on the trace of the solution. We also consider the linearized equations around a perturbation of a constant vortex sheet, and we show that these linearized equations with variable coefficients obey the same energy estimate with loss of one derivative w.r.t. the source terms.

In a second paper [20] we consider the nonlinear problem and prove the existence of supersonic compressible vortex sheets solutions. To prove our result we first extend the energy estimate to Sobolev norms by application of the L^2-estimate to tangential derivatives and combination with an a priori estimate for normal derivatives obtained by the energy method from a vorticity-type equation.

The new estimate shows the loss of one derivative with respect to the source terms, and the loss of three derivatives with respect to the coefficients. The loss is fixed, and we can thus solve the nonlinear problem by a Nash-Moser iteration scheme. Recall that the Nash-Moser procedure was already used to construct other types of waves for multidimensional systems of conservation laws, see, e.g., [2, 25]. However, our Nash-Moser procedure is not completely standard, since the tame estimate for the linearized equations will be obtained under certain nonlinear constraints on the state about which we linearize. We thus need to make sure that these constraints are satisfied at each iteration step. The rest of the present paper is devoted to the presentation of these results.

In [20] we also show how a similar analysis yields the existence of weakly stable shock waves in isentropic gas dynamics, and the existence of weakly stable liquid/vapor phase transitions.

4. The nonlinear equations in a fixed domain

The interface $\Sigma := \{x_2 = \varphi(t, x_1)\}$ is an unknown of the problem. We first straighten the unknown front in order to work in a fixed domain. Let us introduce the change of variables

$$(\tau, y_1, y_2) \to (t, x_1, x_2),$$
$$(t, x_1) = (\tau, y_1), \quad x_2 = \Phi(\tau, y_1, y_2),$$

where

$$\Phi : \{(\tau, y_1, y_2) : y_2 > 0\} \to \mathbb{R},$$

is a smooth function such that

$$\partial_{y_2} \Phi(\tau, y_1, y_2) \geq \kappa > 0, \quad \Phi(\tau, y_1, 0) = \varphi(t, x_1).$$

We define the new unknowns

$$(\rho_\sharp^+, \mathbf{u}_\sharp^+)(\tau, y_1, y_2) := (\rho, \mathbf{u})(\tau, y_1, \Phi(\tau, y_1, y_2)),$$
$$(\rho_\sharp^-, \mathbf{u}_\sharp^-)(\tau, y_1, y_2) := (\rho, \mathbf{u})(\tau, y_1, \Phi(\tau, y_1, -y_2)).$$

The functions $\rho_\sharp^\pm, \mathbf{u}_\sharp^\pm$ are smooth on the fixed domain $\{y_2 > 0\}$. For convenience, we drop the $\sharp$ index and only keep the $+$ and $-$ exponents. Then, we again write (t, x_1, x_2) instead of (τ, y_1, y_2).

Let us denote $\mathbf{u}^\pm = (v^\pm, u^\pm)$. The existence of compressible vortex sheets amounts to proving the existence of smooth solutions to the following first order system:

$$\partial_t \rho^\pm + v^\pm \partial_{x_1} \rho^\pm + (u^\pm - \partial_t \Phi^\pm - v^\pm \partial_{x_1} \Phi^\pm) \frac{\partial_{x_2} \rho^\pm}{\partial_{x_2} \Phi^\pm}$$

$$+ \rho^\pm \partial_{x_1} v^\pm + \rho^\pm \frac{\partial_{x_2} u^\pm}{\partial_{x_2} \Phi^\pm} - \rho^\pm \frac{\partial_{x_1} \Phi^\pm}{\partial_{x_2} \Phi^\pm} \partial_{x_2} v^\pm = 0,$$

$$\partial_t v^\pm + v^\pm \partial_{x_1} v^\pm + (u^\pm - \partial_t \Phi^\pm - v^\pm \partial_{x_1} \Phi^\pm) \frac{\partial_{x_2} v^\pm}{\partial_{x_2} \Phi^\pm}$$

$$+ \frac{p'(\rho^\pm)}{\rho^\pm} \partial_{x_1} \rho^\pm - \frac{p'(\rho^\pm)}{\rho^\pm} \frac{\partial_{x_1} \Phi^\pm}{\partial_{x_2} \Phi^\pm} \partial_{x_2} \rho^\pm = 0, \qquad (6)$$

$$\partial_t u^\pm + v^\pm \partial_{x_1} u^\pm + (u^\pm - \partial_t \Phi^\pm - v^\pm \partial_{x_1} \Phi^\pm) \frac{\partial_{x_2} u^\pm}{\partial_{x_2} \Phi^\pm}$$

$$+ \frac{p'(\rho^\pm)}{\rho^\pm} \frac{\partial_{x_2} \rho^\pm}{\partial_{x_2} \Phi^\pm} = 0,$$

in the fixed domain $\{x_2 > 0\}$, where

$$\Phi^\pm(t, x_1, x_2) := \Phi(t, x_1, \pm x_2),$$

both defined on the half-space $\{x_2 > 0\}$.

The equations are not sufficient to determine the unknowns $U^\pm := (\rho^\pm, v^\pm, u^\pm)$ and $\Phi^\pm$. In fact, the change of variables is only requested to map Σ to $\{x_2 = 0\}$ and is arbitrary outside Σ. In order to simplify the transformed equations of motion we may prescribe that $\Phi^{\pm}$ solve the *eikonal* equations

$$\partial_t \Phi^{\pm} + v^{\pm} \partial_{x_1} \Phi^{\pm} - u^{\pm} = 0 \tag{7}$$

in the domain $\{x_2 > 0\}$.

This choice has the advantage that the boundary matrix of the system for $U^{\pm}$ has constant rank in the whole space domain $\{x_2 \geq 0\}$, and not only at the boundary.

The equations for $U^{\pm}$ are only coupled through the boundary conditions

$$
\begin{aligned}
\Phi^+ &= \Phi^- = \varphi, \\
(v^+ - v^-)\,\partial_{x_1}\varphi - (u^+ - u^-) &= 0, \\
\partial_t \varphi + v^+ \partial_{x_1}\varphi - u^+ &= 0, \\
\rho^+ - \rho^- &= 0,
\end{aligned}
\tag{8}
$$

on the fixed boundary $\{x_2 = 0\}$, which are obtained from (4). We will also consider the initial conditions

$$(\rho^{\pm}, v^{\pm}, u^{\pm})_{|t=0} = (\rho_0^{\pm}, v_0^{\pm}, u_0^{\pm})(x_1, x_2), \quad \varphi_{|t=0} = \varphi_0(x_1), \tag{9}$$

in the space domain $\mathbb{R}_+^2 = \{x_1 \in \mathbb{R}, x_2 > 0\}$.

Thus, compressible vortex sheet solutions should solve (6), (7), (8), (9).

There exist many simple solutions of (6), (7), (8) that correspond (for the Euler equations (1) in the original variables) to stationary rectilinear vortex sheets:

$$
(\rho, \mathbf{u}) = \begin{cases} (\overline{\rho}, \overline{v}, 0), & \text{if } x_2 > 0, \\ (\overline{\rho}, -\overline{v}, 0), & \text{if } x_2 < 0, \end{cases}
$$

where $\overline{\rho}, \overline{v} \in \mathbb{R}$, $\overline{\rho} > 0$. Up to Galilean transformations, every rectilinear vortex sheet has this form. In the straightened variables, these stationary vortex sheets correspond to the following smooth (stationary) solution to (6), (7), (8):

$$\overline{U}^{\pm} \equiv \begin{pmatrix} \overline{\rho} \\ \pm \overline{v} \\ 0 \end{pmatrix}, \quad \overline{\Phi}^{\pm}(t, x) \equiv \pm x_2, \quad \overline{\varphi} \equiv 0. \tag{10}$$

In this paper, we shall assume $\overline{v} > 0$, but the opposite case can be dealt with in the same way.

The following theorem is our main result: for the nonlinear problem (6), (7), (8), (9) of supersonic compressible vortex sheets we prove the existence of solutions close enough to the piecewise constant solution (10).

Theorem 1. [18, 20] *Let $T > 0$, and let $\mu \in \mathbb{N}$, with $\mu \geq 6$. Assume that the stationary solution defined by (10) satisfies the "supersonic" condition:*

$$\overline{v} > \sqrt{2}\, c(\overline{\rho}). \tag{11}$$

Assume that the initial data $(U_0^{\pm}, \varphi_0)$ have the form

$$U_0^{\pm} = \overline{U}^{\pm} + \dot{U}_0^{\pm},$$

with $\dot{U}_0^\pm \in H^{2\mu+3/2}(\mathbb{R}_+^2)$, $\varphi_0 \in H^{2\mu+2}(\mathbb{R})$, and that they satisfy sufficient compatibility conditions. Assume also that $(\dot{U}_0^\pm, \varphi_0)$ have a compact support. Then, there exists $\delta > 0$ such that, if $\|\dot{U}_0^\pm\|_{H^{2\mu+3/2}(\mathbb{R}_+^2)} + \|\varphi_0\|_{H^{2\mu+2}(\mathbb{R})} \le \delta$, then there exists a solution $U^\pm = \overline{U}^\pm + \dot{U}^\pm, \Phi^\pm = \pm x_2 + \dot{\Phi}^\pm, \varphi$ of (6), (7), (8), (9), on the time interval $[0,T]$. This solution satisfies $(\dot{U}^\pm, \dot{\Phi}^\pm) \in H^\mu(]0,T[\times\mathbb{R}_+^2)$, and $\varphi \in H^{\mu+1}(]0,T[\times\mathbb{R})$.

For the compatibility conditions as for all the other details we refer the reader to [18, 20].

The rest of the paper is organized as follows: in Section 5 we introduce the linearized equations around a perturbation of the piecewise constant solution (10) and state the basic a priori L^2 estimate while in Section 6 we describe the main steps of its proof. In 7 we give a tame a priori estimate in Sobolev spaces for the solution of the linearized problem.

In Section 8, we reduce the nonlinear problem (6), (7), (8), (9), to another nonlinear system with zero initial data; then we describe the Nash-Moser iteration scheme that will be used to solve this reduced problem.

5. The L^2-energy estimate for the linearized problem

We introduce the linearized equations around a perturbation of the piecewise constant solution (10). More precisely, let us consider the functions

$$U_{r,l} = \overline{U}^\pm + \dot{U}_{r,l}(t, x_1, x_2),$$

$$\Phi_{r,l} = \pm x_2 + \dot{\Phi}_{r,l}(t, x_1, x_2),$$

where

$$\dot{U}_{r,l} \in W^{2,\infty}(\Omega), \quad \dot{\Phi}_{r,l} \in W^{3,\infty}(\Omega),$$

$(U_{r,l}, \Phi_{r,l})$ satisfy (7), (8), and the perturbations $\dot{U}_{r,l}$ and $\dot{\Phi}_{r,l}$ have compact support.

Let us consider the linearized equations around $U_{r,l}, \Phi_{r,l}$ with solutions denoted by $U_\pm, \Psi_\pm$. The equations take a simpler form by the introduction of the new unknowns (cfr. [2])

$$\dot{U}_+ := U_+ - \frac{\Psi_+}{\partial_{x_2}\Phi_r}\partial_{x_2}U_r, \quad \dot{U}_- := U_- - \frac{\Psi_-}{\partial_{x_2}\Phi_l}\partial_{x_2}U_l. \tag{12}$$

Then the equations are diagonalized and transformed to an equivalent form with constant (singular) boundary matrix.

Denote the new unknowns by $W^\pm$. The linearized equations are then equivalent to

$$\mathcal{N}_r W^+ := \mathbf{A}_0^r \partial_t W^+ + \mathbf{A}_1^r \partial_{x_1} W^+ + \mathbf{I}_2 \partial_{x_2} W^+ + \mathbf{A}_0^r \mathbf{C}^r W^+ = F^+,$$
$$\mathcal{N}_l W^- := \mathbf{A}_0^l \partial_t W^- + \mathbf{A}_1^l \partial_{x_1} W^- + \mathbf{I}_2 \partial_{x_2} W^- + \mathbf{A}_0^l \mathbf{C}^l W^- = F^-, \tag{13}$$

with suitable matrices $\mathbf{A}_j^{r,l} = \mathbf{A}_j^{r,l}(U_{r,l}, \Phi_{r,l}), \mathbf{C}^{r,l} = \mathbf{C}^{r,l}(U_{r,l}, \Phi_{r,l})$, and boundary matrix

$$\mathbf{I}_2 := \text{diag}\,(0, 1, 1).$$

We have

$$\mathbf{A}_j^{r,l} \in W^{2,\infty}(\Omega), \quad \mathbf{C}^{r,l} \in W^{1,\infty}(\Omega).$$

In view of the results in [2, 25], in (13) we have dropped the zero order terms in Ψ_+, Ψ_-. The linearized boundary conditions are

$$\Psi_{+|_{x_2=0}} = \Psi_{-|_{x_2=0}} = \psi,$$
$$\mathbf{b}\,\nabla\psi + \mathbf{M}\,U_{|_{x_2=0}} = g,$$

with suitable matrices $\mathbf{b} = \mathbf{b}(U_{r,l}), \mathbf{M} = \mathbf{M}(\nabla\varphi)$ and where $U = (U_+, U_-)^T$, $\nabla\psi = (\partial_t\psi, \partial_{x_1}\psi)^T$ and $g = (g_1, g_2, g_3)^T$. Introducing $W^{\pm}$ the linearized boundary conditions become equivalent to

$$\Psi_+ = \Psi_- = \psi,$$
$$\mathcal{B}(W^{\mathrm{nc}}, \psi) := \mathbf{b}\,\nabla\psi + \mathbf{\check{b}}\,\psi + \widetilde{\mathbf{M}}W_{|_{x_2=0}} = g. \tag{14}$$

Here $W = (W^+, W^-)^T$, and

$$\mathbf{b}(U_{r,l}) \in W^{2,\infty}(\mathbb{R}^2),$$
$$\mathbf{\check{b}}(\partial_{x_2} U_{r,l}, \nabla\varphi, \partial_{x_2}\Phi_{r,l}) \in W^{1,\infty}(\mathbb{R}^2),$$
$$\widetilde{\mathbf{M}}(U_{r,l}, \nabla\varphi, \nabla\Phi_{r,l}) \in W^{2,\infty}(\mathbb{R}^2).$$

Observe that the boundary conditions involve both ψ and W. Moreover, the matrix $\mathbf{M}$ only acts on the *noncharacteristic* part $W^{\mathrm{nc}} := (W_2^+, W_3^+, W_2^-, W_3^-)$ of the vector W.

Our first goal is to obtain an L^2 a priori estimate of the solution to the linearized problem (13),(14). Let us define

$$\Omega := \{(t, x_1, x_2) \in \mathbb{R}^3 \text{ s.t. } x_2 > 0\} = \mathbb{R}^2 \times \mathbb{R}^+.$$

The boundary $\partial\Omega = \{x_2 = 0\}$ is identified to $\mathbb{R}^2$. Define also

$$H_\gamma^s = H_\gamma^s(\mathbb{R}^2) := \{u \in \mathcal{D}'(\mathbb{R}^2) \text{ s.t. } \exp(-\gamma t)u \in H^s(\mathbb{R}^2)\},$$

equipped with the norm

$$\|u\|_{H_\gamma^s} := \|\exp(-\gamma t)u\|_{H^s(\mathbb{R}^2)}.$$

Define similarly the space $H_\gamma^k(\Omega)$. The space $L^2(\mathbb{R}^+; H_\gamma^s(\mathbb{R}^2))$ is equipped with the norm

$$\|v\|_{L^2(H_\gamma^s)}^2 := \int_0^{+\infty} \|v(\cdot, x_2)\|_{H_\gamma^s(\mathbb{R}^2)}^2\, dx_2.$$

In the sequel, the variable in $\mathbb{R}^2$ is (t, x_1) while x_2 is the variable in $\mathbb{R}^+$.

Our first result is the following (here we denote : $\mathcal{N} = (\mathcal{N}_r, \mathcal{N}_l)$).

Theorem 2. [18] *Assume that the particular solution defined by* (10) *satisfies* (11), *that* $(U_{r,l}, \Phi_{r,l})$ *satisfy* (7), (8), *and that the perturbation* $(\dot{U}_{r,l}, \dot{\Phi}_{r,l})$ *is sufficiently small in* $W^{2,\infty}(\Omega) \times W^{3,\infty}(\Omega)$ *and has compact support. Then, for all* $\gamma \geq 1$ *large enough and for all* $(W, \psi) \in H^2_\gamma(\Omega) \times H^2_\gamma(\mathbb{R}^2)$, *the following estimate holds:*

$$\gamma \|W\|^2_{L^2_\gamma(\Omega)} + \|W^{\mathrm{nc}}{}_{|x_2=0}\|^2_{L^2_\gamma} + \|\psi\|^2_{H^1_\gamma}$$
$$\leq C \left(\frac{1}{\gamma^3} \|\mathcal{N}W\|^2_{L^2(H^1_\gamma)} + \frac{1}{\gamma^2} \|\mathcal{B}(W^{\mathrm{nc}}, \psi)\|^2_{H^1_\gamma} \right). \tag{15}$$

Observe that there is the loss of one (tangential) derivative for W with respect to the source terms, but no loss of derivatives for the front function ψ (as in Majda's work [39, 40] on shock waves). Since the problem is characteristic, only the trace of the noncharacteristic part of the solution may be controlled at the boundary. The loss of control regards the tangential velocity.

6. Proof of Theorem 2

We describe the main steps of the proof of the above theorem.

(1) *Paralinearization of the equations.*

    Using the paradifferential calculus of Bony [8] and Meyer [47], we substitute in the equations the paradifferential operators (w.r. to the tangential variables (t, x_1)) and obtain a system of ordinary differential equations with derivatives in x_2 and symbols instead of derivatives in (t, x_1). This step essentially reduces to the constant coefficient case.

(2) *Elimination of the front.*

    The projected boundary condition onto a suitable subspace of the frequency space gives an elliptic equation of order one for the front ψ. This property is a key point in our work since it allows to *eliminate* the unknown front and to consider a standard boundary value problem with a symbolic boundary condition (this ellipticity property is also crucial in Majda's analysis on shock waves [39, 40]). One obtains an estimate of the form

$$\|\psi\|^2_{H^1_\gamma} \leq C \left(\frac{1}{\gamma^2} \|\mathcal{B}(W^{\mathrm{nc}}, \psi)\|^2_{H^1_\gamma} + \|W^{\mathrm{nc}}{}_{|x_2=0}\|^2_{L^2_\gamma} \right)$$
$$+ \; error \; terms,$$

with no loss of regularity with respect to the source terms. In view of (15), it is enough to estimate W.

(3) *Problem with reduced boundary conditions.*

    The projection of the boundary condition onto the orthogonal subspace gives a boundary condition involving only W^{nc}, i.e., without involving ψ. Thus we are left with the (paradifferential version of the) linear problem for W

$$\mathcal{N}_r W^+ = \mathbf{A}_0^r \, \partial_t W^+ + \mathbf{A}_1^r \, \partial_{x_1} W^+ + \mathbf{I}_2 \, \partial_{x_2} W^+ + \mathbf{A}_0^r \, \mathbf{C}^r \, W^+ = F^+, \quad x_2 > 0,$$

$$\mathcal{N}_l W^- = \mathbf{A}_0^l \, \partial_t W^- + \mathbf{A}_1^l \, \partial_{x_1} W^- + \mathbf{I}_2 \, \partial_{x_2} W^- + \mathbf{A}_0^l \, \mathbf{C}^l \, W^- = F^-, \quad x_2 > 0,$$

$$\mathbf{\Pi} \, \widetilde{\mathbf{M}} \, W_{|x_2=0} = \mathbf{\Pi} \, g, \qquad x_2 = 0, \tag{16}$$

where $\mathbf{\Pi}$ denotes the suitable projection operator.

For this problem the boundary is characteristic with constant multiplicity, as in the analysis of Majda and Osher [43]. Differently from [43], our problem satisfies a Kreiss-Lopatinskii condition in the weak sense and not uniformly. In fact, the Lopatinskii determinant associated to the boundary condition vanishes at some points. Recalling that the uniform Kreiss-Lopatinskii condition is a necessary and sufficient condition for the L^2 estimate with no loss of derivatives, the failure of the uniform Kreiss-Lopatinskii condition yields necessarily a loss of derivatives with respect to the source terms.

The proof of the main energy estimate is based on the construction of a *degenerate Kreiss' symmetrizer*. We add the technics of Majda and Osher [43] for the analysis of characteristic boundaries to Coulombel's technic [14, 15] for the analysis of the singularities near the wave numbers where the Lopatinskii condition fails.

In order to explain the main ideas, let us consider for simplicity the linearization around the piecewise constant solution (10).

Then, instead of (16), we have a problem of the form ($\widehat{W} = \widehat{W}(\delta, \eta)$ is the Fourier transform in (t, x_1))

$$(\tau \mathcal{A}_0 + i\eta \mathcal{A}_1)\widehat{W} + \mathcal{A}_2 \frac{d\widehat{W}}{dx_2} = 0, \quad x_2 > 0,$$

$$\beta(\tau, \eta)\widehat{W}^{\mathrm{nc}} = \widehat{h}, \quad x_2 = 0, \tag{17}$$

where $\tau = \delta + i\eta$ and where $\mathcal{A}_0, \mathcal{A}_1, \mathcal{A}_2$ are matrices with constant coefficients. Because of the characteristic boundary, the two first equations do not involve differentiation with respect to the normal variable x_2:

$$(\tau + iv_r\eta) \, \widehat{W}_1^+ - ic^2\eta \, \widehat{W}_2^+ + ic^2\eta \, \widehat{W}_3^+ = 0,$$

$$(\tau + iv_l\eta) \, \widehat{W}_1^- - ic^2\eta \, \widehat{W}_2^- + ic^2\eta \, \widehat{W}_3^- = 0.$$

For $\mathrm{Re} \, \tau > 0$, we obtain an expression for $\widehat{W}_1^+$ and $\widehat{W}_1^-$ that we plug into the other equations in (17). This operation yields a system of O.D.E. of the form:

$$\frac{d\widehat{W}^{\mathrm{nc}}}{dx_2} = \mathcal{A}(\tau, \eta) \, \widehat{W}^{\mathrm{nc}}, \quad x_2 > 0,$$

$$\beta(\tau, \eta)\widehat{W}^{\mathrm{nc}}(0) = \widehat{h}, \qquad x_2 = 0.$$

By microlocalization, the analysis is performed locally in the neighborhood of points (τ, η) with $\mathrm{Re} \, \tau \geq 0$. In points with $\mathrm{Re} \, \tau > 0$ the matrix $\mathcal{A}(\tau, \eta)$ is regular and the Lopatinskii determinant doesn't vanish; therefore in the neighborhood of those points we can construct a classical Kreiss' symmetrizer. This symmetrizer would yield an L^2 estimate with no loss of derivatives. When $\mathrm{Re}\,\tau = 0$ we find points of the following type:

1) Points where $\mathcal{A}(\tau, \eta)$ is diagonalizable and the Lopatinskii condition is satisfied.

In these points the analysis is the same as for the interior points with $\mathrm{Re}\,\tau > 0$. Therefore we can construct a classical Kreiss' symmetrizer. This symmetrizer would yield an L^2 estimate with no loss of derivatives.

2) Points where $\mathcal{A}(\tau, \eta)$ is diagonalizable and the Lopatinskii condition breaks down.

The points where the Lopatinskii determinant vanishes correspond to critical speeds which are exactly the speeds of the kink modes in [4]. Since the Lopatinskii determinant has simple roots, it behaves like $\gamma = \mathrm{Re}\,\tau$ uniformly in a neighborhood of the points. Using this fact and the diagonalizability of $\mathcal{A}(\tau, \eta)$ we construct a degenerate Kreiss' symmetrizer; this yields an L^2 estimate with loss of one derivative.

3) Points where $\mathcal{A}(\tau, \eta)$ is not diagonalizable. In those points, the Lopatinskii condition is satisfied.

Differently from the other cases we construct a suitable non-diagonal symmetrizer. This case doesn't yield a loss of derivatives.

4) Poles of $\mathcal{A}$. At those points, the Lopatinskii condition is satisfied.

The matrix $\mathcal{A}(\tau, \eta)$ is not smoothly diagonalizable. Consequently, Majda and Osher [43] construction of a symmetrizer in this case involves a singularity in the symmetrizer. We avoid this singularity and construct a smooth symmetrizer by working on the original system (16).

In the end, we consider a partition of unity to patch things together and we get the degenerate Kreiss' symmetrizer used in order to derive the energy estimate.

7. Linearized stability in Sobolev norms

Our second result concerns the well-posedness in Sobolev norm. In view of the future application to the initial boundary value problem, we consider functions defined up to time T. Let us set

$$\begin{aligned} \Omega_T &:= \{(t, x_1, x_2) \in \mathbb{R}^3 \text{ s.t. } -\infty < t < T, \, x_2 > 0\}, \\ \omega_T &:= \{(t, x_1, x_2) \in \mathbb{R}^3 \text{ s.t. } -\infty < t < T, \, x_2 = 0\}. \end{aligned} \tag{18}$$

Theorem 3. [20] *Let $T > 0$ and $m \in \mathbb{N}$. Assume that*
(i) *the particular solution $\overline{U}^\pm$ defined by (10) satisfies (11),*
(ii) *$(\overline{U}^\pm + \dot{U}_{r,l}, \pm x_2 + \dot{\Phi}_{r,l})$ satisfies (7) and (8),*
(iii) *the perturbation $(\dot{U}_{r,l}, \dot{\Phi}_{r,l}) \in H_\gamma^{m+3}(\Omega_T)$ has compact support and is sufficiently small in $H^6(\Omega_T)$.*

Then there exist some constants $C > 0$ and $\gamma \geq 1$ such that, if $(F_\pm, g) \in H^{m+1}(\Omega_T) \times H^{m+1}(\omega_T)$ vanish in the past (i.e., for $t < 0$), then there exists a unique solution $(W^\pm, \psi) \in H^m(\Omega_T) \times H^{m+1}(\omega_T)$ to (13), (14) that vanishes in the past. Moreover the following estimate holds:

$$\|W\|_{H_\gamma^m(\Omega_T)} + \|W^{nc}_{|x_2=0}\|_{H_\gamma^m(\omega_T)} + \|\psi\|_{H_\gamma^{m+1}(\omega_T)} \leq C \Big\{ \|F\|_{H_\gamma^{m+1}(\Omega_T)} \tag{19}$$

$$+ \|g\|_{H_\gamma^{m+1}(\omega_T)} + \Big(\|F\|_{H_\gamma^4(\Omega_T)} + \|g\|_{H_\gamma^4(\omega_T)} \Big) \|(\dot{U}_{r,l}, \dot{\Phi}_{r,l})\|_{H_\gamma^{m+3}(\Omega_T)} \Big\}.$$

Observe that there is the loss of *one* derivative for W with respect to the source terms, and the loss of *three* derivatives with respect to the coefficients. Again we have no loss of derivatives for the front function ψ (as in Majda's work [39, 40] on shock waves).

For the forthcoming analysis of the nonlinear problem by a Nash-Moser procedure it's important to observe that (19) is a *"tame estimate"* (roughly speaking: linear in high norms which are multiplied by low norms).

Proof. We describe the main steps of the proof of the above theorem.

1) *Estimate of tangential derivatives.* The tangential derivatives $\partial_t^h \partial_{x_1}^k W$ and the front function ψ are estimated by differentiation along tangential directions of the equations and application of the L^2 energy estimate given in Theorem 2. We obtain

$$\sqrt{\gamma} \|W\|_{L^2(H_\gamma^m)} + \|W^{nc}_{|x_2=0}\|_{H_\gamma^m(\omega_T)} + \|\psi\|_{H_\gamma^{m+1}(\omega_T)} \tag{20}$$

$$\leq \frac{C}{\gamma} \Big\{ \|F\|_{L^2(H_\gamma^{m+1})} + \|g\|_{H_\gamma^{m+1}(\omega_T)} + \|W\|_{W^{1,\infty}(\Omega_T)} \|(\dot{U}_{r,l}, \nabla \dot{\Phi}_{r,l}\|_{H_\gamma^{m+2}(\Omega_T)}$$

$$+ \Big(\|W^{nc}_{|x_2=0}\|_{L^\infty(\omega_T)} + \|\psi\|_{W^{1,\infty}(\omega_T)} \Big) \|(\dot{U}_{r,l}, \partial_{x_2} \dot{U}_{r,l}, \nabla \dot{\Phi}_{r,l})_{|x_2=0}\|_{H^{m+1}(\omega_T)} \Big\},$$

where $\|\cdot\|_{L^2(H_\gamma^m)}$ denotes the norm of $L^2(\mathbb{R}^+; H_\gamma^m(\omega_T))$.

2) *Estimate of the linearized vorticity.* Consider the original non linear equations. On both sides of the interface the solution is smooth and satisfies

$$\rho(\partial_t \mathbf{u} + (\mathbf{u} \cdot \nabla)\mathbf{u}) + \nabla p(\rho) = 0.$$

Hence the vorticity $\xi := \partial_{x_1} u - \partial_{x_2} v$ satisfies on both sides

$$\partial_t \xi + \mathbf{u} \cdot \nabla \xi + \xi \nabla \cdot \mathbf{u} = 0.$$

Recalling that the interface is a streamline and that there is continuity of the normal velocity across the interface, this suggests the possibility of estimates of the vorticity on either part of the front. This leads to introduce the "linearized vorticity"

$$\dot{\xi}_\pm := \partial_{x_1} \dot{u}_\pm - \frac{1}{\partial_{x_2} \Phi_{r,l}} (\partial_{x_1} \Phi_{r,l} \partial_{x_2} \dot{u}_\pm + \partial_{x_2} \dot{v}_\pm).$$

Then

$$\partial_t \dot{\xi}_+ + v_r \partial_{x_1} \dot{\xi}_+$$
$$= \partial_{x_1} \mathcal{F}_2^+ - \frac{1}{\partial_{x_2} \Phi_r}(\partial_{x_1} \Phi_r \partial_{x_2} \mathcal{F}_2^+ + \partial_{x_2} \mathcal{F}_1^+) + \Lambda_1^r \cdot \partial_{x_1} \dot{U}_+ + \Lambda_2^r \cdot \partial_{x_2} \dot{U}_+ \,,$$

$$\partial_t \dot{\xi}_- + v_l \partial_{x_1} \dot{\xi}_-$$
$$= \partial_{x_1} \mathcal{F}_2^- - \frac{1}{\partial_{x_2} \Phi_l}(\partial_{x_1} \Phi_l \partial_{x_2} \mathcal{F}_2^- + \partial_{x_2} \mathcal{F}_1^-) + \Lambda_1^l \cdot \partial_{x_1} \dot{U}_- + \Lambda_2^l \cdot \partial_{x_2} \dot{U}_- \,,$$

$$(21)$$

where $\Lambda_{1,2}^{r,l}$ are $\mathcal{C}^\infty$ functions of $(\dot{U}_{r,l}, \nabla \dot{U}_{r,l}, \nabla \dot{\Phi}_{r,l}, \nabla^2 \dot{\Phi}_{r,l})$ and where $\mathcal{F}_{1,2}^\pm$ are $\mathcal{C}^\infty$ functions of $U_{r,l}, \nabla \Phi_{r,l}$ and depend linearly on $F^\pm$ and $W^\pm$.

A standard energy argument may be applied to (21). In fact we may observe that, if we take any derivative ∂^α of (21), multiply by $\partial^\alpha \xi^\pm$ and integrate over Ω_T, then the usual integrations by parts give no boundary terms. We obtain the estimate

$$\gamma \|\dot{\xi}_\pm\|_{H_\gamma^{m-1}(\Omega_T)} \le C \Big\{ \|F\|_{H_\gamma^m(\Omega_T)} + \|F\|_{L^\infty(\Omega_T)} \|\nabla \dot{\Phi}_{r,l}\|_{H_\gamma^m(\Omega_T)}$$
$$+ \|W\|_{H_\gamma^m(\Omega_T)} + \|W\|_{W^{1,\infty}(\Omega_T)} \big(\|\dot{U}_{r,l}\|_{H^{m+1}(\Omega_T)} + \|\nabla \dot{\Phi}_{r,l}\|_{H^m(\Omega_T)} \big) \Big\}.$$

$$(22)$$

3) *Estimate of normal derivatives.* We have

$$\partial_{x_2} W_1^\pm = \frac{1}{\langle \partial_{x_1} \Phi_{r,l} \rangle^2}$$
$$\times \Big\{ \partial_{x_2} \Phi_{r,l} (\partial_{x_1} \dot{u}_\pm - \dot{\xi}_\pm) - \partial_{x_1} \Phi_{r,l} (\partial_{x_2} T_{r,l} W^\pm)_3 - (\partial_{x_2} T_{r,l} W^\pm)_2 \Big\},$$

where $T_{r,l} = T(U_{r,l}, \Phi_{r,l})$ denotes a suitable invertible matrix such that $W^\pm = T_{r,l}^{-1} \dot{U}_\pm$ (recall that $\dot{U}_\pm$ is the unknown defined in (12). The above equality shows that we may estimate $\partial_{x_2} W_1^\pm$ by the previous steps. The estimate of normal derivatives $\partial_{x_2} W^{nc}$ of the noncharacteristic part of the solution follows directly from the equations:

$$\mathbf{I}_2 \, \partial_{x_2} W^\pm = F^\pm - \mathbf{A}_0^{r,l} \, \partial_t W^\pm - \mathbf{A}_1^{r,l} \, \partial_{x_1} W^\pm - \mathbf{A}_0^{r,l} \, \mathbf{C}^{r,l} \, W^\pm \,,$$

since

$$\mathbf{I}_2 := \text{diag} \,(0, 1, 1), \quad W^{nc} := (W_2^+, W_3^+, W_2^-, W_3^-).$$

We obtain for $k = 1, \ldots, m$

$$\|\partial_{x_2}^k W\|_{L^2(H_\gamma^{m-k})}$$
$$\le C \Big\{ \|F\|_{H_\gamma^{m-1}(\Omega_T)} + \|\dot{\xi}_\pm\|_{H_\gamma^{m-1}(\Omega_T)} + \|\dot{\xi}_\pm\|_{L^\infty(\Omega_T)} \|\nabla \dot{\Phi}_{r,l}\|_{H_\gamma^{m-1}(\Omega_T)}$$
$$+ \|W\|_{L^\infty(\Omega_T)} \|(\dot{U}_{r,l}, \nabla \dot{\Phi}_{r,l})\|_{H_\gamma^m(\Omega_T)} + \|W\|_{L^2(H_\gamma^m)} + \|W\|_{H_\gamma^{m-1}(\Omega_T)} \Big\}.$$

$$(23)$$

By a combination of (20), (22) and (23) we finally obtain (19). The existence of the solution of the linear problem (13), (14) is a consequence of the well-posedness result of [16].   □

8. The Nash-Moser iterative scheme

8.1. Preliminary steps

We reduce the nonlinear problem (6), (7), (8), (9), to a new problem with solution vanishing in the past. We proceed as follows.

1) Given initial data $U_0^\pm = \overline{U}^\pm + \dot{U}_0^\pm$, $\dot{U}_0^\pm \in H^{2\mu+3/2}(\mathbb{R}_+^2)$, and $\varphi_0 \in H^{2\mu+2}(\mathbb{R})$, $\dot{U}_0^\pm$ and φ_0 with compact support and small enough, there exist an approximate "solution" U^a, Φ^a, φ^a, such that $U^a - \overline{U} = \dot{U}^a \in H^{2\mu+2}(\Omega)$, $\Phi^{a\pm} \mp x_2 = \dot{\Phi}^{a\pm} \in H^{2\mu+3}(\Omega)$, $\varphi^a \in H^{2\mu+5/2}(\omega)$, and such that

$$\partial_t^j \mathbb{L}(U^a, \Phi^a)_{|t=0} = 0, \qquad \text{for } j = 0, \dots, 2\mu, \tag{24}$$

$$\partial_t \Phi^a + v^a \, \partial_{x_1} \Phi^a - u^a = 0, \tag{25}$$

$$\varphi^a = \Phi_{|x_2=0}^{a+} = \Phi_{|x_2=0}^{a-}, \tag{26}$$

$$\mathbb{B}(U_{|x_2=0}^a, \varphi^a) = 0. \tag{27}$$

The functions $\dot{U}^a, \dot{\Phi}^{a\pm}, \varphi^a$ satisfy a suitable a priori estimate and may be taken with compact supports.

2) We write the equations (6), (8) for $U = (U^+, U^-), \Phi = (\Phi^+, \Phi^-)$ in the form

$$\mathbb{L}(U, \Phi) = 0, \qquad \mathbb{B}(U_{|x_2=0}, \varphi) = 0,$$

and introduce

$$\begin{cases} f^a := -\mathbb{L}(U^a, \Phi^a), & t > 0, \\ f^a := 0, & t < 0. \end{cases}$$

Because $\dot{U}^a \in H^{2\mu+2}(\Omega)$, and $\dot{\Phi}^a \in H^{2\mu+3}(\Omega)$, (24) yields $f^a \in H^{2\mu+1}(\Omega)$.

3) For all real number $T > 0$, we let Ω_T^+, and ω_T^+ denote the sets

$$\omega_T^+ :=]0, T[\times \mathbb{R}, \quad \Omega_T^+ :=]0, T[\times \mathbb{R} \times]0, +\infty[= \omega_T^+ \times \mathbb{R}^+.$$

Given the approximate solution (U^a, Φ^a) and the function f^a, then $(U, \Phi) = (U^a, \Phi^a) + (V, \Psi)$ is a solution on Ω_T^+ of (6), (7), (8), (9), if $V = (V^+, V^-), \Psi = (\Psi^+, \Psi^-)$ satisfy the following system:

$$\mathcal{L}(V, \Psi) = f^a, \qquad\qquad\qquad\qquad \text{in } \Omega_T,$$
$$\mathcal{E}(V, \Psi) := \partial_t \Psi + (v^a + v) \, \partial_{x_1} \Psi - u + v \, \partial_{x_1} \Phi^a = 0, \qquad \text{in } \Omega_T,$$
$$\Psi_{|x_2=0}^+ = \Psi_{|x_2=0}^- =: \psi, \qquad\qquad\qquad \text{on } \omega_T, \tag{28}$$
$$\mathcal{B}(V_{|x_2=0}, \psi) = 0, \qquad\qquad\qquad\qquad \text{on } \omega_T,$$
$$(V, \Psi) = 0, \qquad\qquad\qquad\qquad\qquad\quad \text{for } t < 0,$$

where

$$\mathcal{L}(V, \Psi) := \mathbb{L}(U^a + V, \Phi^a + \Psi) - \mathbb{L}(U^a, \Phi^a),$$
$$\mathcal{B}(V_{|x_2=0}, \psi) := \mathbb{B}(U^a_{|x_2=0} + V_{|x_2=0}, \varphi^a + \psi). \tag{29}$$

We note that $(V, \Psi) = 0$ satisfy (28) for $t < 0$, because $f^a = 0$ for $t < 0$, and $\mathbb{B}(U^a_{|x_2=0}, \varphi^a) = 0$ for all $t \in \mathbb{R}$. Therefore the initial nonlinear problem on Ω_T^+ is now substituted for a problem on Ω_T. The initial data (9) are absorbed into the equations by the introduction of the approximate solution (U^a, Φ^a, φ^a), and the problem has to be solved in the class of functions vanishing in the past (i.e., for $t < 0$), which is exactly the class of functions in which we have a well-posedness result for the linearized problem, see Theorem 3.

4) We solve problem (28) by a Nash-Moser type iteration. This method requires a family of smoothing operators. For $T > 0$, $s \geq 0$, and $\gamma \geq 1$, we let

$$\mathcal{F}^s_\gamma(\Omega_T) := \left\{ u \in H^s_\gamma(\Omega_T), \quad u = 0 \text{ for } t < 0 \right\}.$$

The definition of $\mathcal{F}^s_\gamma(\omega_T)$ is entirely similar.

Proposition 1. *Let $T > 0$, $\gamma \geq 1$, and let $M \in \mathbb{N}$, with $M \geq 4$. There exists a family $\{S_\theta\}_{\theta \geq 1}$ of operators*

$$S_\theta: \quad \mathcal{F}^3_\gamma(\Omega_T) \times \mathcal{F}^3_\gamma(\Omega_T) \longrightarrow \bigcap_{\beta \geq 3} \mathcal{F}^\beta_\gamma(\Omega_T) \times \mathcal{F}^\beta_\gamma(\Omega_T),$$

and a constant $C > 0$ (depending on M), such that

$$\|S_\theta U\|_{H^\beta_\gamma(\Omega_T)} \leq C \, \theta^{(\beta-\alpha)_+} \|U\|_{H^\alpha_\gamma(\Omega_T)}, \quad \forall \alpha, \beta \in \{1, \dots, M\},$$
$$\|S_\theta U - U\|_{H^\beta_\gamma(\Omega_T)} \leq C \, \theta^{\beta-\alpha} \|U\|_{H^\alpha_\gamma(\Omega_T)}, \quad 1 \leq \beta \leq \alpha \leq M,$$
$$\left\|\frac{d}{d\theta} S_\theta U\right\|_{H^\beta_\gamma(\Omega_T)} \leq C \, \theta^{\beta-\alpha-1} \|U\|_{H^\alpha_\gamma(\Omega_T)}, \quad \forall \alpha, \beta \in \{1, \dots, M\}.$$

Moreover,

(i) *if $U = (u^+, u^-)$ satisfies $u^+ = u^-$ on ω_T, then $S_\theta u^+ = S_\theta u^-$ on ω_T,*
(ii) *the following estimate holds:*

$$\|(S_\theta u^+ - S_\theta u^-)_{|x_2=0}\|_{H^\beta_\gamma(\omega_T)} \leq C \, \theta^{(\beta+1-\alpha)_+} \|(u^+ - u^-)_{|x_2=0}\|_{H^\alpha_\gamma(\omega_T)},$$
$$\forall \alpha, \beta \in \{1, \dots, M\}.$$

There is another family of operators, still denoted S_θ, that acts on functions that are defined on the boundary ω_T, and that enjoy the above properties with the norms $\|\cdot\|_{H^\alpha_\gamma(\omega_T)}$.

In our case it appears to be convenient the choice $M := 2\mu + 3$.

8.2. Description of the iterative scheme

The iterative scheme starts from $V_0 = 0, \Psi_0 = 0, \psi_0 = 0$. Assume that V_k, Ψ_k, ψ_k are already given for $k = 0, \dots, n$ and verify

$$(V_k, \Psi_k, \psi_k) = 0, \qquad \text{for } t < 0,$$
$$(\Psi_k^+)_{|x_2=0} = (\Psi_k^-)_{|x_2=0} = \psi_k, \quad \text{on } \omega_T.$$

Given $\theta_0 \geq 1$, let us set $\theta_n := (\theta_0^2 + n)^{1/2}$, and consider the smoothing operators S_{θ_n}. Let us set

$$V_{n+1} = V_n + \delta V_n, \qquad \Psi_{n+1} = \Psi_n + \delta \Psi_n, \qquad \psi_{n+1} = \psi_n + \delta \psi_n. \tag{30}$$

We introduce the decomposition ($\mathcal{L}$ is defined in (29))

$$\begin{aligned}
\mathcal{L}(V_{n+1}, \Psi_{n+1}) - \mathcal{L}(V_n, \Psi_n) &= \mathbb{L}(U^a + V_{n+1}, \Phi^a + \Psi_{n+1}) - \mathbb{L}(U^a + V_n, \Phi^a + \Psi_n) \\
&= \mathbb{L}'(U^a + V_n, \Phi^a + \Psi_n)(\delta V_n, \delta \Psi_n) + e_n' \\
&= \mathbb{L}'(U^a + S_{\theta_n} V_n, \Phi^a + S_{\theta_n} \Psi_n)(\delta V_n, \delta \Psi_n) + e_n' + e_n'',
\end{aligned}$$

where e_n' denotes the usual "quadratic" error of Newton's scheme, and e_n'' the "substitution" error, due to the regularization of the state where the operator is calculated.

Thanks to the properties of the smoothing operators, we have $(S_{\theta_n} \Psi_n^+)_{|x_2=0} = (S_{\theta_n} \Psi_n^-)_{|x_2=0}$ and we denote $\psi_n^\sharp$ the common trace of these two functions. With this notation, we have

$$\begin{aligned}
&\mathcal{B}((V_{n+1})_{|x_2=0}, \psi_{n+1}) - \mathcal{B}((V_n)_{|x_2=0}, \psi_n) \\
&= \mathbb{B}'((U^a + V_n)_{|x_2=0}, \varphi^a + \psi_n)((\delta V_n)_{|x_2=0}, \delta \psi_n) + \tilde{e}_n' \\
&= \mathbb{B}'((U^a + S_{\theta_n} V_n)_{|x_2=0}, \varphi^a + \psi_n^\sharp)((\delta V_n)_{|x_2=0}, \delta \psi_n) + \tilde{e}_n' + \tilde{e}_n'',
\end{aligned}$$

where $\tilde{e}_n'$ denotes the "quadratic" error, and $\tilde{e}_n''$ the "substitution" error.

The inversion of the operator $(\mathbb{L}', \mathbb{B}')$ requires the linearization around a state satisfying the constraints (7), (8). We thus need to introduce a smooth modified state, denoted $V_{n+1/2}, \Psi_{n+1/2}, \psi_{n+1/2}$, that satisfies the above-mentioned constraints, see [20] for details of the construction. Accordingly, we introduce the decompositions

$$\begin{aligned}
&\mathcal{L}(V_{n+1}, \Psi_{n+1}) - \mathcal{L}(V_n, \Psi_n) \\
&= \mathbb{L}'(U^a + V_{n+1/2}, \Phi^a + \Psi_{n+1/2})(\delta V_n, \delta \Psi_n) + e_n' + e_n'' + e_n''',
\end{aligned}$$

and

$$\begin{aligned}
&\mathcal{B}((V_{n+1})_{|x_2=0}, \psi_{n+1}) - \mathcal{B}((V_n)_{|x_2=0}, \psi_n) \\
&= \mathbb{B}'((U^a + V_{n+1/2})_{|x_2=0}, \varphi^a + \psi_{n+1/2})((\delta V_n)_{|x_2=0}, \delta \psi_n) + \tilde{e}_n' + \tilde{e}_n'' + \tilde{e}_n''',
\end{aligned}$$

where $e_n''', \tilde{e}_n'''$ denote the second "substitution" errors. The final step is the introduction of the "good unknown":

$$\delta \dot{V}_n := \delta V_n - \delta \Psi_n \frac{\partial_{x_2}(U^a + V_{n+1/2})}{\partial_{x_2}(\Phi^a + \Psi_{n+1/2})}. \tag{31}$$

This leads to

$$\mathcal{L}(V_{n+1}, \Psi_{n+1}) - \mathcal{L}(V_n, \Psi_n) = \mathbb{L}'_e(U^a + V_{n+1/2}, \Phi^a + \Psi_{n+1/2})\delta\dot{V}_n$$
$$+ e'_n + e''_n + e'''_n + \frac{\delta\Psi_n}{\partial_{x_2}(\Phi^a + \Psi_{n+1/2})} \partial_{x_2}\left\{\mathbb{L}(U^a + V_{n+1/2}, \Phi^a + \Psi_{n+1/2})\right\}, \quad (32)$$

and

$$\mathcal{B}((V_{n+1})_{|x_2=0}, \psi_{n+1}) - \mathcal{B}((V_n)_{|x_2=0}, \psi_n)$$
$$= \mathbb{B}'_e((U^a + V_{n+1/2})_{|x_2=0}, \varphi^a + \psi_{n+1/2})((\delta\dot{V}_n)_{|x_2=0}, \delta\psi_n) + \tilde{e}'_n + \tilde{e}''_n + \tilde{e}'''_n, \quad (33)$$

Here $\mathbb{L}'_e \delta\dot{V}$ denotes the "effective" linear operator obtained by linearizing $\mathbb{L}\delta V$, substituting the good unknown $\delta\dot{V}$ in place of the unknown δV and neglecting the zero order term in $\delta\Psi$, see (13). Similarly, $\mathbb{B}'_e$ is the operator obtained from linearization of the boundary conditions and the introduction of the good unknown.

For the sake of brevity we set

$$D_{n+1/2} := \frac{1}{\partial_{x_2}(\Phi^a + \Psi_{n+1/2})} \partial_{x_2}\left\{\mathbb{L}(U^a + V_{n+1/2}, \Phi^a + \Psi_{n+1/2})\right\},$$

$$\mathbb{B}'_{n+1/2} := \mathbb{B}'_e\left((U^a + V_{n+1/2})_{|x_2=0}, \varphi^a + \psi_{n+1/2}\right).$$

Let us also set

$$e_n := e'_n + e''_n + e'''_n + D_{n+1/2}\,\delta\Psi_n,$$

$$\tilde{e}_n := \tilde{e}'_n + \tilde{e}''_n + \tilde{e}'''_n.$$

The iteration proceeds as follows. Given

$$
\begin{array}{llll}
V_0 := 0, & \Psi_0 := 0, & \psi_0 := 0, & \\
f_0 := S_{\theta_0} f^a, & g_0 := 0, & E_0 := 0, & \tilde{E}_0 := 0, \\
V_1, \ldots, V_n, & \Psi_1, \ldots, \Psi_n, & \psi_1, \ldots, \psi_n, & \\
f_1, \ldots, f_{n-1}, & g_1, \ldots, g_{n-1}, & & \\
e_0, \ldots, e_{n-1}, & \tilde{e}_0, \ldots, \tilde{e}_{n-1}, & &
\end{array}
$$

we first compute for $n \geq 1$

$$E_n := \sum_{k=0}^{n-1} e_k, \qquad \tilde{E}_n := \sum_{k=0}^{n-1} \tilde{e}_k.$$

These are the accumulated errors at the step n. Then we compute f_n, and g_n from the equations:

$$\sum_{k=0}^{n} f_k + S_{\theta_n} E_n = S_{\theta_n} f^a, \qquad \sum_{k=0}^{n} g_k + S_{\theta_n} \tilde{E}_n = 0, \quad (34)$$

and we solve the linear problem

$$\mathbb{L}'_e(U^a + V_{n+1/2}, \Phi^a + \Psi_{n+1/2})\,\delta\dot{V}_n = f_n \qquad \text{in } \Omega_T \,,$$

$$\mathbb{B}'_{n+1/2}((\delta\dot{V}_n)_{|x_2=0}, \delta\psi_n) = g_n \qquad \text{on } \omega_T \,, \qquad (35)$$

$$\delta\dot{V}_n = 0, \quad \delta\psi_n = 0 \qquad \text{for } t < 0 \,,$$

finding $(\delta\dot{V}_n, \delta\psi_n)$. Now we need to construct $\delta\Psi_n = (\delta\Psi_n^+, \delta\Psi_n^-)$ that satisfies $(\delta\Psi_n^\pm)_{|x_2=0} = \delta\psi_n$. Using the explicit expression of the boundary conditions in (35), we first note that $\delta\psi_n$ solves the equation:

$$\partial_t \delta\psi_n + (v^{a+} + v^+_{n+1/2})_{|x_2=0}\, \partial_{x_1}\delta\psi_n$$

$$+ \left\{ \partial_{x_1}(\varphi^a + \psi_{n+1/2}) \frac{\partial_{x_2}(v^{a+} + v^+_{n+1/2})_{|x_2=0}}{\partial_{x_2}(\Phi^{a+} + \Psi^+_{n+1/2})_{|x_2=0}} - \frac{\partial_{x_2}(u^{a+} + u^+_{n+1/2})_{|x_2=0}}{\partial_{x_2}(\Phi^{a+} + \Psi^+_{n+1/2})_{|x_2=0}} \right\} \delta\psi_n$$

$$+ \partial_{x_1}(\varphi^a + \psi_{n+1/2})\,(\delta\dot{v}_n^+)_{|x_2=0} - (\delta\dot{u}_n^+)_{|x_2=0} = g_{n,2} \,, \quad (36)$$

and the equation

$$\partial_t \delta\psi_n + (v^{a-} + v^-_{n+1/2})_{|x_2=0}\, \partial_{x_1}\delta\psi_n$$

$$+ \left\{ \partial_{x_1}(\varphi^a + \psi_{n+1/2}) \frac{\partial_{x_2}(v^{a-} + v^-_{n+1/2})_{|x_2=0}}{\partial_{x_2}(\Phi^{a-} + \Psi^-_{n+1/2})_{|x_2=0}} - \frac{\partial_{x_2}(u^{a-} + u^-_{n+1/2})_{|x_2=0}}{\partial_{x_2}(\Phi^{a-} + \Psi^-_{n+1/2})_{|x_2=0}} \right\} \delta\psi_n$$

$$+ \partial_{x_1}(\varphi^a + \psi_{n+1/2})\,(\delta\dot{v}_n^-)_{|x_2=0} - (\delta\dot{u}_n^-)_{|x_2=0} = g_{n,2} - g_{n,1} \,. \quad (37)$$

We shall thus define $\delta\Psi_n^+, \delta\Psi_n^-$ as the solutions to the following equations:

$$\partial_t \delta\Psi_n^+ + (v^{a+} + v^+_{n+1/2})\, \partial_{x_1}\delta\Psi_n^+$$

$$+ \left\{ \partial_{x_1}(\Phi^{a+} + \Psi^+_{n+1/2}) \frac{\partial_{x_2}(v^{a+} + v^+_{n+1/2})}{\partial_{x_2}(\Phi^{a+} + \Psi^+_{n+1/2})} - \frac{\partial_{x_2}(u^{a+} + u^+_{n+1/2})}{\partial_{x_2}(\Phi^{a+} + \Psi^+_{n+1/2})} \right\} \delta\Psi_n^+$$

$$+ \partial_{x_1}(\Phi^{a+} + \Psi^+_{n+1/2})\,\delta\dot{v}_n^+ - \delta\dot{u}_n^+ = \mathcal{R}_T g_{n,2} + h_n^+ \,, \quad (38)$$

and

$$\partial_t \delta\Psi_n^- + (v^{a-} + v^-_{n+1/2})\, \partial_{x_1}\delta\Psi_n^-$$

$$+ \left\{ \partial_{x_1}(\Phi^{a-} + \Psi^-_{n+1/2}) \frac{\partial_{x_2}(v^{a-} + v^-_{n+1/2})}{\partial_{x_2}(\Phi^{a-} + \Psi^-_{n+1/2})} - \frac{\partial_{x_2}(u^{a-} + u^-_{n+1/2})}{\partial_{x_2}(\Phi^{a-} + \Psi^-_{n+1/2})} \right\} \delta\Psi_n^-$$

$$+ \partial_{x_1}(\Phi^{a-} + \Psi^-_{n+1/2})\,\delta\dot{v}_n^- - \delta\dot{u}_n^- = \mathcal{R}_T(g_{n,2} - g_{n,1}) + h_n^- \,. \quad (39)$$

In (38), and (39), the source terms $h_n^\pm$ have to be chosen suitably. First we require that $h_n^\pm$ vanish on the boundary ω_T, and in the past, so that the unique smooth solutions to (38), and (39) will vanish in the past, and will satisfy the continuity condition $(\delta\Psi_n^\pm)_{|x_2=0} = \delta\psi_n$. In order to compute the source terms $h_n^\pm$, we use a decomposition that is similar to (32) for the operator $\mathcal{E}$ (defined in (28)). We have:

$$\mathcal{E}(V_{n+1}, \Psi_{n+1}) - \mathcal{E}(V_n, \Psi_n) = \mathcal{E}'(V_{n+1/2}, \Psi_{n+1/2})(\delta V_n, \delta\Psi_n) + \hat{e}_n' + \hat{e}_n'' + \hat{e}_n''', \quad (40)$$

where $\hat{e}_n'$ is the "quadratic" error, $\hat{e}_n''$ is the first "substitution" error, and $\hat{e}_n'''$ is the second "substitution" error. We denote

$$\hat{e}_n := \hat{e}_n' + \hat{e}_n'' + \hat{e}_n''', \quad \hat{E}_n := \sum_{k=0}^{n-1} \hat{e}_k.$$

Using the good unknown (31), and omitting the $\pm$ superscripts, we compute

$$\mathcal{E}'(V_{n+1/2}, \Psi_{n+1/2})(\delta V_n, \delta\Psi_n) = \partial_t \delta\Psi_n + (v^a + v_{n+1/2})\,\partial_{x_1}\delta\Psi_n$$
$$+ \left\{ \partial_{x_1}(\Phi^a + \Psi_{n+1/2}) \frac{\partial_{x_2}(v^a + v_{n+1/2})}{\partial_{x_2}(\Phi^a + \Psi_{n+1/2})} - \frac{\partial_{x_2}(u^a + u_{n+1/2})}{\partial_{x_2}(\Phi^a + \Psi_{n+1/2})} \right\} \delta\Psi_n$$
$$+ \partial_{x_1}(\Phi^a + \Psi_{n+1/2})\,\delta\dot{v}_n - \delta\dot{u}_n.$$

Consequently, (38) and (40) yield

$$\mathcal{E}(V_{n+1}^+, \Psi_{n+1}^+) - \mathcal{E}(V_n^+, \Psi_n^+) = \mathcal{R}_T g_{n,2} + h_n^+ + \hat{e}_n^+.$$

Summing these relations, and using $\mathcal{E}(V_0^+, \Psi_0^+) = 0$, we get

$$\mathcal{E}(V_{n+1}^+, \Psi_{n+1}^+) = \mathcal{R}_T\left(\sum_{k=0}^n g_{k,2}\right) + \sum_{k=0}^n h_k^+ + \hat{E}_{n+1}^+$$
$$= \mathcal{R}_T\left(\mathcal{E}((V_{n+1}^+)_{|x_2=0}, \psi_{n+1}) - \tilde{E}_{n+1,2}\right) + \sum_{k=0}^n h_k^+ + \hat{E}_{n+1}^+,$$

where in the last equality, we have summed (33) and used the relation

$$\left(\mathcal{B}((V_{n+1})_{|x_2=0}, \psi_{n+1})\right)_2 = \mathcal{E}((V_{n+1}^+)_{|x_2=0}, \psi_{n+1}).$$

The previous relations lead to the following definition of the source term h_n^+:

$$\sum_{k=0}^n h_k^+ + S_{\theta_n}\left(\hat{E}_n^+ - \mathcal{R}_T \tilde{E}_{n,2}\right) = 0.$$

The definition of h_n^- is entirely similar:

$$\sum_{k=0}^n h_k^- + S_{\theta_n}\left(\hat{E}_n^- - \mathcal{R}_T \tilde{E}_{n,2} + \mathcal{R}_T \tilde{E}_{n,1}\right) = 0.$$

Once $\delta\Psi_n$ is computed, the function δV_n is obtained from (31), and the functions $V_{n+1}, \Psi_{n+1}, \psi_{n+1}$ are obtained from (30).

Finally, we compute $e_n, \hat{e}_n, \tilde{e}_n$ from

$$\mathcal{L}(V_{n+1}, \Psi_{n+1}) - \mathcal{L}(V_n, \Psi_n) = f_n + e_n,$$

$$\mathcal{E}(V_{n+1}^+, \Psi_{n+1}^+) - \mathcal{E}(V_n^+, \Psi_n^+) = \mathcal{R}_T g_{n,2} + h_n^+ + \hat{e}_n^+,$$

$$\mathcal{E}(V_{n+1}^-, \Psi_{n+1}^-) - \mathcal{E}(V_n^-, \Psi_n^-) = \mathcal{R}_T(g_{n,2} - g_{n,1}) + h_n^- + \hat{e}_n^-, \tag{41}$$

$$\mathcal{B}((V_{n+1})_{|x_2=0}, \psi_{n+1}) - \mathcal{B}((V_n)_{|x_2=0}, \psi_n) = g_n + \tilde{e}_n.$$

To compute V_1, Ψ_1, ψ_1 we only consider steps (35), (38), (39), (41) for $n = 0$.
Adding (41) from 0 to N, and combining with (34) gives

$$\mathcal{L}(V_{N+1}, \Psi_{N+1}) - f^a = (S_{\theta_N} - I)f^a + (I - S_{\theta_N})E_N + e_N,$$

$$\mathcal{E}(V_{N+1}^+, \Psi_{N+1}^+) = \mathcal{R}_T\Big(\mathcal{E}((V_{N+1}^+)_{|x_2=0}, \psi_{N+1})\Big)$$
$$+ (I - S_{\theta_N})(\hat{E}_N^+ - \mathcal{R}_T \tilde{E}_{N,2}) + \hat{e}_N^+ - \mathcal{R}_T \tilde{e}_{N,2},$$

$$\mathcal{E}(V_{N+1}^-, \Psi_{N+1}^-) = \mathcal{R}_T\Big(\mathcal{E}((V_{N+1}^-)_{|x_2=0}, \psi_{N+1})\Big)$$
$$+ (I - S_{\theta_N})(\hat{E}_N^- - \mathcal{R}_T(\tilde{E}_{N,2} - \tilde{E}_{N,1}))$$
$$+ \hat{e}_N^- - \mathcal{R}_T(\tilde{e}_{N,2} - \tilde{e}_{N,1}),$$

$$\mathcal{B}((V_{N+1})_{|x_2=0}, \psi_{N+1}) = (I - S_{\theta_N})\tilde{E}_N + \tilde{e}_N.$$

Because $S_{\theta_N} \to I$ as $N \to +\infty$, and since we expect $(e_N, \hat{e}_N, \tilde{e}_N) \to 0$, we will formally obtain the solution of the problem (28) from $\mathcal{L}(V_{N+1}, \Psi_{N+1}) \to f^a$, $\mathcal{B}((V_{N+1})_{|x_2=0}, \psi_{N+1}) \to 0$, and $\mathcal{E}(V_{N+1}, \Psi_{N+1}) \to 0$.

The rigorous proof of convergence follows from a priori estimates of V_k, Ψ_k, ψ_k proved by induction for every k. In the limit we obtain a solution (V, Ψ) on Ω_T of (28), vanishing in the past, which yields that $(U, \Phi) = (U^a, \Phi^a) + (V, \Psi)$ is a solution on Ω_T^+ of (6), (7), (8), (9). This concludes the proof of Theorem 1.

References

[1] R. Agemi. The initial boundary value problem for inviscid barotropic fluid motion. *Hokkaido Math. J.* 10: 156–182, 1981.

[2] S. Alinhac. Existence d'ondes de raréfaction pour des systèmes quasi-linéaires hyperboliques multidimensionnels. *Commun. Partial Diff. Eqs.*, 14(2):173–230, 1989.

[3] S. Alinhac, P. Gérard. *Opérateurs pseudo-différentiels et théorème de Nash-Moser.* InterEditions/Editions du CNRS, Paris, Meudon, 1991.

[4] M. Artola, A. Majda. Nonlinear development of instabilities in supersonic vortex sheets. I. The basic kink modes. *Phys. D*, 28(3):253–281, 1987.

[5] H. Beirão da Veiga. On the barotropic motion of compressible perfect fluids. *Ann. Sc. Norm. Sup. Pisa*, 8: 317–351, 1981.

[6] G. Birkhoff. Helmholtz and Taylor instability. In *Hydrodynamics Instability*, Proc. Symp. Appl. Math., A.M.S., Providence, RI, 13: 55–76, 1962.

[7] A. Blokhin, Y. Trakhinin. Stability of strong discontinuities in fluids and MHD. In *Handbook of Mathematical Fluid Dynamics, Vol. I*, pages 545–652, North-Holland, 2002.

[8] J.-M. Bony. Calcul symbolique et propagation des singularités pour les équations aux dérivées partielles non linéaires. *Ann. Sci. École Norm. Sup. (4)*, 14(2):209–246, 1981.

[9] R.E. Caflish, O.F. Orellana. Long time existence for a slightly perturbed vortex sheet. *Commun. Pure Appl. Math.*, 39:807–838, 1986.

[10] R.E. Caflish, O.F. Orellana. Singular solutions and ill-posedness for the evolution of vortex sheets. *SIAM J. Math. Anal.*, 20:293–307, 1989.

[11] J.-Y. Chemin. *Fluides parfaits incompressibles*. Astérisque (Vol. 230), Paris, 1995.

[12] A. Corli. Asymptotic analysis of contact discontinuities. *Ann. Mat. Pura Appl. (4)*, 173:163–202, 1997.

[13] A. Corli, M. Sablé-Tougeron. Stability of contact discontinuities under perturbations of bounded variation. *Rend. Sem. Mat. Univ. Padova*, 97:35–60, 1997.

[14] J.-F. Coulombel. Weak stability of nonuniformly stable multidimensional shocks. *SIAM J. Math. Anal.*, 34(1):142–172, 2002.

[15] J.-F. Coulombel. Weakly stable multidimensional shocks. *Annales de l'IHP, Analyse Non Linéaire*, 21(4):401–443, 2004.

[16] J.-F. Coulombel. Well-posedness of hyperbolic initial boundary value problems. *J. Math. Pures Appl.*, 84(6):786–818, 2005.

[17] J.-F. Coulombel, A. Morando. Stability of contact discontinuities for the nonisentropic Euler equations. *Ann. Univ. Ferrara Sez. VII (N.S.)*, 50:79–90, 2004.

[18] J.-F. Coulombel, P. Secchi. The stability of compressible vortex sheets in two space dimensions. *Indiana Univ. Math. J.*, 53:941–1012, 2004.

[19] J.-F. Coulombel, P. Secchi. On the transition to instability for compressible vortex sheets. *Proc. Roy. Soc. Edinburgh*, 134A:885–892, 2004.

[20] J.-F. Coulombel, P. Secchi. Nonlinear compressible vortex sheets in two space dimensions, submitted.

[21] J.M. Delort. Existence de nappes de tourbillon en dimension deux. *J. Amer. Math. Soc.*, 4:553–586, 1991.

[22] J. Duchon, R. Robert. Global vortex sheet solutions of Euler equations in the plane. *J. Diff. Eqns.*, 73:215–224, 1988.

[23] L.C. Evans, S. Muller. Hardy spaces and the two-dimensional Euler equations with nonnegative vorticity. *J. Amer. Math. Soc.*, 1:199–219, 1994.

[24] J.A. Fejer, J. W. Miles. On the stability of a plane vortex sheet with respect to three-dimensional disturbances. *J. Fluid Mech.*, 15:335–336, 1963.

[25] J. Francheteau, G. Métivier. *Existence de chocs faibles pour des systèmes quasi-linéaires hyperboliques multidimensionnels*. Astérisque (Vol. 268), Paris, 2000.

[26] M. Grassin Global smooth solutions to Euler equations for a perfect gas. *Indiana Univ. Math. J.*, 47:1397–1432, 1998.

[27] M. Grassin, D. Serre Existence de solutions globales et régulières aux équations d'Euler pour un gaz parfait isentropique. *C. R. Acad. Sci. Paris Sér. I Math.*, 325(7):721–726, 1997.

[28] E. Harabetian. A convergent series expansion for hyperbolic systems of conservation laws. *Trans. Amer. Math. Soc.*, 294 (1986), no. 2, 383–424.

[29] D. Hateau. Instabilité des feuilles de tourbillon avec conduction de chaleur. *C. R. Acad. Sci. Paris Sér. I Math.*, 330(7):629–633, 2000.

[30] T. Kato. The Cauchy problem for quasi-linear symmetric hyperbolic systems. *Arch. Rat. Mech. Anal.* 58:181–205, 1975.

[31] R. Krasny. On singularity formation in a vortex sheet by the point-vortex approximation. *J. Fluid Mech.* 167: 65–93, 1986.

[32] R. Krasny. Computing vortex sheet motion. In *Proc. Int. Congress Math., Kyoto 1990* Math. Soc. Japan, Tokyo, 2: 1573–1583, 1991.

[33] H.-O. Kreiss. Initial boundary value problems for hyperbolic systems. *Commun. Pure Appl. Math.*, 23:277–298, 1970.

[34] P.D. Lax. Hyperbolic systems of conservation laws. II. *Commun. Pure Appl. Math.*, 10:537–566, 1957.

[35] G. Lebeau. Régularité du problème de Kelvin-Helmholtz pour l'équation d'Euler 2d. *ESAIM Control Optim. Calc. Var.*, 8:801–825, 2002.

[36] H. Lindblad. Well-posedness for the motion of an incompressible liquid with free surface boundary. *Annals of Math.*, 162 no. 1, 2005.

[37] H. Lindblad. Well-posedness for the motion of a compressible liquid with free surface boundary. *Commun. Math. Phys.*, to appear.

[38] T.-L. Liu, T. Yang. Compressible Euler equations with vacuum. *J. Diff. Eqns.* 137: 223–237, 1997.

[39] A. Majda. The stability of multidimensional shock fronts. *Mem. Amer. Math. Soc.*, 41(275):iv+95, 1983.

[40] A. Majda. The existence of multidimensional shock fronts. *Mem. Amer. Math. Soc.*, 43(281):v+93, 1983.

[41] A. Majda. Remarks on weak solutions for vortex sheets with a distinguished sign. *Indiana Univ. Math. J.*, 42:921–939, 1993.

[42] A. Majda, A. Bertozzi. *Vorticity and incompressible flow.* Cambridge University Press, Cambridge, 2002.

[43] A. Majda, S. Osher. Initial-boundary value problems for hyperbolic equations with uniformly characteristic boundary. *Commun. Pure Appl. Math.*, 28(5):607–675, 1975.

[44] C. Marchioro, M. Pulvirenti. *Mathematical theory of incompressible nonviscous fluids.* Springer, 1994.

[45] D.I. Meiron, G.R. Baker, S.A. Orszag. Analytic structure of vortex sheet dynamics. Part I. *J. Fluid Mech.*, 114: 283, 1982.

[46] G. Métivier. Stability of multidimensional shocks. In *Advances in the theory of shock waves*, PNDEA, Birkhäuser, 25–103, 2001.

[47] Y. Meyer. Remarques sur un théorème de J.-M. Bony. *Rend. Circ. Mat. Palermo (2)*, suppl. 1:1–20, 1981.

[48] J.W. Miles. On the disturbed motion of a plane vortex sheet. *J. Fluid Mech.*, 4:538–552, 1958.

[49] D.W. Moore. The spontaneous appearance of a singularity in the shape of an evolving vortex sheet. *Proc. Roy. Soc. London Ser. A*, 365: 105–119, 1979.

[50] V. Scheffer. An inviscid flow with compact support in space-time. *J. Geom. Anal.* 3: 343–401, 1993.

[51] D. Serre. Solutions classiques globales des équations d'Euler pour un fluid parfait compressible. *Ann. Inst. Fourier*, 47:139–153, 1997.

[52] D. Serre. *Systems of conservation laws.* 2. Cambridge Univ. Press, Cambridge, 2000.

[53] A. Shnirelman. On the non-uniqueness of weak solutions of the Euler equations. *Commun. Pure Appl. Math.* L: 1261–1286, 1997.

[54] T. Sideris. Formation of singularities in three-dimensional compressible flow. *Commun. Math. Phys.* 10: 475–485, 1985.

[55] C. Sulem, P.L. Sulem, C. Bardos, U. Frisch. Finite time analyticity for the two and three-dimensional Kevin-Helmholtz instability. *Commun. Math. Phys.*, 80:485–516, 1981.

[56] P. Woodward. Simulation of the Kelvin-Helmholtz instability of a supersonic slip surface with a piecewise parabolic method. L.L.L. preprint, 1984.

[57] S. Wu. Recent progress in mathematical analysis of vortex sheets I.C.M. 2002, Vol. III, 233–242.

[58] T. Yang, C.J. Zhu. Non-existence of global smooth solutions to symmetrizable non-linear hyperbolic systems. *Proc. R. Soc. Edinb.* 133A: 719–728, 2003.

Paolo Secchi
Dipartimento di Matematica
Facoltà di Ingegneria
Via Valotti 9
I-25123 Brescia, Italy
e-mail: `paolo.secchi@ing.unibs.it`

Analysis and Simulation of Fluid Dynamics

Advances in Mathematical Fluid Mechanics, 229–246

© 2006 Birkhäuser Verlag Basel/Switzerland

Existence and Stability of Compressible and Incompressible Current-Vortex Sheets

Yuri Trakhinin

Abstract. Recent author's results in the investigation of current-vortex sheets (MHD tangential discontinuities) are surveyed. A sufficient condition for the neutral stability of planar compressible current-vortex sheets is first found for a general case of the unperturbed flow. In astrophysical applications, this condition can be considered as the sufficient condition for the stability of the heliopause, which is modelled by an ideal compressible current-vortex sheet and caused by the interaction of the supersonic solar wind plasma with the local interstellar medium (in some sense, the heliopause is the boundary of the solar system). The linear variable coefficients problem for nonplanar compressible current-vortex sheets is studied as well. Since the tangential discontinuity is characteristic, the functional setting is provided by the anisotropic weighted Sobolev spaces. The a priori estimate deduced for this problem is a necessary step to prove the local-in-time existence of current-vortex sheet solutions of the nonlinear equations of ideal compressible MHD. Analogous results are obtained for incompressible current-vortex sheets. In the incompressibility limit the sufficient stability condition found for compressible current-vortex sheets describes exactly the half of the whole parameter domain of linear stability of planar discontinuities in ideal incompressible MHD.

1. Introduction

We consider the equations of magnetohydrodynamics (MHD) governing the motion of an ideal (inviscid and perfectly conducting) compressible fluid. In the nonconservative form the MHD equations read (see, e.g., [11, 13]):

$$\frac{1}{\rho c^2}\frac{dp}{dt} + \operatorname{div}\mathbf{v} = 0\,, \qquad \rho\frac{d\mathbf{v}}{dt} - (\mathbf{H}, \nabla\mathbf{H}) + \nabla q = 0\,,$$

$$\frac{d\mathbf{H}}{dt} - (\mathbf{H}, \nabla)\mathbf{v} + \mathbf{H}\operatorname{div}\mathbf{v} = 0\,, \qquad \frac{dS}{dt} = 0\,. \tag{1}$$

This research was supported under EPSRC research grants number GR/R79753/01 and GR/S96609/01.

Here $\rho = \rho(t, \mathbf{x})$, $\mathbf{v} = \mathbf{v}(t, \mathbf{x}) = (v_1, v_2, v_3)$, $\mathbf{H} = \mathbf{H}(t, \mathbf{x}) = (H_1, H_2, H_3)$, $p = p(t, \mathbf{x})$, $S = S(t, \mathbf{x})$ are the density, the fluid velocity, the magnetic field, the pressure, and the entropy respectively, $q = p + |\mathbf{H}|^2/2$ is the total pressure, $c^2 = \partial p/\partial \rho$ is the square of the sound velocity, t is the time, $\mathbf{x} = (x_1, x_2, x_3)$ are space variables, and $d/dt = \partial_t + (\mathbf{v}, \nabla)$, $\partial_t = \partial/\partial t$, $\partial_j = \partial/\partial x_j$. With a state equation of medium, $p = p(\rho, S)$, we can consider (1) as a closed system for the vector of unknowns $\mathbf{U} = \mathbf{U}(t, \mathbf{x}) = (p, \mathbf{v}, \mathbf{H}, S)$. Moreover, system (1) should be supplemented by the divergent constraint

$$\operatorname{div} \mathbf{H} = 0, \tag{2}$$

that is just an additional requirement on the initial data $\mathbf{U}(0, \mathbf{x}) = \mathbf{U}_0(\mathbf{x})$. As is known, system (1) is written in the form

$$A_0(\mathbf{U})\mathbf{U}_t + \sum_{k=1}^{3} A_k(\mathbf{U})\mathbf{U}_{x_k} = 0, \tag{3}$$

with symmetric matrices A_α, and system (3) is symmetric hyperbolic if the following natural assumptions (hyperbolicity conditions) hold:

$$\rho > 0, \quad c^2 > 0. \tag{4}$$

Let $\Gamma(t) = \{x_1 - f(t, \mathbf{x}') = 0\}$ be a smooth hypersurface in $\mathbb{R} \times \mathbb{R}^3$ ($\mathbf{x}' = (x_2, x_3)$ are tangential coordinates). We assume that $\Gamma(t)$ is a surface of tangential discontinuity [13] (*current-vortex sheet*) for solutions of the MHD system. This is the type of contact discontinuities for which the normal component of the magnetic field is zero on $\Gamma(t)$, and the tangential components of the velocity and the magnetic field may undergo any jump on $\Gamma(t)$:

$$H_{\mathrm{N}}^{\pm} = 0, \quad [\mathbf{v}_\tau] \neq 0, \quad [\mathbf{H}_\tau] \neq 0$$

Here

$$H_{\mathrm{N}} = (\mathbf{H}, \mathbf{N}), \quad v_{\mathrm{N}} = (\mathbf{v}, \mathbf{N}), \quad \mathbf{v}_\tau = (v_{\tau_1}, v_{\tau_2}), \quad \mathbf{H}_\tau = (H_{\tau_1}, H_{\tau_2}),$$

$$v_{\tau_i} = (\mathbf{v}, \boldsymbol{\tau}_i), \quad H_{\tau_i} = (\mathbf{H}, \boldsymbol{\tau}_i), \quad \boldsymbol{\tau}_1 = (f_{x_2}, 1, 0), \quad \boldsymbol{\tau}_2 = (f_{x_3}, 0, 1);$$

$\mathbf{N} = (1, -f_{x_2}, -f_{x_3})$ is a space normal vector to $\Gamma(t)$, $[g] = g^+ - g^-$ denotes the jump for every regularly discontinuous function g with corresponding values behind ($g^+ := g|_{x_1 - f(t, \mathbf{x}') \to +0}$) and ahead ($g^- := g|_{x_1 - f(t, \mathbf{x}') \to -0}$) of the discontinuity front Γ. For current-vortex sheets, the general MHD Rankine-Hugoniot conditions (see, e.g., [11, 13, 4, 23]) are satisfied in the following way:

$$f_t = v_{\mathrm{N}}^{\pm}, \quad H_{\mathrm{N}}^{\pm} = 0, \quad [q] = 0. \tag{5}$$

The initial boundary value problem for system (1) in the domains $\Omega^{\pm}(t) := \{x_1 \gtrless f(t, \mathbf{x}')\}$ with the boundary conditions (5) on the hypersurface $\Gamma(t)$ is a free boundary value problem. Indeed, the function $f(t, \mathbf{x}')$ defining Γ is one of the unknowns of problem (1), (5) with the corresponding initial data

$$f(0, \mathbf{x}') = f_0(\mathbf{x}'), \quad \mathbf{x}' \in \mathbb{R}^2; \quad \mathbf{U}(0, \mathbf{x}) = \mathbf{U}_0(\mathbf{x}), \quad \mathbf{x} \in \Omega^{\pm}(0). \tag{6}$$

It is worth to note that for problem (1), (5), (6) the divergent constraint (2) as well as the boundary conditions

$$H_N^+ = 0, \quad H_N^- = 0 \tag{7}$$

can be regarded as the restrictions only on the initial data (6). This fact was not formally proved in [23] but this can be done by analogy with the proof in [24] for the case of incompressible MHD (see below).

Definition 1.1. *A piecewise smooth vector-function* $\mathbf{U}(t, \mathbf{x})$ *is called current-vortex sheet solution of the MHD equations* (1) *if there exists a smooth hypersurface* Γ *such that* $\mathbf{U}$ *is a classical solution of* (1) *on either side of* Γ *and conditions* (5) *hold at each point of* Γ.

To prove the existence of current-vortex sheet solutions for the MHD equations it needs to reply on the following question: does there exist a solution $(\mathbf{U}, f)$ of problem (1), (5), (6)? Because of the general properties of hyperbolic conservation laws it is natural to expect only the local-in-time existence of current-vortex sheet solutions. In this connection, the question on the nonlinear Lyapunov's stability of current-vortex sheet has no sense.

At the same time, the study of the linearized stability of current-vortex sheet solutions is not only a necessary step to prove local-in-time existence but also is of independent interest in connection with various astrophysical applications (see, e.g., [18, 3, 19]). Piecewise constant solutions of (1) satisfying (5) on a *planar* discontinuity are a simplest case of current-vortex sheet solutions. In astrophysics and geophysics the linear stability of a planar compressible current-vortex sheet is usually interpreted as the macroscopic stability of the *heliopause* (see, e.g., [19] and references therein). The model of heliopause was suggested in [3], and the heliopause is in fact a current-vortex sheet separating the interstellar plasma compressed at the bow shock from the solar wind plasma compressed at the termination shock wave. That is, the heliopause is the model for the boundary of the solar system.

We can show that for the constant coefficients linearized problem for planar current-vortex sheets (see Sect. 3) the uniform Kreiss-Lopatinski condition [12, 16] is never satisfied [23]. That is, planar current-vortex sheets are never uniformly stable and can be only neutrally (*weakly*) stable or violently unstable. In the 1960–90's, in a number of works (see [19] and references therein) motivated by astrophysical applications (in particular, by applications to the heliopause) the linear stability of planar compressible current-vortex sheets was examined by the normal modes analysis. But, before the recent result in [23] neither stability nor instability conditions were found for a general case of the unperturbed flow. The main difficulty in the normal modes analysis is connected with the fact that the Lopatinski determinant is generically reduced to an algebraic equation of the tenth degree depending on seven dimensionless parameters and one more inner parameter determining the wave vector (see [23]). Moreover, the squaring was applied under the reduction of the Lopatinski determinant to this algebraic equation and, therefore, it can introduce spurious roots. For all these reasons both the analytical analysis and the *full* numerical study of the Lopatinski determinant are unacceptable for finding the Lopatinski condition for compressible current-vortex sheets. The alternative energy method suggested in [23] has first enabled to find sufficient conditions for their weak stability.

Unlike the case of compressible current-vortex sheets, for planar current-sheets in *incompressible* MHD the linear stability conditions can be straightforwardly found [22, 2]. At the same time, the question on the local-in-time existence of incompressible current-vortex sheets remains open. First results in this direction were recently obtained in [24].

A current-vortex sheet solution of the system of ideal incompressible MHD

$$\frac{d\mathbf{v}}{dt} - (\mathbf{H}, \nabla\mathbf{H}) + \nabla q = 0, \quad \frac{d\mathbf{H}}{dt} - (\mathbf{H}, \nabla)\mathbf{v} = 0, \quad \mathrm{div}\,\mathbf{v} = 0 \qquad (8)$$

is determined as a piecewise smooth solution $\mathbf{U} = (\mathbf{v}, \mathbf{H})$ of (8) being a classical solution of (8) on either side of a smooth hypersurface Γ and satisfying the jump conditions (5) at each point of Γ. Here the magnetic field is measured in Alfvén velocity units and the pressure p was divided by the density ρ ($\rho \equiv \mathrm{const} > 0$). Other notations are the same as in (1). Generically, for current-vortex sheets the density can be piecewise constant. But, since this gives no trouble, we suppose for simplicity that it is the same constant ($\rho^+ = \rho^- = \rho$) on either side of Γ. We can show (see [24]) that for the free boundary value problem (8), (5), (6) the divergent constraint (2) and the boundary conditions (7) can be regarded as the restrictions only on the initial data (6).

2. A "secondary" symmetrization of the compressible MHD equations

The crucial role in obtaining the a priori estimate for the linearized problem associated to (1), (5), (6) (see Sect. 3) is played by a new symmetric form [23] of the MHD equations that is a kind of "secondary" symmetrization of the symmetric system (3). Using the linear analog of this symmetrization one can get a conserved energy integral for the constant coefficients linearized problem, $dI(t)/dt = 0$ (see Sect. 3), that implies the desired a priori estimate for planar compressible current-vortex sheets, provided that $I(t) > 0$. The last inequality gives a sufficient condition of the linear stability of a planar compressible current-vortex sheet (the macroscopic stability of the heliopause).

The mentioned "secondary" symmetrization is performed as follows. In view of the divergent constraint (2), system (3) implies

$$P A_0 \mathbf{U}_t + \sum_{k=1}^{3} P A_k \mathbf{U}_{x_k} + \mathbf{R}\,\mathrm{div}\mathbf{H} = 0, \qquad (9)$$

where the matrix $P = P(\mathbf{U})$ and the vector $\mathbf{R} = \mathbf{R}(\mathbf{U})$ are yet arbitrary. If we choose

$$
P = \begin{pmatrix}
1 & \dfrac{\lambda H_1}{\rho c^2} & \dfrac{\lambda H_2}{\rho c^2} & \dfrac{\lambda H_3}{\rho c^2} & 0 & 0 & 0 & 0 \\
\lambda H_1 \rho & 1 & 0 & 0 & -\rho\lambda & 0 & 0 & 0 \\
\lambda H_2 \rho & 0 & 1 & 0 & 0 & -\rho\lambda & 0 & 0 \\
\lambda H_3 \rho & 0 & 0 & 1 & 0 & 0 & -\rho\lambda & 0 \\
0 & -\lambda & 0 & 0 & 1 & 0 & 0 & 0 \\
0 & 0 & -\lambda & 0 & 0 & 1 & 0 & 0 \\
0 & 0 & 0 & -\lambda & 0 & 0 & 1 & 0 \\
0 & 0 & 0 & 0 & 0 & 0 & 0 & 1
\end{pmatrix}, \quad
\mathbf{R} = -\lambda \begin{pmatrix}
1 \\ 0 \\ 0 \\ 0 \\ H_1 \\ H_2 \\ H_3 \\ 0
\end{pmatrix}
$$

(the function $\lambda = \lambda(\mathbf{U})$ is arbitrary), then system (9) is again symmetric:

$$
\mathcal{A}_0(\mathbf{U})\mathbf{U}_t + \sum_{k=1}^{3} \mathcal{A}_k(\mathbf{U})\mathbf{U}_{x_k} = 0, \tag{10}
$$

where

$$
\mathcal{A}_0 = P A_0 = \begin{pmatrix}
\dfrac{1}{\rho c^2} & \dfrac{\lambda H_1}{c^2} & \dfrac{\lambda H_2}{c^2} & \dfrac{\lambda H_3}{c^2} & 0 & 0 & 0 & 0 \\
\dfrac{\lambda H_1}{c^2} & \rho & 0 & 0 & -\rho\lambda & 0 & 0 & 0 \\
\dfrac{\lambda H_2}{c^2} & 0 & \rho & 0 & 0 & -\rho\lambda & 0 & 0 \\
\dfrac{\lambda H_3}{c^2} & 0 & 0 & \rho & 0 & 0 & -\rho\lambda & 0 \\
0 & -\rho\lambda & 0 & 0 & 1 & 0 & 0 & 0 \\
0 & 0 & -\rho\lambda & 0 & 0 & 1 & 0 & 0 \\
0 & 0 & 0 & -\rho\lambda & 0 & 0 & 1 & 0 \\
0 & 0 & 0 & 0 & 0 & 0 & 0 & 1
\end{pmatrix};
$$

for the concrete form of the matrices $\mathcal{A}_k$ we refer to [23]. Note that $\lambda\sqrt{\rho}$ is a dimensionless value, and for $\lambda = 0$ system (10) coincides with (3).

The symmetric system (10) is hyperbolic if $\mathcal{A}_0 > 0$ (this also guarantees that $\det P \neq 0$). Direct calculations show that the last condition is satisfied if inequalities (4) hold together with the additional requirement

$$
\rho\lambda^2 < \frac{1}{1 + c_A^2/c^2}, \tag{11}
$$

where $c_A = |\mathbf{H}|/\sqrt{\rho}$. Of course, the hyperbolicity conditions for system (10) are much more restrictive than the usual natural assumptions (4). It should be also noted that condition (11) guarantees the equivalence of systems (1) and (10) on smooth solutions provided that $\lambda(\mathbf{U})$ is a smooth function of its variables (components of the vector $\mathbf{U}$).

Proposition 2.1. *Let the hyperbolicity conditions* (4) *and* (11) *hold for systems* (1) *and* (10) *respectively, and the initial data for these systems satisfy the divergent constraint* (2). *Let* $\lambda = \lambda(\mathbf{U}) : \mathbb{R}^8 \to \mathbb{R}$ *is a smooth enough function of their arguments. Assume* $[0, T]$ *is a time interval on which both hyperbolic systems* (1) *and* (10) *have a unique classical solution. Then classical solutions of the Cauchy problems for systems* (1) *and* (10) *coincide on the interval* $[0, T]$.

For the proof of Proposition 2.1 we refer to [23]. In principle, analogous assertion could be proved for current-vortex sheet solutions of the MHD system.

The "incompressible" counterpart of symmetrization (10) reads [24]:

$$\mathcal{A}_0(\mathbf{U})\mathbf{U}_t + \sum_{k=1}^{3} \mathcal{A}_k(\mathbf{U})\mathbf{U}_{x_k} + \mathbf{b} \otimes \nabla q = 0, \tag{12}$$

where

$$\mathcal{A}_0 = \begin{pmatrix} 1 & -\lambda \\ -\lambda & 1 \end{pmatrix} \otimes I_3, \quad \mathcal{A}_k = \begin{pmatrix} v_k + \lambda H_k & -H_k - \lambda v_k \\ -H_k - \lambda v_k & v_k + \lambda H_k \end{pmatrix} \otimes I_3,$$

$$\mathbf{b} = \begin{pmatrix} 1 \\ -\lambda \end{pmatrix};$$

$\lambda = \lambda(\mathbf{U})$ is an arbitrary function, I_j is the unit matrix of order j. System (12) is equivalent to (8) if $\det \mathcal{A}_0 \neq 0$ and coincides with (8) if $\lambda = 0$. Moreover, the matrix $\mathcal{A}_0$ is positive definite if $|\lambda| < 1$. In [24] the "symmetric" form (12) of system (8) plays the crucial role for obtaining "linearized" a priori estimates for problem (8), (5), (6). The condition $|\lambda| < 1$ being satisfied on either side of a planar current-vortex sheet defines exactly the half of the whole parameter domain [22, 2] of linear stability of incompressible current-vortex sheets (see Sect. 3).

3. Linear stability of planar current-vortex sheets

To work in fixed domains instead of the domains $\Omega^{\pm}(t)$ we make, as usual (see, e.g., [16]), the following change of variables in $\mathbb{R} \times \mathbb{R}^3$:

$$\tilde{t} = t, \quad \tilde{x}_1 = x_1 - f(t, \mathbf{x}'), \quad \tilde{\mathbf{x}}' = \mathbf{x}'. \tag{13}$$

Then, $\widetilde{\mathbf{U}}(\tilde{t}, \tilde{\mathbf{x}}) := \mathbf{U}(t, \mathbf{x})$ is a smooth vector-function for $\tilde{\mathbf{x}} \in \mathbb{R}^3_{\pm}$, and problem (1), (5), (6) is reduced to the following problem (we omit tildes to simplify the notation):

$$L(\mathbf{U}, \mathbf{F})\mathbf{U} = 0 \quad \text{in } [0, T] \times (\mathbb{R}^3_+ \cup \mathbb{R}^3_-), \tag{14}$$

$$v_N^{\pm} = f_t, \quad H_N^{\pm} = 0, \quad [q] = 0 \quad \text{on } [0, T] \times \{x_1 = 0\} \times \mathbb{R}^2, \tag{15}$$

$$\mathbf{U}|_{t=0} = \mathbf{U}_0 \quad \text{in } \mathbb{R}^3_+ \cup \mathbb{R}^3_-, \quad f|_{t=0} = f_0 \quad \text{in } \mathbb{R}^2. \tag{16}$$

Here

$$L = L(\mathbf{U}, \mathbf{F}) = A_0(\mathbf{U})\partial_t + A_\nu(\mathbf{U}, \mathbf{F})\partial_1 + A_2(\mathbf{U})\partial_2 + A_3(\mathbf{U})\partial_3,$$
$$\mathbf{F} = \mathbf{F}(t, \mathbf{x}') = (f_t, f_{x_2}, f_{x_3});$$

$$A_\nu = A_\nu(\mathbf{U}, \mathbf{F}) = \sum_{\alpha=0}^{3} \nu_\alpha A_\alpha = A_1(\mathbf{U}) - f_t A_0(\mathbf{U}) - \sum_{k=2}^{3} f_{x_k} A_k(\mathbf{U})$$

is the *boundary matrix*,

$$\boldsymbol{\nu} = (\nu_0, \ldots, \nu_n) = (-f_t, \mathbf{N}) \quad \text{is the space-time normal vector to } \Gamma(t).$$

Since the boundary matrix A_ν is singular at $x_1 = 0$ (see [23]), compressible current-vortex sheets are *characteristic* discontinuities. Note also that the boundary matrix is of constant rank 2 on the boundary $x_1 = 0$. Unlike [8], where 2D compressible vortex sheets were analyzed, we make here the straightening of variables that is standard for shock waves [16]. The change of variables (13) is quite suitable for studying problem (14)–(16) by the *energy method*. However, for applying, as in [8], Kreiss' symmetrizer technique it needs to make another change of variables [10, 8]. For the change of variable used in [10, 8] the boundary matrix will have constant rank in the whole spaces $\mathbb{R}_\pm^3$, but not only on the boundary $x_1 = 0$.

Let $(\widehat{\mathbf{U}}(t, \mathbf{x}), \hat{f}(t, \mathbf{x}'))$ be a given vector-function, where $\widehat{\mathbf{U}} = (\hat{p}, \hat{\mathbf{v}}, \widehat{\mathbf{H}}, \widehat{S})$ is supposed to be smooth for $\mathbf{x} \in \mathbb{R}_\pm^3$. Then the linearization of (14)–(16) results in the following variable coefficients problem for determining small perturbations $(\delta \mathbf{U}, \delta f)$ (below we drop δ):

$$L(\widehat{\mathbf{U}}, \widehat{\mathbf{F}})\mathbf{U} + \widehat{C}\mathbf{U} = \{L(\widehat{\mathbf{U}}, \widehat{\mathbf{F}})f\}\widehat{\mathbf{U}}_{x_1} \quad \text{in } [0, T] \times (\mathbb{R}_+^3 \cup \mathbb{R}_-^3), \qquad (17)$$

$$v_N^\pm = f_t + \hat{v}_2^\pm f_{x_2} + \hat{v}_3^\pm f_{x_3}, \quad H_N^\pm = \widehat{H}_2^\pm f_{x_2} + \widehat{H}_3^\pm f_{x_3}, \quad [q] = 0 \quad \text{if } x_1 = 0, \ (18)$$

and the initial data for the perturbation $(\mathbf{U}, f)$ coincide with (16). Here, $\widehat{\mathbf{F}} = (\hat{f}_t, \hat{f}_{x_2}, \hat{f}_{x_2})$, $v_N = (\mathbf{v}, \widehat{\mathbf{N}})$, $H_N = (\mathbf{H}, \widehat{\mathbf{N}})$, $q = p + (\widehat{\mathbf{H}}, \mathbf{H})$, $\widehat{\mathbf{N}} = (1, -\hat{f}_{x_2}, -\hat{f}_{x_2})$, etc. The matrix $\widehat{C} = \widehat{C}(\widehat{\mathbf{U}}, \widehat{\mathbf{U}}_t, \nabla \widehat{\mathbf{U}}, \widehat{\mathbf{F}})$ is determined as follows:

$$\widehat{C}\mathbf{U} = (\mathbf{U}, \nabla_u A_0(\widehat{\mathbf{U}}))\widehat{\mathbf{U}}_t + (\mathbf{U}, \nabla_u A_\nu(\widehat{\mathbf{U}}, \widehat{\mathbf{F}}))\widehat{\mathbf{U}}_{x_1} + \sum_{k=2}^{3}(\mathbf{U}, \nabla_u A_k(\widehat{\mathbf{U}}))\widehat{\mathbf{U}}_{x_k},$$

$(\mathbf{U}, \nabla_u) := \sum_{i=1}^{8} u_i \partial/\partial u_i$, $(u_1, \ldots, u_8) := (p, \mathbf{v}, \mathbf{H}, S)$. Problem (17), (18), (16) is the *genuine linearization* of (14)–(16) in the sense that we keep all the lower order terms in (17).

It should be noted that the differential operator in system (17) is a first order operator in f. This fact can give some trouble in the application of the energy method to (17), (18), (16). To avoid this difficulty we make the change of unknowns (see [1])

$$\bar{\mathbf{U}} = \mathbf{U} - f\widehat{\mathbf{U}}_{x_1}. \qquad (19)$$

In terms of the "good unknown" (19) problem (17), (18) takes the form

$$L(\widehat{\mathbf{U}}, \widehat{\mathbf{F}})\bar{\mathbf{U}} + \widehat{C}\bar{\mathbf{U}} + f\partial_1\{L(\widehat{\mathbf{U}}, \widehat{\mathbf{F}})\widehat{\mathbf{U}}\} = 0 \quad \text{in } [0, T] \times (\mathbb{R}^3_+ \cup \mathbb{R}^3_-), \qquad (20)$$

$$\begin{cases} \bar{v}_N^\pm = f_t + \hat{v}_2^\pm f_{x_2} + \hat{v}_3^\pm f_{x_3} - (\hat{v}_N)_{x_1}^\pm f, \\ \bar{H}_N^\pm = \widehat{H}_2^\pm f_{x_2} + \widehat{H}_3^\pm f_{x_3} - (\widehat{H}_N)_{x_1}^\pm f, \\ [\bar{q}] = -f[\hat{q}_{x_1}] \hspace{4cm} \text{if } x_1 = 0, \end{cases} \qquad (21)$$

where $\bar{v}_N = (\bar{\mathbf{v}}, \widehat{\mathbf{N}})$, $(\hat{v}_N)_{x_1}^\pm = (\hat{v}_N)_{x_1}|_{x_1 \to \pm 0}$, $\bar{q} = \bar{p} + (\widehat{\mathbf{H}}, \bar{\mathbf{H}})$, etc.

For the successful application of the energy method to (20), (21) it would be enough if the operator in (20) had not involved first order terms in f (zero order terms in f give no trouble while applying the energy method). Therefore, without loss of generality we can drop the term $f\partial_1\{L(\widehat{\mathbf{U}}, \widehat{\mathbf{F}})\widehat{\mathbf{U}}\}$ as well as the term $\widehat{C}\bar{\mathbf{U}}$ appearing in (20). As the result, the linearized equations associated to (14), (15) and obtained by dropping the lower order terms in (20) read:

$$L(\widehat{\mathbf{U}}, \widehat{\mathbf{F}})\mathbf{U} = \mathbf{f} \quad \text{in } [0, T] \times (\mathbb{R}^n_+ \cup \mathbb{R}^n_-), \qquad (22)$$

$$\begin{pmatrix} f_t + \hat{v}_2^+ f_{x_2} + \hat{v}_3^+ f_{x_3} - (\hat{v}_N)_{x_1}^+ f - v_N^+ \\ f_t + \hat{v}_2^- f_{x_2} + \hat{v}_3^- f_{x_3} - (\hat{v}_N)_{x_1}^- f - v_N^- \\ \widehat{H}_2^+ f_{x_2} + \widehat{H}_3^+ f_{x_3} - (\widehat{H}_N)_{x_1}^+ f - H_N^+ \\ \widehat{H}_2^- f_{x_2} + \widehat{H}_3^- f_{x_3} - (\widehat{H}_N)_{x_1}^- f - H_N^- \\ [q] + f[\hat{q}_{x_1}] \end{pmatrix} = \mathbf{g} \quad \text{if } x_1 = 0. \qquad (23)$$

Here we introduce the source terms $\mathbf{f}(t, \mathbf{x}) = \mathbf{f}^\pm(t, \mathbf{x})$ for $\mathbf{x} \in \mathbb{R}^n_\pm$ and $\mathbf{g}(t, \mathbf{x}')$ to make the interior equations and the boundary conditions inhomogeneous (this is needed to attack the nonlinear problem).

For planar discontinuities $\hat{f}(t, \mathbf{x}')$ is a linear function:

$$\hat{f}(t, \mathbf{x}') = \sigma t + (\boldsymbol{\sigma}', \mathbf{x}'), \quad \boldsymbol{\sigma} = (\sigma, \boldsymbol{\sigma}') \in \mathbb{R}^3. \qquad (24)$$

Without loss of generality we can suppose that $\boldsymbol{\sigma} = 0$. For the case of a piecewise constant solution,

$$\widehat{\mathbf{U}} = \begin{cases} \widehat{\mathbf{U}}^+, & x_1 > 0, \\ \widehat{\mathbf{U}}^-, & x_1 < 0, \end{cases}$$

equations (22), (23) have constant ("frozen") coefficients:

$$\widehat{A}_0^\pm \mathbf{U}_t + \sum_{k=1}^3 \widehat{A}_k^\pm \mathbf{U}_{x_k} = 0 \quad \text{if } \mathbf{x} \in \mathbb{R}^3_\pm, \qquad (25)$$

$$\begin{cases} f_t = v_1^\pm - \hat{v}_2^\pm f_{x_2} - \hat{v}_3^\pm f_{x_3}, \\ H_1^\pm = \widehat{H}_2^\pm f_{x_2} + \widehat{H}_3^\pm f_{x_3}, \\ [q] = 0 \hspace{3cm} \text{if } x_1 = 0, \end{cases} \qquad (26)$$

where $\widehat{A}_\alpha^\pm := A_\alpha(\widehat{\mathbf{U}}^\pm)$; $q = p + (\widehat{\mathbf{H}}^\pm, \mathbf{H})$ for $\mathbf{x} \in \mathbb{R}^3_\pm$. Since the constant coefficients linearized problem for compressible current-vortex sheets is of independent interest in connection with astrophysical applications mentioned in Section 1, we do not introduce in (25), (26) artificial source terms. At the same time, the a priori estimates proved in [23] for problem (25), (26) (see below) can be easily generalized to the case of inhomogeneous problem (see also the next section, where the variable coefficients inhomogeneous problem (22), (23) is considered).

We can show that compressible current-vortex sheets cannot be *uniformly stable*, i.e., the following important proposition is true [23].

Proposition 3.1. *For the initial boundary value problem* (25), (26), (16), *the uniform Kreiss-Lopatinski condition is never satisfied.*

Since problem (25), (26), (16) is a hyperbolic problem with characteristic boundary, there appears a loss of control on derivatives in the normal (x_1-)direction. Therefore, in the theorem below we use the following "nonsymmetric" Sobolev norms for solutions of (25), (26), (16):

$$\|\mathbf{U}(t)\|^2_{\widetilde{W}^1_2(\mathbb{R}^3_\pm)} = \|\mathbf{U}_\mathrm{n}(t)\|^2_{W^1_2(\mathbb{R}^3_\pm)} + \|\mathbf{U}_\mathrm{tan}(t)\|^2_{W^{1,\mathrm{tan}}_2(\mathbb{R}^3_\pm)},$$

where $\mathbf{U}_\mathrm{n} = (q, v_1, H_1)$, $\mathbf{U}_\mathrm{tan} = (v_2, v_3, H_2, H_3, S)$,

$$\|(\cdot)(t)\|^2_{W^{1,\mathrm{tan}}_2(\mathbb{R}^3_\pm)} = \|(\cdot)(t)\|^2_{L_2(\mathbb{R}^3_\pm)} + \|(\cdot)_{x_2}(t)\|^2_{L_2(\mathbb{R}^3_\pm)} + \|(\cdot)_{x_3}(t)\|^2_{L_2(\mathbb{R}^3_\pm)}.$$

Theorem 3.1. *If* $\widehat{\mathbf{H}}^+ \times \widehat{\mathbf{H}}^- \neq 0$, $[\hat{\mathbf{v}}] \neq 0$, *and*

$$|[\hat{\mathbf{v}}]| < |\sin(\varphi^+ - \varphi^-)| \min\left\{\frac{\gamma^+}{|\sin\varphi^-|}, \frac{\gamma^-}{|\sin\varphi^+|}\right\}, \tag{27}$$

where

$$\gamma^\pm = \frac{\hat{c}^\pm \hat{c}^\pm_\mathrm{A}}{\sqrt{(\hat{c}^\pm)^2 + (\hat{c}^\pm_\mathrm{A})^2}}, \quad \cos\varphi^\pm = \frac{([\hat{\mathbf{v}}], \widehat{\mathbf{H}}^\pm)}{|[\hat{\mathbf{v}}]|\,|\widehat{\mathbf{H}}^\pm|},$$

then, for Problem (25), (26), (16), *the Lopatinski condition is satisfied and the a priori estimates*

$$\|\mathbf{U}(t)\|_{\widetilde{W}^1_2(\mathbb{R}^3_+)} + \|\mathbf{U}(t)\|_{\widetilde{W}^1_2(\mathbb{R}^3_-)}$$
$$\leq C_1 \left\{\|\mathbf{U}_0\|_{\widetilde{W}^1_2(\mathbb{R}^3_+)} + \|\mathbf{U}_0\|_{\widetilde{W}^1_2(\mathbb{R}^3_-)}\right\}, \tag{28}$$

$$\|f(t)\|_{W^1_2(\mathbb{R}^2)} \leq \|f_0\|_{L_2(\mathbb{R}^2)} + C_2 \left\{\|\mathbf{U}_0\|_{\widetilde{W}^1_2(\mathbb{R}^3_+)} + \|\mathbf{U}_0\|_{\widetilde{W}^1_2(\mathbb{R}^3_-)}\right\} \tag{29}$$

hold for any $t \in (0, T)$. *Here* T *is a positive constant;* C_1 *and* $C_2(T)$ *are positive constants independent of the initial data* (16).

If $\widehat{\mathbf{H}}^+ \times \widehat{\mathbf{H}}^- = 0$, $\widehat{\mathbf{H}}^\pm \times [\hat{\mathbf{v}}] = 0$, $[\hat{\mathbf{v}}] \neq 0$, *and*

$$|[\hat{\mathbf{v}}]| < \max\left\{\max\{\gamma^+, \gamma^-\}, 2\min\{\gamma^+, \gamma^-\}\right\} \tag{30}$$

the a priori estimate (28) *holds as well, but for the function* $f(t, \mathbf{x}')$ *we have the weaker estimate*

$$\|f(t)\|_{L_2(\mathbb{R}^2)} \leq \|f_0\|_{L_2(\mathbb{R}^2)} + C_3 \left\{\|\mathbf{U}_0\|_{\widetilde{W}^1_2(\mathbb{R}^3_+)} + \|\mathbf{U}_0\|_{\widetilde{W}^1_2(\mathbb{R}^3_-)}\right\}, \tag{31}$$

where $C_3(T)$ *is a positive constant independent of* (16).

For current sheets, i.e., for the case $[\hat{v}] = 0$ *the Lopatinski condition is always satisfied and estimate* (28) *takes place. Furthermore, estimate* (29) *holds if* $\widehat{\mathbf{H}}^+ \times \widehat{\mathbf{H}}^- \neq 0$, *otherwise we have the weaker estimate* (31). *For current sheets the case* $\widehat{\mathbf{H}}^+ \times \widehat{\mathbf{H}}^- = 0$ *corresponds to the transition to violent instability* (*the Lopatinski condition is violated if* $\widehat{\mathbf{H}}^+ \times \widehat{\mathbf{H}}^- = 0$ *and* $|[\hat{v}]|/\hat{c}^+ \ll 1$).

For the detailed proof of Theorem 3.1 we refer to [23]. The crucial point in the proof is that (25) implies a symmetric system which is the linearization of system (10) about the piecewise constant solution. This symmetric system accompanied with the boundary conditions (26) has the conserved integral

$$I(t) = \int_{\mathbb{R}^3_+} (\widehat{\mathcal{A}}_0^+ \mathbf{U}, \mathbf{U}) \, d\mathbf{x} + \int_{\mathbb{R}^3_-} (\widehat{\mathcal{A}}_0^- \mathbf{U}, \mathbf{U}) \, d\mathbf{x}$$

for a certain choice of the constants $\lambda^\pm = \lambda(\widehat{\mathbf{U}}^\pm)$, where $\widehat{\mathcal{A}}_0^\pm = \mathcal{A}_0(\widehat{\mathbf{U}}^\pm)$ (see (10)). In view of (11), for the most general case $\widehat{\mathbf{H}}^+ \times \widehat{\mathbf{H}}^- \neq 0$, $[\hat{v}] \neq 0$, the hyperbolicity conditions $\widehat{\mathcal{A}}_0^\pm > 0$ give the stability condition (27), that can be interpreted as the sufficient condition of the macroscopic stability of the *heliopause*. Note also that the process of deducing the a priori estimates (28), (29) can be formalized by introducing the notations of *dissipative p-symmetrizers* [25]. In fact, for problem (25), (26) the dissipative (but *not* strictly dissipative [25]) 0-symmetrizer

$$\mathbb{S} = \left\{ P(\widehat{\mathbf{U}}^+), P(\widehat{\mathbf{U}}^-), \mathbf{R}(\widehat{\mathbf{U}}^+), \mathbf{R}(\widehat{\mathbf{U}}^-) \right\}$$

(cf. (9)) has been constructed [23].

For *incompressible* current-vortex sheets, the variable coefficients linearized problem is the initial boundary value problem for the system

$$L(\widehat{\mathbf{U}}, \widehat{\mathbf{F}})\mathbf{U} + \mathbf{e} \otimes \nabla_f q = \mathbf{f}, \quad \operatorname{div} \mathbf{u} = 0 \quad \text{in } [0, T] \times (\mathbb{R}^n_+ \cup \mathbb{R}^n_-), \qquad (32)$$

with the boundary conditions (23) and the initial data (16), where

$$\mathbf{e} = (1, 0), \quad \nabla_f = \widehat{\mathbf{N}} \partial_1 + \mathbf{e}_2 \partial_2 + \mathbf{e}_3 \partial_3, \quad \mathbf{e}_k = (0, \delta_{2k}, \delta_{3k}), \quad \mathbf{u} = (v_N, v_2, v_3),$$

$A_\alpha := \mathcal{A}_\alpha|_{\lambda=0}$ (see (12)), and other notations are the same as for compressible fluid.

The constant coefficients linearized problem for planar incompressible current-vortex sheets is the problem for the system

$$\mathbf{U}_t + \sum_{k=2}^{3} \widehat{A}_k^\pm \mathbf{U}_{x_k} + \mathbf{e} \otimes \nabla q = 0, \quad \operatorname{div} \mathbf{v} = 0, \qquad \mathbf{x} \in \mathbb{R}^3_\pm, \qquad (33)$$

with the boundary conditions (26) and the initial data (16). The necessary and sufficient conditions for the nonexistence of Hadamard-type ill-posedness examples for problem (33), (26) (*linear stability* of a planar current-vortex sheet) were found in [22, 2] (see also [17] for the 2D case):

$$|[\hat{v}]|^2 < 2 \left\{ |\widehat{\mathbf{H}}^+|^2 + |\widehat{\mathbf{H}}^-|^2 \right\}, \quad \left\{ |\widehat{\mathbf{H}}^+ \times [\hat{v}]|^2 + |\widehat{\mathbf{H}}^- \times [\hat{v}]|^2 \right\} \leq 2|\widehat{\mathbf{H}}^+ \times \widehat{\mathbf{H}}^-|^2.$$
$$(34)$$

Using the "symmetric" form (12), a priori estimates for problem (33), (26), (16) were obtained in [24] for a part of the parameter domain (34).

Theorem 3.2. *If* $\widehat{\mathbf{H}}^+ \times \widehat{\mathbf{H}}^- \neq 0$, $[\hat{\mathbf{v}}] \neq 0$, *and*

$$|[\hat{\mathbf{v}}]| < |\sin(\varphi^+ - \varphi^-)| \min\left\{ \frac{|\mathbf{H}^+|}{|\sin\varphi^-|}, \frac{|\mathbf{H}^-|}{|\sin\varphi^+|} \right\}, \tag{35}$$

where $\varphi^{\pm}$ *are the same as in Theorem 3.1, then the a priori estimates* (28), (29), *and*

$$\|\nabla q(t)\|_{L_2(\mathbb{R}^3_+)} + \|\nabla q(t)\|_{L_2(\mathbb{R}^3_-)} \leq C_4 \left\{ \|\mathbf{U}_0\|_{\widetilde{W}^1_2(\mathbb{R}^3_+)} + \|\mathbf{U}_0\|_{\widetilde{W}^1_2(\mathbb{R}^3_-)} \right\} \tag{36}$$

hold for any $t \in (0, T)$. *If* $\widehat{\mathbf{H}}^+ \times \widehat{\mathbf{H}}^- = 0$, $\widehat{\mathbf{H}}^{\pm} \times [\hat{\mathbf{v}}] = 0$, $[\hat{\mathbf{v}}] \neq 0$, *and*

$$|[\hat{\mathbf{v}}]| < \max\left\{ \max\{|\mathbf{H}^+|, |\mathbf{H}^-|\}, 2\min\{|\mathbf{H}^+|, |\mathbf{H}^-|\} \right\}, \tag{37}$$

then the a priori estimates (28), (36) *hold as well, but for the function* $f(t, \mathbf{x}')$ *we have the weaker estimate* (31).

For current sheets, i.e., for the case $[\hat{\mathbf{v}}] = 0$, *the a priori estimates* (28), (36) *always take place. Furthermore, estimate* (29) *holds if* $\widehat{\mathbf{H}}^+ \times \widehat{\mathbf{H}}^- \neq 0$, *otherwise we have the weaker estimate* (31).

For the detailed proof of Theorem 3.2 we refer to [24]. Note that, in view of (34), the particular case $\widehat{\mathbf{H}}^+ \times \widehat{\mathbf{H}}^- = 0$ corresponds to transition to violent instability (ill-posedness). For a general case when $\widehat{\mathbf{H}}^+ \times \widehat{\mathbf{H}}^- \neq 0$, we can show that the first inequality in (34) is redundant, and the second inequality in (34) can be rewritten as follows:

$$|[\hat{\mathbf{v}}]| \leq \frac{\sqrt{2}\,|\widehat{\mathbf{H}}^+|\,|\widehat{\mathbf{H}}^-|\,|\sin(\varphi^+ - \varphi^-)|}{\sqrt{|\widehat{\mathbf{H}}^+|^2 \sin^2\varphi^+ + |\widehat{\mathbf{H}}^-|^2 \sin^2\varphi^-}}. \tag{38}$$

In the *incompressibility limit*, $\hat{c}^{\pm} = \infty$, the sufficient stability condition (27) is reduced to inequality (35) (for simplicity we consider the case $\hat{\rho}^+ = \hat{\rho}^-$). Moreover, if we introduce the dimensionless parameters

$$x = \frac{|[\hat{\mathbf{v}}]|^2 \sin^2\varphi^+}{|\widehat{\mathbf{H}}^-|^2 \sin^2(\varphi^+ - \varphi^-)}, \qquad y = \frac{|[\hat{\mathbf{v}}]|^2 \sin^2\varphi^-}{|\widehat{\mathbf{H}}^+|^2 \sin^2(\varphi^+ - \varphi^-)},$$

then in the xy-plane inequalities (35) and (38) determine the domains

$$D_1 = \{x > 0,\ y > 0,\ \max\{x, y\} < 1\} \quad \text{and} \quad D_2 = \{x > 0,\ y > 0,\ x + y \leq 2\}$$

respectively. It is clear that mes $D_1 = (1/2)$ mes D_2, and the domain $D_2 \backslash D_1$ is an "interlayer" between the domain D_1 of well-posedness (cf. Theorem 3.2) and the domain of ill-posedness: $\{x > 0,\ y > 0,\ x + y > 2\}$.

As regards a possibility to prove a priori estimates for the "interlayer" domain $D_2 \backslash D_1$, it seems that this could be done by adapting Kreiss's symmetrizer technique [12] to the *nonhyperbolic* problem (33), (26), (16). In this connection, there appears an interesting problem on obtaining "exponentially weighted" a priori estimates for nonhyperbolic initial boundary value problems like (33), (26), (16).

4. The variable coefficients analysis

Since the original nonlinear problem (14)–(16) is a *free boundary value problem*, to prove local-in-time existence by standard fixed-point argument we should gain the "additional derivative" for the front perturbation f (see [4, 16]). As we can see from (28), (29), we do not have an estimate for the second derivatives of f provided that $\mathbf{U} \in \widetilde{W}_2^1(\mathbb{R}_+^3) \cap \widetilde{W}_2^1(\mathbb{R}_-^3)$. This is because the uniform Lopatinski condition is violated and, therefore, it is principally impossible that the boundary conditions were *strictly* dissipative. And so, we have the loss of the trace $(\mathbf{U}_n^+, \mathbf{U}_n^-)$ in a high norm, i.e., we are not able to include the norm $\|\mathbf{U}_n^+\|_{W_2^1(\partial\mathbb{R}_+^3)} + \|\mathbf{U}_n^-\|_{W_2^1(\partial\mathbb{R}_-^3)}$ in estimate (28) (to estimate f in a high norm we need only the "noncharacteristic part" of the trace of $(\mathbf{U}^+, \mathbf{U}^-)$ whereas the loss of the trace for the "characteristic" unknown $\mathbf{U}_{\mathrm{tan}}$ even in a lower norm does not give any trouble).

Since even for constant coefficients we have a loss of one derivative for f, the standard scheme of proving the local-in-time existence theorem for the original nonlinear problem does not work for current-vortex sheets. In this connection, another possibility to attack the nonlinear problem is to use the so-called Nash-Moser technique (for hyperbolic problems see [1, 15, 10] and references therein). To apply the Nash-Moser method for compressible current-vortex sheets it needs to carry out an accurate analysis of the corresponding variable coefficients linearized problem. At the same time, after performing the change of unknowns (19) the lower order terms in the interior equations (20) can be neglected because they give no trouble while deducing a priori estimates. Thus, it is enough to study problem (22), (23). The main difficulties in the variable coefficients analysis are connected with lower order terms in the boundary conditions (23). Analogous remarks take place for the nonhyperbolic problem (32), (23) for incompressible fluid.

For the basic state $(\widehat{\mathbf{U}}, \hat{f})$ (it can be, in particular, an exact current-vortex sheet solution to the MHD equations), we assume that

$$\widehat{\mathbf{U}} \in X_4([0,T], \mathbb{R}_+^3) \cap X_4([0,T], \mathbb{R}_-^3), \quad \hat{f} \in W_2^5([0,T] \times \mathbb{R}^2), \qquad (39)$$

where

$$X_k([0,T], \mathbb{R}_\pm^3) := \bigcap_{j=0}^k C^j([0,T], W_2^{k-j}(\mathbb{R}_\pm^3)), \quad \|\cdot\|_{X_k} = \max_{t\in[0,T]} \sum_{j=0}^k \|\partial_t^j(\cdot)(t)\|_{W_2^{k-j}}^2.$$

Then, in view of Sobolev's imbedding, we have

$$\widehat{\mathbf{U}} \in W_\infty^2([0,T] \times \mathbb{R}_+^3) \cap W_\infty^2([0,T] \times \mathbb{R}_-^3), \quad \widehat{\mathbf{F}} \in W_\infty^2([0,T] \times \mathbb{R}^2).$$

In view of (39), there exists a constant $M > 0$ such that

$$\|\hat{f}\|_{W_2^5([0,T]\times\mathbb{R}^2)} + \sum_\pm \|\widehat{\mathbf{U}}\|_{X_4([0,T],\mathbb{R}_\pm^3)} \leq M.$$

For the constant coefficients linearized problem, in the a priori estimate (28) we have the so-called loss of derivatives in the normal direction to the boundary. But, for constant coefficients it was enough to use usual Sobolev norms. At the same time, in the variable coefficients analysis we have to require a little bit more regularity for solutions. The natural functional setting is provided by the anisotropic weighted Sobolev space $W_2^{m,\sigma}$ ($:= H_*^m$; see, e.g., [21] and references therein). Following [21], we now give the definition of the spaces $W_2^{m,\sigma}(\mathbb{R}_\pm^3)$. Let $\sigma(x_1) \in C^\infty(\mathbb{R}_+) \cap C^\infty(\mathbb{R}_-)$ is a monotone increasing function for $x_1 > 0$ and monotone decreasing for $x_1 < 0$ such that $\sigma(x_1) = |x_1|$ in a neighborhood of the origin and $\sigma(x_1) = 1$ for $|x_1|$ large enough. Let us introduce the so-called conormal derivative

$$\partial_*^\alpha = (\sigma(x_1)\partial_1)^{\alpha_1}\partial_2^{\alpha_2}\partial_3^{\alpha_3} .$$

Then, given $m \geq 1$, the function space $W_2^{m,\sigma}(\Omega)$ ($\Omega = \mathbb{R}_+^3$ or $\Omega = \mathbb{R}^3$) is defined as the set of functions $u \in L_2(\Omega)$ such that $\partial_*^\alpha \partial_1^k u \in L_2(\Omega)$ if $|\alpha| + 2k \leq m$. The space $W_2^{m,\sigma}(\Omega)$ is normed by

$$\|u\|_{W_2^{m,\sigma}(\Omega)}^2 = \sum_{|\alpha|+2k\leq m} \|\partial_*^\alpha \partial_1^k u\|_{L_2(\Omega)}^2 .$$

For solutions of problem (22), (23) we use also the norms

$$\||\mathbf{U}(t)\||_{\widetilde{W}_2^{m,\sigma}(\mathbb{R}_\pm^3)}^2 = \||\mathbf{U}(t)\||_{W_2^{m,\sigma}(\mathbb{R}_\pm^3)}^2 + \||\partial_1\mathbf{U}_{\mathrm{n}}(t)\||_{W_2^{m-1,\sigma}(\mathbb{R}_\pm^3)}^2 ,$$

where $\||(\cdot)(t)\||_{W_2^{k,\sigma}}^2 = \sum_{j=0}^k \|\partial_t^j(\cdot)(t)\|_{W_2^{k-j,\sigma}}^2$; $\mathbf{U}_{\mathrm{n}} = (q, v_{\mathrm{N}}, H_{\mathrm{N}})$ is the "noncharacteristic part" of $\mathbf{U}$. We are now in a position to formulate the main result from [23] obtained for the variable coefficients linearized problem for compressible current-vortex sheets. Note that unlike [23] we formulate the theorem below for the inhomogeneous problem (22), (23) (arguments in [23] can be easily extended to the case when $\mathbf{f} \neq 0$ and $\mathbf{g} \neq 0$).

Theorem 4.1. *Let the basic state $(\widehat{\mathbf{U}}, \hat{f})$ satisfy assumptions (39), the Rankine-Hugoniot conditions (5), and the hyperbolicity conditions (4). Let there also exist a positive constant δ such that*

$$|\hat{\mathbf{h}}^+(t, \mathbf{x}') \times \hat{\mathbf{h}}^-(t, \mathbf{x}')| \geq \delta > 0 \tag{40}$$

for all $t \in [0, T]$, $\mathbf{x}' \in \mathbb{R}^2$ and the condition

$$r(t, \mathbf{x}) < b(t, \mathbf{x}) \tag{41}$$

holds for all $t \in [0, T]$ at each point $\mathbf{x} \in \overline{\mathbb{R}_\pm^3}$ such that $\hat{\mathbf{u}}^+(t, \mathbf{x}') \neq \hat{\mathbf{u}}^-(t, \mathbf{x}')$, where

$$\hat{\mathbf{h}} = (\widehat{H}_{\mathrm{N}}, \widehat{H}_2, \widehat{H}_3), \quad \hat{\mathbf{u}} = (\hat{v}_{\mathrm{N}} - \hat{f}_t, \hat{v}_2, \hat{v}_3),$$

$$r(t, \mathbf{x}) = \sqrt{\hat{\rho}(1 + \hat{c}_{\mathrm{A}}^2/\hat{c}^2)}, \quad b(t, \mathbf{x}) = \begin{cases} b^+(t, \mathbf{x}') & \text{if } x_1 > 0, \\ b^-(t, \mathbf{x}') & \text{if } x_1 < 0; \end{cases}$$

$$b^\pm(t, \mathbf{x}') = \frac{|\hat{\mathbf{h}}^\pm| \, |\sin(\varphi^+ - \varphi^-)|}{|[\hat{\mathbf{u}}]| \, |\sin\varphi^\mp|}, \quad \cos\varphi^\pm(t, \mathbf{x}') = \frac{([\hat{\mathbf{u}}], \hat{\mathbf{h}}^\pm)}{|[\hat{\mathbf{u}}]| \, |\hat{\mathbf{h}}^\pm|} .$$

Then, for problem (22), (23) *the a priori estimate*

$$\sum_{\pm} \|\|\mathbf{U}(t)\|\|_{\widetilde{W}_2^{1,\sigma}(\mathbb{R}_{\pm}^3)} + \|f\|_{W_2^1([0,T]\times\mathbb{R}^2)}$$

$$\leq C\Big\{\sum_{\pm}\Big\{\|\mathbf{f}^{\pm}\|_{W_2^1([0,T]\times\mathbb{R}_{\pm}^3)} + \|\|\mathbf{U}_0\|\|_{\widetilde{W}_2^{1,\sigma}(\mathbb{R}_{\pm}^3)}\Big\} \tag{42}$$

$$+ \|g\|_{W_2^2([0,T]\times\mathbb{R}^2)} + \|f_0\|_{W_2^1(\mathbb{R}^2)}\Big\}$$

holds for any $t \in [0,T]$. *Here* $C = C(T,M)$ *is a positive constant independent of the initial data* $(\mathbf{U}_0, f_0)$.

For the detailed proof of Theorem 4.1 we refer to [23]. Inequality (41) appearing in this theorem is the analogue of the stability condition (27) for variable coefficients. The a priori estimate (42) is, in some sense, a base estimate, and to attack the original nonlinear problem we should deduce estimates in higher order norms.

Corollary 4.1. *Let all the assumptions of Theorem* 4.1 *be satisfied and*

$$(\widehat{\mathbf{U}}, \widehat{f}) \in \Big\{\bigcap_{\pm}\bigcap_{j=0}^s W_{\infty}^j([0,T], W_2^{s-j,\sigma}(\mathbb{R}_{\pm}^3))\Big\} \times W_2^{s+1}([0,T]\times\mathbb{R}^2)$$

for some $s \geq 8$. *Let* $1 \leq m \leq s$. *Then, for problem* (22), (23) *the a priori estimate*

$$\sum_{\pm}\|\|\mathbf{U}(t)\|\|_{\widetilde{W}_2^{m,\sigma}(\mathbb{R}_{\pm}^3)} + \|f\|_{W_2^m([0,T]\times\mathbb{R}^2)}$$

$$\leq C\Big\{\sum_{\pm}\Big\{\|\mathbf{f}^{\pm}\|_{W_2^{m,\sigma}([0,T]\times\mathbb{R}_{\pm}^3)} + \|\|\mathbf{U}_0\|\|_{\widetilde{W}_2^{m,\sigma}(\mathbb{R}_{\pm}^3)}\Big\}$$

$$+ \|g\|_{W_2^{m+1}([0,T]\times\mathbb{R}^2)} + \|f_0\|_{W_2^m(\mathbb{R}^2)}\Big\}$$

holds for any $t \in [0,T]$, *where* $\|\mathbf{f}^{\pm}\|^2_{W_2^{m,\sigma}([0,T]\times\mathbb{R}_{\pm}^3)} = \int_0^T \|\|\mathbf{f}^{\pm}(t)\|\|^2_{W_2^{m,\sigma}(\mathbb{R}_{\pm}^3)}dt$.

To prove Corollary 4.1 it needs to take into account the imbeddings $W_2^{s,\sigma} \subset W_2^{[s/2]} \subset W_2^4$ (cf. (39)) and use the standard technique based on the application of the Gagliardo-Nirenberg inequalities.

For incompressible current-vortex sheets, the technique of obtaining a priori estimates for problem (32), (23) is similar to that used in [23] for compressible current-vortex sheets. However, there is, of course, an essential principal difference from the "hyperbolic" energy method utilized in [23]. As is known, for the system of incompressible MHD the total pressure q is an "elliptic" unknown. Hence, the differentiation of system (32) with respect to t cannot help to prove the energy a priori estimate. Instead of this, roughly speaking, we estimate ∇q through spatial derivatives of $\mathbf{U}$ and then, from system (32), we obtain an estimate for $\mathbf{U}_t$. Actually, since in our case the boundary is a strong discontinuity, for the function q we have an elliptic boundary value problem that is like *diffraction problems* [14]. It is true that the estimate for ∇q can be obtained directly from system (32) (see [24] for details).

Let us now formulate the corresponding theorem proved in [24]. Let

$$\widehat{\mathbf{U}} \in \bigcap_{k=0}^{1} C^k\left([0,T], \bigcap_{\pm} W_2^{4-k}(\mathbb{R}_{\pm}^3)\right), \quad \hat{f} \in C\left([0,T], W_2^4(\mathbb{R}^2)\right). \tag{43}$$

Moreover, we assume also that

$$\nabla \hat{q} \in C\left([0,T], W_\infty^1(\mathbb{R}_+^3) \cap W_\infty^1(\mathbb{R}_-^3)\right) \tag{44}$$

and

$$\nabla \hat{q} \in \bigcap_{j=0}^{1} C\left([0,T]; W_2^j\left(\mathbb{R}_+, W_\infty^{1-j}(\mathbb{R}^2)\right) \cap W_2^j\left(\mathbb{R}_-, W_\infty^{1-j}(\mathbb{R}^2)\right)\right). \tag{45}$$

Theorem 4.2. *Let the basic state $(\widehat{\mathbf{U}}, \hat{f})$ satisfy assumptions (43)–(45). Let also conditions (40) and (41) be fulfilled with $r(t, \mathbf{x}) \equiv 1$ (other notations are the same as in Theorem 4.1). Then, for problem (32), (23) the a priori estimate*

$$\sum_{\pm} \left\{ \|\|\mathbf{U}(t)\|\|_{\widetilde{W}_2^{1,\sigma}(\mathbb{R}_\pm^3)} + \|f\|_{W_2^1([0,T]\times\mathbb{R}^2)} + \|\nabla q\|_{L_2([0,T]\times\mathbb{R}_\pm^3)} \right\}$$

$$\leq C \left\{ \sum_{\pm} \left\{ \|\mathbf{f}^\pm\|_{W_2^1([0,T]\times\mathbb{R}_\pm^3)} + \|\|\mathbf{U}_0\|\|_{\widetilde{W}_2^{1,\sigma}(\mathbb{R}_\pm^3)} \right\} \right. \tag{46}$$

$$\left. + \|\mathbf{g}\|_{W_2^2([0,T]\times\mathbb{R}^2)} + \|f_0\|_{W_2^1(\mathbb{R}^2)} \right\}$$

holds for any $t \in [0, T]$.

As for problem (22), (23) (see Corollary 4.1), we can obtain estimates in higher order norms for problem (32), (23). Note also that the a priori estimates (42) and (46) imply the *uniqueness* of solutions of the original nonlinear problems for compressible and incompressible current-vortex sheets respectively (see [24] for standard arguments).

5. Concluding remarks

In [23], the sufficient condition (27) for the neutral stability of planar compressible current-vortex sheets is first found for a general case of the unperturbed flow. This condition can be interpreted as the sufficient condition of the macroscopic stability of the heliopause [3, 19]. In the incompressibility limit this condition describes exactly the half of the whole parameter domain of stability. The variable coefficients linearized problem for nonplanar compressible current-vortex sheets has been studied as well. Since the current-vortex sheet is a characteristic discontinuity, there appears a loss of derivatives in the normal direction to the discontinuity front, and the natural functional setting is provided by the anisotropic weighted Sobolev spaces [21]. Furthermore, since the uniform Lopatinski condition is not satisfied, we have a loss of one derivative for the front perturbation as well as for the source term in the boundary conditions.

To prove the local-in-time existence of current-vortex sheets solutions for compressible MHD it needs, first, to prove the existence of solutions of the variable coefficients linearized problem (22), (23) in the functional spaces indicated in the estimate in Corollary 4.1. To this end, it is necessary to generalize recent results from [5] to the case of boundary conditions with variable coefficients, when the boundary conditions are not strictly dissipative. Note that in [5] earlier results from [20] for linear symmetric hyperbolic problems with characteristic boundary were extended to nonhomogeneous problems. Then, it seems that we can construct solutions of the nonlinear problem by the Nash-Moser method.

At the same time, one can suggest an alternative programme based on Kreiss' symmetrizer technique and paradifferential calculus (and the Nash-Moser method as well). Such a programme is now being realized for weakly stable shock waves and 2D vortex sheets (see [6, 7, 8]). Indeed, the energy method applied in [23] for the constant coefficients problem for compressible current-vortex sheets can be considered as an indirect test of the Kreiss-Lopatinski condition. Then, we should construct Kreiss' symmetrizer for this problem and follow the arguments from [6, 7, 8] for variable coefficients.

As regards incompressible current-vortex sheets [24], the nonhyperbolic problem for them has, of course, some peculiarities. But, in principle, with appropriate modifications the above remarks for compressible current-vortex sheets can be made for this problem as well.

At last, we note that for astrophysical applications it would be extremely important to generalize the sufficient stability condition (27) to improved heliopause models like, for example, the multifluid neutral MHD model newly developed by astrophysicists (see [9] and references therein). For possible future results, it is especially important that this MHD model is a system of hyperbolic balance laws.

Acknowledgment

The author thanks Konstantin Ilin and Vladimir Vladimirov for their kind hospitality during his stay at the Hull Institute of Mathematical Sciences and Applications of the University of Hull.

References

[1] ALINHAC S. Existence d'ondes de raréfaction pour des systèmes quasi-linéaires hyperboliques multidimensionnels. *Comm. Partial Differential Equations* **14**, 173–230.

[2] AXFORD W.I. Note on a problem of magnetohydrodynamic stability. *Can. J. Phys.* **40** (1962), 654–655.

[3] BARANOV V.B., KRASNOBAEV K.V., KULIKOVSKY A.G. A model of interaction of the solar wind with the interstellar medium. *Sov. Phys. Dokl.* **15** (1970), 791–793.

[4] BLOKHIN A., TRAKHININ YU. Stability of strong discontinuities in fluids and MHD. In: *Handbook of mathematical fluid dynamics*, vol. 1, S. Friedlander and D. Serre, eds., North-Holland, Amsterdam, 2002, pp. 545–652.

[5] CASELLA E., SECCHI P., TREBESCHI P. Non-homogeneous linear symmetric hyperbolic systems with characteristic boundary. Preprint, 2005.

[6] COULOMBEL J.-F. Weakly stable multidimensional shocks. *Ann. Inst. H. Poincaré Anal. Non Linéaire* **21** (2004), 401–443.

[7] COULOMBEL J.-F. Well-posedness of hyperbolic initial boundary value problems. *J. Math. Pures Appl.* (9) **84** (2005), 786–818.

[8] COULOMBEL J.-F., SECCHI P. The stability of compressible vortex sheets in two space dimensions. *Indiana Univ. Math. J.* **53** (2004), 941–1012.

[9] FLORINSKI V., POGORELOV N.V., ZANK G.P., WOOD B.E., COX D.P. On the possibility of a strong magnetic field in the local interstellar medium. *Astrophys. J,* **604** (2004), 700–706.

[10] FRANCHETEAU J., MÉTIVIER G. *Existence de chocs faibles pour des systèmes quasi-linéaires hyperboliques multidimensionnels.* Astérisque, no. 268, Soc. Math. France, Paris, 2000.

[11] JEFFREY A., TANIUTI T. *Non-linear wave propagation. With applications to physics and magnetohydrodynamics.* Academic Press, New York, London, 1964.

[12] KREISS H.-O. Initial boundary value problems for hyperbolic systems. *Commun. Pure and Appl. Math.* **23** (1970), 277–296.

[13] KULIKOVSKY A.G., LYUBIMOV G.A. *Magnetohydrodynamics.* Addison-Wesley, Massachusets, 1965.

[14] LADYZHENSKAYA O.A. *The boundary value problems of mathematical physics.* Springer-Verlag, New York, 1985.

[15] LI D. Rarefaction and shock waves for multi-dimensional hyperbolic conservation laws. *Commun. Partial Differ. Equations* **16** (1991), 425–450.

[16] MAJDA A. *The stability of multi-dimensional shock fronts.* Mem. Amer. Math. Soc. **41**(275), 1983.

[17] MICHAEL D.H. The stability of a combined current and vortex sheet in a perfectly conducting fluid. *Proc. Cambridge Philos. Soc.* **51** (1955), 528–532.

[18] PARKER E.N. Dynamical properties of stellar coronas and stellar winds. III. The dynamics of coronal streamers. *Astrophys. J.* **139** (1964), 690–709.

[19] RUDERMAN M.S., FAHR H.J. The effect of magnetic fields on the macroscopic instability of the heliopause. II. Inclusion of solar wind magnetic fields. *Astron. Astrophys.* **299** (1995), 258–266.

[20] SECCHI P. Linear symmetric hyperbolic systems with characteristic boundary. *Math. Methods Appl. Sci.* **18** (1995), 855–870.

[21] SECCHI P. Some properties of anisotropic Sobolev spaces. *Arch. Math.* **75** (2000), 207–216.

[22] SYROVATSKIJ S.I. The stability of tangential discontinuities in a magnetohydrodynamic medium. *Z. Eksperim. Teoret. Fiz.* **24** (1953), 622–629 (in Russian).

[23] TRAKHININ YU. On existence of compressible current-vortex sheets: variable coefficients linear analysis. *Arch. Rational Mech. Anal.*, to appear.

[24] TRAKHININ YU. On the existence of incompressible current-vortex sheets: study of a linearized free boundary value problem. *Math. Methods Appl. Sci.* **28** (2005), 917–945.

[25] TRAKHININ YU. Dissipative symmetrizers of hyperbolic problems and their applications to shock waves and characteristic discontinuities, *submitted*.

Yuri Trakhinin
Sobolev Institute of Mathematics
Koptyuga pr.4
630090 Novosibirsk, Russia

and

Department of Mathematics
University of Hull
Cottingham Road
Hull, HU6 7RX, UK
e-mail: trakhin@math.nsc.ru; y.l.trakhinin@hull.ac.uk